Molecular and Cellular Biology of Filamentous Fungi

The Practical Approach Series

Related **Practical Approach** Series Titles

Functional Genomics
Bioinformatics: sequence, structure and databanks
Protein Purification Techniques

Please see the **Practical Approach** series website at
http://www.oup.co.uk/pas
for full contents lists of all Practical Approach titles.

Molecular and Cellular Biology of Filamentous Fungi

A Practical Approach

Edited by

Nick Talbot

University of Exeter, School of Biological Sciences, Washington Singer Laboratories, Perry Road, Exeter EX4 4QG, UK

OXFORD
UNIVERSITY PRESS

This book has been printed digitally and produced in a standard specification in order to ensure its continuing availability

OXFORD
UNIVERSITY PRESS

Great Clarendon Street, Oxford OX2 6DP

Oxford University Press is a department of the University of Oxford.
It furthers the University's objective of excellence in research, scholarship,
and education by publishing worldwide in

Oxford New York

Auckland Cape Town Dar es Salaam Hong Kong Karachi
Kuala Lumpur Madrid Melbourne Mexico City Nairobi
New Delhi Shanghai Taipei Toronto
With offices in
Argentina Austria Brazil Chile Czech Republic France Greece
Guatemala Hungary Italy Japan South Korea Poland Portugal
Singapore Switzerland Thailand Turkey Ukraine Vietnam

Published in the United States
by Oxford University Press Inc., New York

Reprinted 2007

ISBN 978-0-19-963837-6

Preface

The study of fungal biology has changed dramatically in recent years as the techniques for isolating and characterizing genes have become easier and more information has steadily accumulated regarding fundamental process such as cell cycle control, hyphal growth, septation and cell wall development, and fungal pathogenicity. The advent of genome-level analysis, following the completion of the genome sequences of the yeasts *Saccharomyces cerevisiae*, *Schizosaccharomyces pombe*, and *Candida albicans* has also changed the research landscape enormously, and the probability that filamentous fungal genomes will soon be available in public databases is an exciting prospect. For those of us who have spent the early parts of their careers identifying and studying single genes from fungi, the change in the way fungal biology will be investigated in the future, will be very great, but also very welcome. For the first time, there is the real prospect of studying the integrated action of many genes during developmental and physiological process and producing a far greater understanding of fungal biology than has hitherto been possible.

With this new research landscape, however, come significant challenges. An entire generation of fungal research scientists have been trained primarily in molecular biology, and more specifically in cloning and sequencing. Although such tools will remain the cornerstone of biological investigation for some time to come, there is a critical need to carry out multi-disciplinary investigations of fungal biology. It was set against this background that I conceived the idea for this book. I wanted to bring together the diversity of tools that are necessary for contemporary study of fungal biology into a single volume. Although, assimilating the theoretical background and wealth of literature in all of the necessary disciplines would have been an impossible job, I considered that this was of less importance than providing simple, well-explained laboratory protocols of the relevant procedures. After, all it is at the front line of fungal research that I want this book to be used; by the graduate students and postdoctoral fellows searching for ways to determine the function of genes and proteins they have isolated, and trying by their own experiments to further the understanding of fungal biology.

The book starts with an overview of how to classify and determine the identity of a fungal organism. Chapter 1 describes the important considerations that must be made when attempting to determine inter-generic and inter-species relationships among fungi using molecular tools. The types of analysis that can be carried out to examine genotypic variability within isolates of a particular species are also carefully explained. In Chapter 2, a small number of tested methods for extraction of DNA and RNA from a variety of fungal species are described. Many molecular approaches succeed or fail based simply on the quality of this starting material. In Chapter 3 methods for DNA-mediated transformation are explained, including new *Agrobacterium*-mediated protocols and biolistic methods on introducing DNA into fungi. Chapter 4 explains in detail how to carry out genetic analysis in ascomycete fungi, including advice and guidance on devising mutant selection strategies and carrying out tetrad and random ascospore genetics. The theme continues in Chapter 5, but this time concentrating on basidiomycete fungi and exploring homothallism and heterothallism and giving practical guidance on carrying out genetic analysis in diverse species within this group. Details regarding the use of restriction enzyme-mediated insertional (REMI) mutagenesis are also presented here. In Chapter 6 methods for exploring fungi at the genome level are introduced, including genetic map construction and assignment of molecular markers to chromosomally-defined loci. Here, instruction in chromosomal-sized DNA isolation and BAC library construction is given; critical tools for genomic studies. The focus is changed in Chapter 7 to introduce cell biological methods that can be applied to filamentous fungi. These include analysis of cytoskeletal components, cellular localization of proteins, and fundamental studies regarding cell wall synthesis and septation. In Chapter 8 the cell biological procedures applied to the oomycetes are described. Although not true fungi, these organisms are studied by fungal biologists and exhibit remarkable parallels in cellular organization, especially considering their taxonomic distance from fungi. Chapter 9 describes important procedures for investigating signal transduction processes in fungi and concentrates on methods for characterizing and assaying the activity of protein kinases. Chapter 10 synthesizes the methods required to study fungal biology using electron microscopy. Procedures described here include free-substitution fixation, freeze-fracture, and cryo-scanning electron microscopy. These methods have already contributed much to the understanding of fungal cell biology, but will be of far greater relevance in the post-genomic era. Chapter 11 describes the essential biochemical methods required for investigating fungal biology. Methods for isolating plasma membranes, vacuoles, nuclei, and other organelles are described alongside protein purification procedures and separation techniques. Finally in Chapter 12 the application of immunological techniques is described in detail, including the use of monoclonal antibody technology, ELISA for quantitative analysis of antibody–antigen interactions, and immunolocalization methods.

My hope in assembling this variety of experimental approaches in a single volume is that any researcher studying a fungal organism can use this book as their first reference for starting a new avenue of investigation. Many of the procedures are new, but others would have been more familiar to fungal biologists from the pre-molecular biology years than to the current generation. This is important because there is likely to be an element of re-discovery of important techniques in the post-genomic era. Collectively, I believe there is a powerful combination of experimental tools here that will be broadly applicable.

Finally, and most importantly, I would like to thank Catherine Talbot for proofreading the manuscript and all of the authors for their clear explanations, their hard work, patience, and often their collective sense of humour, in bringing this project to fruition.

Exeter
August 2000

N. J. T.

Contents

Protocol list

Genetic mapping

Staining nuclei and cell walls

Observations of living cells

Protein extraction: importance of protecting state of phosphorylation

Conventional preparation techniques for TEM

Abbreviations

AFLP	amplification fragment length polymorphism
BAC	bacterial artificial chromosome
CHEF	contour-clamped homogenous electric field
CI	chloroform/isoamyl alcohol
CIP	calf intestinal phosphatase
CTAB	cetyltrimethyl ammonium bromide
DAF	DNA amplification fingerprinting
DAS-ELISA	double-antibody-sandwich-ELISA
DEB	1,3-butadiene diepoxide
DEO	1,2,7,8-diepoxyoctane
DGGE	denaturing gradient gel electrophoresis
dH_2O	deionized water
DNase	deoxyribonuclease
DTT	dithiothreitol
EDTA	ethylenediaminetetraacetic acid
ELISA	enzyme-linked immunosorbent assay
EMS	ethyl methane sulfonate
FGSC	Fungal Genetic Stock Center
FITC	fluorescein isothiocyanate
IF	immunofluorescence
ip	intraperitoneal
IP	immunoprecipitation
IPTG	isopropyl-thio-β-D-galactoside
ITS	internal transcribed spacer
LB	Luria Bertani broth
LN_2	liquid nitrogen
mAb	monoclonal antibody
MMS	methyl methane sulfonate
MNNG	1-methyl-3-nitro-1-nitrosoguanidine
4NQO	4-nitroquinoline *N*-oxide
nt	nucleotides
pAb	polyclonal antibody

PAP	potato acid phosphatase
PATAg	periodic acid–thiocarbohydrazide–silver proteinate
PCI	phenol/chloroform/isoamyl alcohol
PCR	polymerase chain reaction
PMSF	phenylmethylsulfonyl fluoride
PTA-ELISA	plate-trapped-antigen-enzyme-linked immunosorbent assay
RAPD	randomly amplified polymorphic DNA
RecA-AC	RecA-assisted Achilles' cleavage
REMI	restriction enzyme-mediated integration
RFLP	restriction fragment length polymorphism
RNase	ribonuclease
RT	room temperature
sc	subcutaneously
SCAR	sequence characterized amplified region
SDS	sodium dodecyl sulfate
SDS–PAGE	SDS polyacrylamide gel electrophoresis
SEM	scanning electron microscopy
SSCP	single-strand conformation polymorphism
SSR	simple sequence repeat
STS	sequence-tagged site
TCH	thiocarbohydrazide
TEM	transmission electron microscopy
TGGE	temperature gradient gel electrophoresis
VNTR	variable number of tandem repeat
X-gal	5-bromo-4-chloro-3-indolyl-β-D-galactoside
YAC	yeast artificial chromosome

Chapter 1
Identification and genotypic characterization of filamentous fungi

Michele D. Piercey-Normore
Department of Botany, University of Manitoba, Winnipeg, MB R3T 2N2, Canada.

Keith N. Egger
College of Science & Management, University of Northern British Columbia, Prince George, BC V2N 4Z9, Canada.

1 Introduction

In the past decade a wide array of techniques, and an often bewildering number of acronyms, have been developed for detecting genetic variation in a variety of organisms, including fungi. It is not within the scope of this chapter to describe all of these techniques. Rather we will attempt to provide the reader with an understanding of the main classes of techniques, strategies for determining which are appropriate to answer the biological question being posed, and point out sources of information to assist in implementing the protocols.

2 A synopsis of genome variation

Despite the high fidelity of the process of DNA replication, the fungal genome is dynamic, with changes constantly occurring. Nucleotide substitutions, insertion and deletion events, rearrangements of gene regions or entire chromosomes, transposition events, movements of genes from organelles to the nucleus, and even acquisitions of genetic elements from other organisms, may take place. Most of these changes are exceedingly rare, but are detectable over the periods of time appropriate to evolutionary analysis. Thus, the fungal genome presents a myriad of DNA regions and features that can be analysed, and within the genome of any fungal species, there are likely to be regions that exhibit the level of resolution appropriate for the question under consideration.

A distinction can be made between coding and non-coding regions of the genome. Coding regions include sequences that encode proteins and structural

RNAs. Non-coding regions include spacer elements between genes, introns, and sequences with no known function. Non-coding regions may or may not be transcribed into RNA. In general, coding regions are more highly conserved (i.e. less variable) than non-coding regions. However, even non-coding sequences may be constrained by the need for maintenance of function, for example gene regulatory regions, or transcribed spacer regions whose ability to form secondary structures is necessary to enable proper RNA processing. Introns are non-coding sequences that are interspersed within coding regions (exons) in most eukaryotic genes. Intron sequences are more variable than the exon sequences that surround them, but their position within homologous genes is often highly conserved. As a result, conserved exon sequences can be used to anchor primers to amplify highly variable introns.

Copy number of sequences in the genome varies. Some sequences are present in relatively low copy number while others are repeated many times in the genome. Most protein-coding genes are present in low copy number, but this is at least partly related to gene function. Proteins or structural RNAs that are needed for high demand metabolic functions or structural elements in the cell tend to occur in high copy number to maximize transcription rates. However, copy number is also influenced by the process which gave rise to the genetic element. Some genetic elements are more susceptible to duplication processes, for example sequences derived from transposable elements may have populated the genome via a process of gene duplication and transposition.

While single copy genes occur, it is not uncommon for genes to occur in families. A gene family may contain multiple copies of identical or nearly identical genes, but it is not unusual to detect divergent gene families that contain similar but non-identical copies. In some cases non-identical copies produce functional variants of the gene product, in other cases they are non-functional gene copies called pseudogenes.

Many multicopy genes are arranged as tandem repeats, which are multiple copies of the same gene arranged head to tail. Tandemly repeated gene copies are usually more similar to each other than would be expected if random mutations were occurring in the individual copies. This tendency towards homogenization of tandem repeats has been called 'concerted evolution' (see ref. 1). Concerted evolution is not well understood (2), but is hypothesized to be due to a combination of several different genetic processes (sometimes referred to as 'molecular drive' *sensu* (3)). This tendency for tandemly repeated sequences to be homogenized makes tandem repeats useful for studies of species, as many of the mechanisms postulated to be responsible for concerted evolution operate most effectively when genetic exchange is occurring.

An example of a high copy number tandemly repeated sequence is the gene repeat that encodes the structural RNAs that are part of the ribosomes. The DNA region that encodes the nuclear ribosomal RNA gene repeat (rDNA) is a popular target for genetic analysis. Each fungal ribosomal rDNA repeat contains at least three genes: the small subunit (SSU or 16S–18S) gene, the 5.8S gene, and the large subunit (LSU or 26S–28S) gene. Between the SSU and 5.8S gene, and the

5.8S and LSU gene, are internal transcribed spacer regions ITS1 and ITS2, respectively. This rDNA gene repeat is transcribed as a single unit, then processed to splice out the spacer regions. Between each transcript is an intergenic spacer (IGS) region that in some basidiomycetes may contain a fourth gene, the 5S RNA, which is transcribed separately. Within a single repeat there are regions that have different rates of evolutionary change, and hence may be used to examine evolutionary histories of groups of taxa that differ in age since divergence. The regions encoding the ribosomal RNA genes (SSU, 5.8S, LSU) are most highly conserved, although there are moderately variable regions, called divergent domains, dispersed throughout the gene regions. The divergent domains are more variable by virtue of the fact that they encode optional structural domains in the ribosomal RNAs that have little impact upon ribosome function. The ITS regions are moderately variable, while the IGS region contains the most highly variable regions in the rDNA. Zimmerman and Dahlberg (4) provide an extensive review of nuclear rDNA.

Originally, DNA was purified by ultracentrifugation in high density gradients (e.g. caesium chloride). In such a gradient, the nuclear DNA generally forms as a single, broad band. However, minor bands representing some fraction of total cellular DNA are also found at different positions within the gradient. These bands represent DNA sequences having similar densities, and became known as 'satellite' DNA. For example, the ribosomal DNA can form a separate band originally referred to as alpha satellites. The name satellite DNA has been retained to refer to similar sequences that occur in high copy number in the genome. The term 'minisatellite' was coined by Jeffreys *et al.* (5) to refer to sequences that occur as tandem arrays containing a core sequence of 15–30 bp. These repeats are mostly located near telomeres of chromosomes and tend to be maintained by recombination processes (6). The term 'microsatellite' has been applied to tandem arrays of simple sequence repeats (SSR) of 2–10 bp in length (7, 8). Microsatellite loci are randomly distributed throughout the genome and are maintained by processes that include replication slippage, rolling circle amplification, and unequal exchange (9). Loci where minisatellites or microsatellites occur are susceptible to variation in the number of tandem repeats of the core sequence, and are referred to as variable number of tandem repeat (VNTR) loci.

Some high copy number sequences are not tandemly repeated, but are dispersed throughout the genome. Many dispersed repeats are derived from transposable elements (9) which have undergone repeated transposition events. Over time, copies mutate and lose their transposition function; ultimately they are lost from the genome via chromosomal rearrangements.

Finally, not all DNAs that form satellites in high density gradients are sequence repeats; some represent extra-nuclear sequences. Mitochondria occur in all fungi, and contain their own small complement of DNA sequences, including genes that encode mitochondrial versions of ribosomal RNAs and proteins involved in electron transport. Mitochondrial genes tend to evolve at faster rates than their nuclear counterparts, which makes them desirable for some types of studies. While mitochondrial sequences in animals have the highly desirable property of being uniparentally inherited (i.e. clonally and without recombination), there is evidence of both biparental inheritance and recombination in mitochondrial genomes of fungi (10).

All of the sequences and genetic elements discussed above can potentially be subjected to molecular analysis. In the next section we will discuss how to decide which gene regions to analyse and which techniques are appropriate for their study.

3 Determining your objectives

Molecular studies will have at least one of two primary objectives: establishing identity, or determining relationship. Establishment of identity asks whether individuals represented by different genetic samples are the same. If samples differ, then this falsifies the hypothesis, but leads to the often more difficult question of whether the differences observed are sufficient for the samples to represent different genetic or taxonomic units. Determining relationship presupposes that there are differences, and attempts to interpret the pattern of variation among individual samples to infer phylogenetic relationships.

There are three approaches one may take to resolving a question of identity or relationship. You may choose to analyse a defined region of DNA at a single locus, such as a ribosomal RNA gene or protein-coding gene. Reasons for this choice may include prior knowledge of gene variation in the taxa under investigation, or availability of the primers for that gene. In this way the researcher can use pre-existing knowledge of genetic variation to select the gene region with the appropriate level of resolution required for the study. Alternatively, the researcher may choose DNA scanning techniques, which arbitrarily target DNA regions from multiple loci distributed throughout the genome. Since targeting is arbitrary, the markers produced will range from non-variable to highly variable, thus permitting the researcher to select the characters appropriate to the resolution required. A third approach, with a more functional aspect, involves examination of gene expression using messenger RNA. This approach could be valuable for examining ecological or phylogenetic differences which are based on gene function.

4 Techniques for genome analysis

4.1 PCR amplification

As a result of the advent of heat stable DNA polymerases that can tolerate multiple heating and cooling cycles (11), the polymerase chain reaction (PCR) has become the standard tool for molecular analysis. An excellent introduction to the polymerase chain reaction, including protocols, troubleshooting tips, and guidelines for primer design can be found in Palumbi (12). Although several factors such as reagents and reaction conditions affect the specificity of PCR, the primary control of specificity of amplification of a PCR product is the primers. In this section, we will discuss aspects of primer selection and design.

4.1.1 Primer specificity

All primers, under some reaction conditions, will amplify a PCR product or products. What controls the specificity of the amplification is the reaction conditions (temperature of the cycling steps, number of cycles, reaction components, template concentration and purity, characteristics of the thermal cycler). Denaturation is an important first step in amplification, since the PCR can only be initiated from a single-strand template. A typical denaturation temperature is 94°C, although some protocols recommend temperatures as low as 92°C. Keep in mind, however, that if the temperature is too low or of too short a duration, PCR failure can result due to incomplete denaturation of the template DNA. The annealing temperature varies with the primer, and is dependent upon primer length and base composition. The T_m is the theoretical temperature at which half of the potential primer sites are bound by the primer, and can be calculated using commonly available computer programs (see *Table 1*). However, a simple rule-of-thumb is that T_m is four times the number of G and C nucleotides plus two times the number of A and T nucleotides (G/C pairs require a higher temperature to denature because they form three hydrogen bonds as opposed to two for A/T pairs). At temperatures higher than T_m fewer primers will be bound to the template, and at temperatures lower than T_m more primers will bind to mismatched sites, resulting in amplification of non-target products.

4.1.2 Primer design

Many primers have been described for amplification of specific regions and specific genes. Some of these are 'universal' primers, designed to match highly conserved DNA regions, and hence amplify from a wide variety of organisms. Others are highly specific, designed to amplify a gene or gene region from a single species, or even a single variant within a species. If the primer is designed for specificity it should produce an amplification product only in the taxa for which it was targeted. If a primer is designed to be universal it will produce an amplification product in many distantly related taxa. In many cases universal primers have been used to amplify across a variable region from which taxon-specific primers are designed.

Designing for universality presents a different strategy than designing for specificity, but the same basic principles of primer design apply. Primers which are designed to amplify non-specifically are typically short (e.g. 10 nucleotides in length for RAPD), while primers that are designed to amplify specific products are typically 18–24 nucleotides (nt). Primers may be longer, but as length increases so does the proportion of incorrect primers that result from errors incorporated during synthesis (12). In the case of longer primers it may be necessary to purify the primers to ensure optimal specificity.

Primers can be any sequence that matches the template of interest, but certain features may impair amplification. Internal repeats may result in formation of primer dimers, amplification artefacts that result when the primer anneals directly to itself or to a complementary primer. As primer dimers are short, they

```
Protein                           SerProGlyArgCysGluGluAspIleSerThr↔LysArgSerLeuTrpGluHisProAlaValGluAspTyrLeu
Sequence #1                 5'-AGCCCTGGAGATGTGAAGAAGATATTTCAACA↔AAAAGGAGTTTGTGGGAGCACCCCGCCGTAGAAGATTATCTT-3'
Sequence #2                 5'-TCCCCAGGTAGGTGCGAGGAAGACATAAGCACC↔AAGCGCTCTTTATGGGAGCACCCTGCGGTGGAAGACTACTTG-3'
Sequence #3                 5'-TCTCCCGGACGGTGCGAGGAGGATATCTCGACG↔AAGAGATCCCTGTGGGAACATCCGGCAGTTGAGGACTACCTG-3'
Consensus                   5'-WSYCCHGGDMGRTGYGARGARGAYATHWSVACV↔AARMGVWSYYTRTGGGARCAYCCVGCVGTDGARGAYTAYYTK-3'
Primer specific for Seq #2      5'-GGTGCGAGGAAGACATAAGC-3'↔  3'-GAGAAATACCCTCGTGGGAC-5'
Complete Consensus          5'-WSNCCNGGNMGNTGYGARGARGAYATHWSNACN↔AARMGNWSNYTNTGGGARCAYCCNGCNGTNGARGAYTAYYTN-3'
Universal Primer               5'-GGTAGGTGTGAGGAGGATAT-3'      ↔        3'-ACCCTTGTGGGTCGTCATCT-5'
                                          1     2  3                                4     5
```

Figure 1 DNA sequences are written 5′ to 3′ and extension takes place from the 3′ end. Only the template strand is shown, not the complementary strand (but note that the reverse primers will correspond to the sequence of the complementary strand). The consensus sequence is shown for the known sample (three) sequences. The complete consensus indicates all possible nucleotides at each position for each amino acid according to the universal genetic code: N = any nucleotide; R = purine (A or G); Y = pyrimidine (C or T); B = not A (i.e. C, G, or T); D = not C (i.e. A, G, or T); H = not G (i.e. A, C, or T); V = not T (i.e. A, C, or G). Shown are two theoretical primers, one that should specifically amplify sequence No. 2, the other a universal primer that should amplify all three known sequences plus other potential sequences with the same amino acid composition. Shaded positions are conserved between the sequences and the primers. Numbered positions with the bases in bold indicate different decisions that were made depending upon whether the primer is designed to be specific or universal. 1, 3: Since the complementary strand can have either A or G in this position, a T:G mismatch between the primer and the template will only slightly affect the yield compared to an C:A mismatch. 2: Since the complementary strand can have C or T in this position, when designing for universality a G:T mismatch is preferable to an A:C mismatch. 4: Since the template strand can have A or G in this position, a T:G mismatch in the universal primer is preferable to a C:A mismatch. 5: Since the template strand can have any base in this position, T:T, T:G, and T:C mismatches are preferable for the universal primer.

amplify very efficiently and reduce the yield of the desired template. Complementary internal sequences may cause the primer to form secondary structures, called hairpin loops, that may also impair the efficiency of the primer. Finally, primers that are predominantly A/T or G/C will operate at extreme annealing temperature optima, thus making it difficult to optimize the annealing temperature for these primer pair combinations. Primers that contain a balanced mixture of the four nucleotides, without internal repeats or complementary regions, tend to function most effectively.

The match of the primer to the template controls the specificity of the primer. The 3′ end of the primer is most critical for proper amplification, as this region must be tightly bound before the DNA polymerase will initiate replication. Mismatches within five nucleotides of the 3′ end have significant effects on primer annealing and often result in failure of the PCR. Mismatches upstream of the 3′ end have a progressively reduced effect on amplification, particularly if they are surrounded by matched bases. However, certain mismatches are preferable to others in this regard. A:G and C:C mismatches reduce yield of PCR product the greatest (about 100-fold), A:A mismatches have less impact (about 20-fold reduction in yield), and T:G, T:T, and T:C pairs have the least impact upon efficiency of annealing (12). Therefore the choice of nucleotide in potentially variable positions can have implications when designing primers for specificity or universality (see *Figure 1*).

Designing primers for non-protein-coding regions (e.g. ribosomal RNA genes) (13) is relatively straightforward. Sequences are aligned across a range of taxa, and nucleotides are matched (in the case of universal primers) or mismatched (in the case of specific primers) to give the desired mix of universality versus specificity. However, designing primers for protein-coding genes offers unique challenges. Nucleotides corresponding to amino acid codons in protein-coding genes vary considerably depending upon position in the codon: the first and second nucleotide positions being mostly conserved for a particular amino acid, with the third position being more variable. In selecting a region to design universal primers for a protein-coding gene, it is useful to select a stretch of seven to nine amino acids that has minimum redundancy in the genetic code. By selecting low redundancy codons (i.e. amino acids encoded by only one or two codons) near the 3′ end of the primer, it is possible to minimize mismatches in this critical region and match at least the last five nucleotides at the 3′ end (see *Figure 1*).

A type of universal primer, the degenerate primer, can be useful with variable template DNA sequences. The degenerate primer is actually a mixture of primers with most nucleotides identical, but varying at some positions. At degenerate sites any combination of two, three, or four different bases may be present at a particular position on a subset of the primers in the mixture. This allows an exact match between at least a subset of primers and the template DNA, permitting amplification of a PCR product. However, depending on the degree of degeneracy, the concentration of the optimal primer may be low, potentially reducing the amount of amplification product. Most primer synthesis facilities can incorporate degenerate positions.

4.2 Analysing variation at a single locus

Within a defined DNA region, variation can be detected based upon differences in electrophoretic mobility of fragments, by sampling nucleotide substitutions within one fragment relative to another (e.g. restriction endonuclease digestion), or by determining base composition of the entire region. Variation in fragment size can be quantified on either agarose or acrylamide gels depending on the magnitude of the differences in sizes. For example, agarose gels can reliably detect length differences of 5% or less depending on type and concentration of agarose used in the gel. Polyacrylamide gels can detect fragments that differ in length by less than 1%, although preparing and analysing acrylamide gels requires more elaborate protocols than using agarose. DNA sequencing reveals the maximum information from a defined fragment, and is becoming less labour-intensive and expensive as automated DNA sequencers proliferate.

4.2.1 RFLP and related techniques

Restriction fragment length polymorphism (RFLP) analysis is based upon the action of restriction endonucleases, enzymes that cleave dsDNA at specific target sequences (usually ranging from 4–6 nt in length). Restriction endonucleases are derived from enzymes in prokaryotic organisms that protect the cell from foreign DNAs.

Traditionally, RFLP analysis was carried out using a restriction endonuclease to cleave total cellular DNA, the fragments were separated by electrophoresis on agarose or acrylamide gel, transferred to nitrocellulose or nylon membrane, and hybridized to a DNA probe, a process called Southern blotting (14). DNA probes can be a cloned fragment from the taxon of interest or a related taxon (e.g. the rDNA of *Neurospora crassa* was used as a probe to study rDNA variation in the genus *Wilcoxina*) (15), or even a repetitive element (e.g. minisatellite probe), such as those used for DNA fingerprinting. Hybridization probes need not be from a known region, anonymous cloned or amplified DNA sequences may also be used as probes to detect variation.

A more recent variation, called PCR-RFLP, involves the use of PCR primers to amplify a DNA fragment, followed by digestion of the fragment with restriction endonucleases, electrophoresis, and visualization by staining with ethidium bromide. This is much easier to perform than Southern blotting, and has become the method of choice for rapid RFLP analysis.

4.2.2 DGGE, TGGE, and SSCP

Techniques such as denaturing gradient gel electrophoresis (DGGE), temperature gradient gel electrophoresis (TGGE), and single-strand conformation polymorphism (SSCP) analysis are also used to differentiate DNA fragments (16). The advantage of these techniques is that they not only separate DNA fragments that differ in length, but also fragments of the same length that differ by one or more base substitutions. DGGE and TGGE depend upon differences in denaturation characteristics as fragments move through a gel with an increasing denaturing gradient (a chemical denaturing agent in DGGE; a temperature gradient in TGGE). As a dsDNA fragment migrates through the denaturing gradient, it reaches a point where it starts to denature, which alters the electrophoretic mobility of the fragment. Since fragments that differ in sequence denature at different points in the gradient, they can be distinguished. SSCP is similar, but detects differences in electrophoretic mobility based upon differences in secondary structure.

4.2.3 Exon primed intron-crossing (EPIC)

The highly conserved position of many eukaryotic introns has allowed the development of PCR primers to conserved DNA sequences in the exons flanking these introns. These primers can be used to amplify across the highly variable intron region. Variation can be screened by RFLP analysis, or by DNA sequencing. This approach has been applied extensively in animals (12), and similar approaches are applicable to fungi.

4.2.4 DNA sequencing

The maximum degree of polymorphism can be detected by sequencing appropriate regions of DNA and identifying which of the four possible nucleotides occurs at each position. DNA can be sequenced either directly from PCR product or from cloned fragments. Sequencing directly from PCR product is straightforward. Cloning DNA fragments is often done if sequence polymorphism is suspected. Techniques for sequencing and cloning are described in detail in Hillis *et al.* (17). In the past, sequencing has been expensive and labour-intensive, requiring the use of radioactively-labelled primers. However, recent advances in sequencing technology has led to less expensive and more efficient methods using fluorescently-labelled dideoxy terminators or primers, and automated equipment that reads the sequence directly into computer software.

DNA sequencing is often used for evaluating relatedness among taxa in phylogenetic analysis. However, it is not only the accuracy of the sequence, but also the alignment of the sequences, that ultimately determines the outcome of the analysis (18, 19). Programs are available for aligning DNA sequences (see *Table 1*), but if the sequences have a large amount of variation there may be many possible alignments producing many, and possibly conflicting, tree topologies. Therefore, choice of alignment algorithm can be critical to the final result. On the other hand, too little variation in sequence will produce numerous unresolved trees, indicating that the region chosen is not sufficiently informative for the taxa under investigation. An in depth review of phylogenetic analysis techniques for DNA sequences is beyond the scope of this chapter. The reader is referred to the excellent book by Hillis *et al.* (20).

4.3 Analysing variation at multiple loci

Genome scanning techniques, which do not require previous knowledge of the particular genome, use arbitrary oligonucleotide primers to detect polymorphic DNA regions at multiple loci. These techniques generally involve anonymous regions, in that the researcher does not know the location or function of the region being amplified/probed before or after the study.

4.3.1 RAPD and related techniques

Randomly amplified polymorphic DNA (RAPD) is a PCR technique which uses single, 10 nt primers to amplify fragments of unknown location and function throughout the genome (21). Arbitrarily primed PCR (AP-PCR) (22) is similar, but uses longer primers with the first few PCR cycles at a low annealing temperature to ensure a high initial degree of non-specific binding. A major problem with RAPD and AP-PCR is that the markers produced behave as dominant characters (i.e. heterozygotes can not be distinguished from one of the homozygotes). This can create problems for estimating gene frequencies.

A variation on RAPD uses two different 10 nt primers, rather than a single primer, which allows the RAPD primers to be used to sequence the fragment in either direction. This protocols is called sequencing with arbitrary primer pairs (SWAPP) (23) and has been used to infer relationships among North American biological species of *Armillaria* (24). Alternatively, longer (i.e. more stable) primers can be designed to amplify and sequence the RAPD fragment specifically. This protocol is called sequence characterized amplified regions (SCARs) or sequence-tagged sites (STSs) (25). SCARs were originally developed from cloned RAPD fragments. Standard length primers (18–24 nt) were designed to be complementary to the 5′ and 3′ ends of the RAPD sequence, and were subsequently used to specifically amplify the RAPD fragment as a single locus. Longer primers allow more stringent PCR conditions to be used with SCAR, which make them more reproducible than RAPD markers. Since two specific primers are used in SCAR they tend to be co-dominant, as opposed to the predominantly dominant markers characteristics of RAPD, and therefore more informative for genetic mapping (26).

4.3.2 DAF

DNA amplification fingerprinting (DAF) (27) uses short arbitrary primers five to six nucleotides in length to produce reproducible banding patterns on agarose or acrylamide gels. DAF does not rely on previous knowledge of genome variation, and is a simple, fast, and sensitive method that may be used in a wide variety of organisms. For example, although soybean has low molecular diversity, DAF was able to detect polymorphisms among highly homozygous cultivars sharing a common ancestry (27). DAF produces a more complex fingerprint profile than some genome scanning techniques, such as RAPD, so polyacrylamide gel electrophoresis and silver staining is required, since agarose gel and ethidium bromide staining will detect only the major fragments. Simple DAF patterns may be suitable for genetic mapping, whereas more complex patterns may be used for genotyping individuals.

4.3.3 AFLP

Amplification fragment length polymorphism (AFLP) (28) is a technique that uses restriction digestion in combination with PCR to produce polymorphic fragments. DNA is first digested with a restriction endonuclease, then fragments are ligated to an oligonucleotide adapter that matches the restriction site. Once ligated, the oligonucleotide adapter sequence plus an arbitrary 3′ end is used to amplify a subset of the fragments by PCR, followed by electrophoresis on a polyacrylamide gel to separate the fragments. This technique produces numerous polymorphic bands, but these markers may be susceptible to clustering from a small region of the genome (29).

4.3.4 VNTR loci

The traditional method of DNA fingerprinting of variable number tandem repeat (VNTR) loci is by hybridization of a minisatellite core sequence to Southern blots of restriction endonuclease digested DNA. This is a powerful tool to detect variation at multiple loci. Protocols for DNA fingerprinting of minisatellite sequences are described in Wells (30). VNTR loci can also be arbitrarily sampled by incorporating sequence repeats into PCR primers. Both minisatellite and microsatellite core sequences (e.g. $[CAC]_5$ or $[GACA]_4$) have been incorporated into PCR primers to enrich amplification products for VNTR loci (26, 31–33).

Although VNTR loci are dispersed throughout the genome, individual VNTR loci containing microsatellite SSR sequences can be detected using primers designed to amplify conserved sequences flanking the VNTR locus. However, unless primers are available for a closely related organism, considerable initial effort is usually required to characterize the sequences surrounding the VNTR locus sufficiently to design primers. Once primers are available, only a few loci need to be sampled to reveal considerable genetic variation, in part because of the relatively high rate of mutation in number of tandem repeats of the SSR sequence, and because the markers are co-dominant (i.e. homozygotes and heterozygotes can be distinguished).

Many of the above techniques can be combined to enhance complexity of marker patterns. For example, He *et al.* (34) combined RAPDs with DGGE to detect high levels of reproducible polymorphisms in wheat. McClelland *et al.* (35) used SSCP combined with AP-PCR to increase detectable polymorphisms in PCR fingerprinting of the mouse genome. A combination of AFLPs and cDNA synthesis was introduced by Bachem *et al.* (36), and applied to parasitic nematodes (37) to isolate differentially expressed genes when the nematode became parasitic. DeSalle and Schierwater (38) provide a comprehensive overview of many combined techniques in current use.

4.4 Analysis of environmental samples

All of the above techniques can be applied when amplifying DNA from pure cultures or samples. However, sometimes mixed PCR samples are obtained, either inadvertently, such as when amplifying mycorrhizal fungi from individual root tips (39), or intentionally, such as when trying to characterize the community from a complex environmental sample. In cases where two or more genets or species have been amplified together, one of two approaches can be taken: separate the co-amplified fragments and analyse them individually, or analyse variation in the population of amplified products directly.

Analysis of multiple PCR products creates challenges in selection of PCR primers. The ability of 'universal' primers to amplify from a wide range of organisms can be a liability when amplifying from environmental samples, such as soil or plant tissues. To avoid amplification of non-fungal organisms (e.g. protists, algae, bacteria) it is necessary to use primers that have been designed to selectively amplify groups of fungi. Gardes and Bruns (40) have developed primers for the ribosomal RNA genes of fungi (ITS1F) and for basidiomycetes (ITS4B). Egger (41) designed two primers that preferentially amplify fungi: NL6Amun was designed based upon consensus sequences for Pezizales, and NL6Bmun from consensus sequences of several basidiomycete groups. However, there are only two nucleotide substitution differences between the two primers, so at annealing temperatures normally used for amplification they show little specificity and will amplify a wide range of fungi.

If the intent is to analysis a population of PCR products, then techniques that produce multiple fragments from each individual should be avoided, as it quickly becomes difficult or impossible to assign fragments to specific genetic variants. Techniques that produce multiple fragments include genome scanning techniques and restriction endonuclease digestion techniques (although a variation is discussed below). It is also not possible to perform DNA sequencing on a mixture of PCR products. If these forms of analysis are desired, then the products must first be separated. This can be accomplished electrophoretically if the number of fragments is small and there are differences in size between samples. In these cases individual bands may be excised from the gel matrix and analysed, although this may not be successful if re-amplification is required. Alternatively, the population of fragments can be cloned into a vector, typically a plasmid, and individual clones sampled to assess the variation within the population of heterogeneous DNA sequences.

Separate analysis of multiple PCR products from each sample can make data collection tedious and time-consuming. Therefore, several approaches have been developed to characterize mixtures of PCR products directly. DGGE, TGGE, and related techniques are appropriate techniques for analysis of mixed PCR products because they detect variation without producing multiple fragments from the PCR products. DGGE analysis of mixtures of PCR products derived from environmental samples is becoming a popular approach in microbiological studies, and has been applied to fungi (42).

Techniques that produce multiple fragments from a single PCR product, such as restriction endonuclease digestion, can produce a 'fingerprint' for a community that can be compared across communities or samples, but it is difficult to determine the number of species in the sample since the number of restriction sites in any PCR product will vary from taxon to taxon, and similar size fragments would be expected to occur in closely-related taxa. A modification to reduce the complexity in these complicated patterns involves the use of automated fragment analysers (many DNA sequencers can also serve as fragment analysers). By tagging one or the other PCR primers with a fluorescent dye (or both primers with a different dye in the case of fragment analysers that can detect multiple fluorescent labels), it is possible to apply restriction endonucleases to the mixture of amplified products and detect only those fragments that extend from the primer to the first restriction site downstream (if the 5′ primer is labelled) or upstream (if the 3′ primer is labelled). Using this approach, called terminal restriction fragment length polymorphism (T-RFLP) (43), one can easily assess differences in mixed populations of PCR products, making direct comparison of population or community samples possible. This approach can also be used to assay variation in specific genes of interest in an environmental sample by using primers specific to conserved regions of the functional gene of interest.

4.5 Analysing expressed genes

By targeting primers to functional genes of interest, it may be possible to correlate diversity with gene function. However, targeting the DNA allows the possibility of amplifying unexpressed gene copies, or even non-functional pseudogenes. Therefore, techniques have been developed to study only expressed genes.

RNA fingerprinting is based on the same concept as genome scanning techniques for DNA, only it is the expressed gene, the messenger RNA (mRNA) transcript, that is targeted. The majority of eukaryotic mRNAs have a string of A nucleotides, called a poly(A) tail, that is added during mRNA processing. This allows the mRNA to be reverse transcribed using a primer comprised of a poly(T) sequence at the 5′ end with only one or two degenerate bases at the 3′ end where extension begins. This primer will anneal to the poly(A) tail in mRNA. After extension the resulting DNA strand will be complementary to the RNA transcript and can be used as a template to produce a complementary DNA (cDNA). Subsequent reverse transcriptase PCR (RT-PCR) may used to generate PCR fragments of the RNA transcript.

Alternatively, an approach using RNA arbitrarily primed PCR (RAP-PCR) uses short (10–18 bp) arbitrary primers to generate first and second strand cDNA synthesis using reverse transcriptase under low stringency conditions (44). A single, short arbitrary primer, similar to that used in DNA fingerprinting, can be used to prime the RNA transcript in place of a poly(T) sequence. After the initial complementary strand is synthesized, the second strand cDNA is produced, and subsequent high stringency PCR conditions are used to generate a PCR product that can be subjected to electrophoresis and visualized on an agarose gel. This amplified fragment is located internally on the RNA template, and includes open reading frames. Once probes such as these have been isolated they have the potential to be used as markers in population, parentage, or phylogenetic studies or as probes to isolate full-length CDNAs.

5 Assumptions

Several assumptions are made with the above techniques. First, it is assumed that markers are heritable and that the mode of inheritance is known. Markers that result from competitive interactions between primer annealing sites in RAPD, or other types of PCR artefacts, for example, may not be heritable. The mode of inheritance is also generally assumed to be Mendelian. While this is the case for nuclear genes, it is not for genes localized in organelles such as mitochondria, which may be inherited uniparentally. However, given that the nuclear genome makes up the vast majority of the sequences in the cell, the assumption of Mendelian inheritance is usually correct.

Even if hybridization or PCR-generated banding patterns are mostly inherited in a Mendelian fashion, the dominant behaviour of some of these markers may be a limitation for population analysis in dikaryotic or diploid fungi. RAPD and AFLP analysis produce dominant markers in which heterozygotes will be indistinguishable from one of the parents, as the recessive allele is characterized by absence of a band. This is disadvantageous in studies where allele frequencies must be estimated since the homozygous dominant condition cannot be distinguished from the heterozygote. Dominant markers are effective to derive population parameters in haploid fungi as gene frequencies can be estimated directly. For dikaryotic or diploid fungi, techniques such as RFLP, amplification of specific microsatellite loci, and SWAPP produce co-dominant markers in which the heterozygote can be readily distinguished.

Reproducibility is another factor to consider when deciding which technique is most suitable for a study. Since competition among primer sites can be a problem when using low stringency PCR conditions, reproducibility is especially important in techniques that use short primers. RAPD markers tends to have poor reproducibility from laboratory to laboratory, or even from thermocycler to thermocycler, since differences can occur due to even slight variation in the PCR conditions. Competition among RAPD fragments is another issue that should be considered. Degree of homology of RAPD fragments varies among different organisms, and RAPD primers can compete with one another, favouring amplification of shorter fragments at the low stringency PCR conditions used in RAPD (45). RFLP by Southern blotting is reproducible if highly stringent hybridization and washing conditions are used, but at low stringency condition non-specific hybridization can occur. All reaction components, reagents, and physical conditions must be carefully controlled to ensure adequate reproducibility.

One of the most fundamental assumptions made about PCR is that the amplified product is the expected product. It is unlikely that specific primers will anneal to and consistently amplify the incorrect fragment in all taxa studied. However, a problem arises if inconsistent annealing to non-specific sequences occurs. If the PCR product being compared are not from the same locus, this violates the assumption that you are comparing homologous characters (i.e. the sequences from which the characters are derived are related by descent in all the taxa under study). For fragment data, the usual assumption that is made is that if fragments have the same electrophoretic mobility, then they are homologous. This can be a dubious assumption, because identical size RAPD markers can result from annealing of RAPD primers to different sites, or restriction sites can be in identical locations but result from different mutations within the recognition sequence of the endonuclease. Equivalency in electrophoretic mobility is a somewhat safer assumption for DGGE and SSCP techniques, as the sequence of the fragment itself affects migration through the gel, therefore matching fragments share more than just length.

It can be problematic if DNA sequences are not related by descent in the taxa being studied. Ideally one is comparing orthologous sequences (i.e. sequences whose ancestry can be traced back to a speciation event). However, it is possible to be comparing paralogous sequences (i.e. sequences whose ancestry can be traced back to a gene duplication event). An example of the occurrence of paralogous genes would be amplification and analysis of a functional gene in one sample, but a non-functional pseudogene from the same gene family in another. These genes could share identical primer sites, but if the gene duplication occurred prior to speciation events, then they will not be tracking species relationships (46). In fact, when using ribosomal RNA genes in phylogenetic analysis one is comparing paralogous sequences, however due to the tendency of tandem repeats to evolve in concert (see previous discussion of concerted evolution), the copies behave as if they were orthologous. Another assumption is that genes are transferred by descent (i.e. from parent to offspring). However, lateral (also called horizontal) gene transfer between species has been observed, and is manifested in gene phylogenies by incongruence between data sets (47). Sequences that are laterally transferred are called xenologous.

Adding to the complexity, homologous genes, whether orthologous, paralogous, or xenologous, may recombine, giving rise to sequences that are a mosaic of sequences with different evolutionary histories. Such recombination can even occur in the PCR reaction mixture, a process called 'PCR jumping' (48), if partially elongated primers denature and reanneal to a different gene homologue in the next cycle! These can be significant problems when trying to determine relationships among taxa.

When amplifying a mixture of PCR products from an environmental sample, the assumption is made that all DNA templates are amplifying with equal efficiency. Although this would be expected if all templates have the identical primer site sequence, mutations in primer sites between different groups will result in different efficiency of amplification for different templates. Therefore, the proportional representation of PCR products within a sample may reflect primer-template matching dynamics rather than differences in proportional representation of the organismal DNAs. Mixtures of templates are also more susceptible to 'PCR jumping' artefacts.

6 Choosing the 'best' strategy

The technique you choose and the region of DNA analysed must be appropriate for resolving differences among your genetic samples. As discussed previously, the genomes of fungi contain regions that range from highly conserved to highly variable. Determining the appropriate technique to target areas with sufficient variation to resolve the problem being studied is a necessary part of designing a molecular study, and therefore depends upon an understanding of the genetic units you wish to study.

6.1 Distinguishing individuals or populations

The most fundamental unit of the fungal thallus is the individual, which must be defined in morphological or genetic terms. An individual genotype may comprise a hypha or patch of mycelium, an asexual structure such as a sclerotium or conidium, or a sexual structure such as an apothecium or basidiocarp. Anderson and Kohn (49) reviewed population studies with soil borne fungi, and provided a definition of a fungal individual based upon the genotype. They defined the fungal individual as a 'fungal genet to include all derivatives of a mitotic cell lineage'. These lineages could originate in either of two ways. In fungi with dikaryotic or diploid mycelia (e.g. basidiomycetes and oomycetes), two genetically unlike gametic nuclei in one thallus create a new genet. In fungi with haploid mycelia (e.g. most ascomycetes and zygomycetes), all vegetative derivatives resulting from the fusion and meiosis of two genetically unlike gametic nuclei creates a new genet. Rayner (50) discussed mycelial individualism and its relevance to fungal–fungal interactions. The lichen thallus, often thought to be an individual, can be heterogeneous in chemistry, and may arise from several propagules or thalli (51). Therefore, 'individual' is not an easily defined category in fungi.

Studies of the distribution of genetic variation among individuals within a species fall into the domain of population genetics, and the molecular technique to be used must be able to detect a high level of variation. Since the level of genetic variation among individuals within a species can be quite small, techniques for differentiating individual genotypes target either highly variable single-locus regions (e.g. EPIC), or rely upon genome scanning techniques to randomly sample the genome for variation. Molecular methods appropriate for detecting population-level variation at multiple loci include RFLPs, RAPDs, AP-PCR, DAF, AFLPs, and VNTR loci. Generally, DAF, RAPD, and AFLP produce more markers than RFLP, and are therefore more useful in studies requiring detection of differences among individuals within populations.

EPIC techniques have been applied to fungi. Aoki and O'Donnell for example (52), analysed nucleotide sequences of both exons and introns of the β-tubulin gene to investigate the phylogenetic relationships within *Fusarium*. This approach may also be useful in studies of lichens, as introns have also been found in lichen-forming fungi, and the presence or absence of the intron can itself be used as a marker for population studies (53).

RAPD patterns can be compared between individuals or species, and are commonly used for population studies (54–57). RAPD markers have been used successfully to distinguish intersterility groups in the *Rhizoctonia solani* complex (58), detect variation among intersterility groups in *Heterobasidion annosum* (59, 60), pathotypes of *Pseudocerosporella herpotrichoides* (61), and races of *Gremmeniella abietina* (62). Genotypes within somatically compatible isolates of *Suillus granulatus* have also been separated using RAPD (57). Although RAPD data are most commonly used for population studies, they have also been used for phylogenetic analysis. RAPD data have been used to estimate genetic distances between taxa (61, 63). Cladistic approaches, using presence or absence of RAPD bands as character data, have been used in bacteria (64) and may be applicable to haploid fungi.

DAF is another technique that is useful to population analysis, and tends to produce more markers than RAPD. For example, RAPD analysis would have required between 30–100 more randomly chosen 10 nt primers to detect the same level of variation as DAF when distinguishing individuals of *Discula destructiva*, a fungal pathogen of dogwood (65).

AFLP analysis is also appropriate for detecting population-level variation. AFLP analysis has been successful in demonstrating a gene-for-gene pathogenicity system in *Leptosphaeria maculans* (66), and to evaluate genetic relatedness among field isolates of the same fungus (67). Julian *et al.* (68) correlated AFLP markers with mating type in *Thanatephorus cucumeris* and used them to study progeny types. Levels of genomic variation were examined using AFLPs among members of the genus *Colletotrichum*, providing evidence to identify species and confirming taxonomic placement of other isolates (69).

Dispersed repeats may also be ideal for population level studies. Dispersed moderately repetitive DNA sequences produce numerous unique fingerprints that can be used to examine linkage analysis and inheritance. Milgroom *et al.* (70) used a nuclear dispersed repeat as a probe to study the fungal pathogen *Cryphonectria parasitica*. The probe produced many unique fingerprints, 33 different fingerprints in 39 isolates probed, with the patterns less complex and easier to score than DNA fingerprint patterns found with minisatellite probes. In another study using a dispersed nuclear repeat probe, Kohli *et al.* (71) examined genetic heterogeneity and long distance dispersal of field populations of *Sclerotinia sclerotiorum*. This dispersed repeat produced a different complex fingerprint profile for 33 of the 36 mycelial compatibility groups tested which allowed them to determine that some clones were distributed over large geographic distances, and that each field population of *S. sclerotiorum* contained more than one clone.

Although minisatellite VNTR loci have shown little variation in fungi (72, 73), they have been used to analyse genetic exchange within and between intersterility groups of *Heterobasidion annosum* (74). Loci containing microsatellites amplified using specific primers produced sufficient polymorphism within the pathogenic fungus *Coccidioides immitis* to be used for diagnostic purposes (75). Meyer *et al.* (31) used short sequence repetitive probes and primers to detect variation in fungi. They found both techniques to be reproducible and useful for differentiating among and within species of fungi.

6.2 Distinguishing closely related species

Species vary enormously in their genetic variability. In part this is due to population genetic factors, such as how the species originated (e.g. founder event versus geographical isolation), mechanisms of genetic exchange, and dispersal mechanisms. However, a major reason that species vary is that they differ in age since divergence. Species that diverged more distantly in the past will, in general, have accumulated more nucleotide substitutions in a given DNA region than species that diverged recently. This must be considered in choosing the region of DNA to study, and the technique to be used. While an ITS spacer sequence from the ribosomal RNA gene cluster may be appropriate for differentiating among individuals of one species, it may be too variable for a group of deeply diverged species, or too homogeneous to separate a group of recently diverged taxa, such as biological species of *Armillaria* (76, 77).

Although no one technique or gene region can be guaranteed to detect variation among all species, some regions have been used extensively to assess species-level differences. One such region is the internal transcribed spacer (ITS) region of the nuclear-encoded ribosomal RNA gene repeat (rDNA). The ITS region has been used extensively in analysis of species differences (see 79). Many studies have utilized the ITS region to examine species level relationships in fungi (reviewed in 78, 79). The large subunit (LSU) of rDNA has also been successful to distinguish among species of *Amanita* (80). For closely related taxa, the intergenic spacer region of the rDNA may be useful. This region contains the most highly variable regions in the rDNA, and has been demonstrated to be useful for separating fungal biological species or strains (76, 77, 81).

Techniques for analysis of these regions range from DNA sequencing to RFLP analysis by Southern blotting. Egger and Fortin (15) applied Southern blotting with a heterologous rDNA probe to differentiate species in the genus *Wilcoxina*. Anderson *et al.* (82) used this approach with three different hybridization probes, including anonymous probes from cloned genomic DNA, to examine polymorphisms in the *Armillaria mellea* complex. Karlsson (83) amplified PCR products from ribosomal nuclear ITS and mitochondrial regions that were used as probes in RFLP analysis in *Heterobasidion annosum*. More recently, Southern blotting has been supplanted by PCR-RFLP studies (e.g. 40, 84–87) or DNA sequencing (80, 88–92).

DGGE and related techniques have also demonstrated their use for distinguishing species. Kowalchuk *et al.* (42) characterized fungi that infect roots of a grass (*Ammophila arenaria*) by using DGGE of specific DNA fragments coding for 18S rDNA. They could resolve differences between species of fungi, but not within a species. SSCP was used to differentiate ectomycorrhizal Russulaceae on two orchids (93), and to screen strains of *Sclerotinia sclerotiorum* for sequence level polymorphisms (94).

6.3 Higher taxa

As the objective turns to comparison of taxa above the species level (e.g. genus, family, order), it is necessary to turn to more highly conserved DNA sequences. It can be difficult to find diagnostic differences between higher taxa, as mutations have been accumulating independently in each lineage within that higher taxon. Therefore, studies of higher taxa are generally attempting to determine relationship rather than identity. Regions that exhibit higher rates of change become less useful over time, since as the time since divergence increases, the probability of multiple changes at any particular nucleotide site increases. This makes it difficult to estimate the total number of state changes at a homologous site, and back mutations may make it appear that no changes have taken place when in fact there have been at least two changes. This accumulation of 'noise' makes determining exact relationships difficult due to the small number of nucleotide choices at each site (i.e. A, C, G, and T). Highly conserved regions are required to examine phylogenetic relationships among higher taxa, such as small and large subunits of rDNA, or exons of protein coding genes. In some examples of studies of relationship among higher taxa, the small subunit gene of the rDNA has been used to infer phylogenetic relationships among ascomycetous fungi (95, 96), basidiomycetes (97, 98), as well as among taxa within the entire fungal kingdom (99, 100).

In cases where there is too much 'noise' or too little variation, it may be profitable to combine data sets from different genes or gene regions. Hibbett *et al.* (101) combined small subunit sequences of nuclear and mitochondrial ribosomal genes to infer a phylogeny of gilled mushrooms and puffballs. In cases where the region being analysed is a protein-coding gene it may be useful to translate the DNA sequence into an amino acid sequence, then compare amino acid substitutions across taxa. For example, Radford (102) used amino acid sequences of orotidine 5′-monophosphate decarboxylase to estimate a phylogeny among fungi, mammals, and bacteria.

7 Databases and WWW resources

Many resources exist on the internet. *Table 1* describes some useful sites for further information.

Table 1 Examples of useful Internet resources for genotypic analysis of fungi

Source	Description
Primer design	
http://research.nwfsc.noaa.gov/protocols/oligoTMcalc.html	Basic calculator for determining primer melting temperatures T_m. Also links to molecular biology protocols.
http://www.basic.nwu.edu/biotools/oligocalc.html	More complex calculator with explanations of mathematical calculations.
http://dmg.nott.ac.uk/molbio/primer.htm	Primer design from protein or DNA sequences.
Primer sources	
http://www.botany.duke.edu/fungi/mycolab/primers.htm	Source of primer sequences for rDNA.
http://plantbio.berkeley.edu/~bruns/primers.html	More primer sequences for rDNA.
Molecular techniques	
http://molbio.info.nih.gov/molbio/desk.html	Many molecular biology links.
http://www.horizonpress.com/gateway	Internet resources for molecular biology, microbiology, and biochemistry. '/protocols.html' is especially informative.
http://www.nwfsc.noaa.gov/protocols.html	Many DNA, RNA, and protein methods available.
Genome databases	
http://rrna.uia.ac.be/	The rRNA web server.
http://www.cme.msu.edu/RDP/html/index.html	Ribosomal Database Project.
http://www.genome.ad.jp/	GenomeNet WWW Server.
http://www.embl-heidelberg.de/	EMBL European Molecular Biology Laboratory.
www.ncbi.nlm.nih.gov	GENBANK genome database and search engines.
Phylogenetic analysis	
http://evolution.genetics.washington.edu/phylip.html	PHYLIP Home page with links to phylogenetic analysis software.
http://herbaria.harvard.edu/treebase/index.html	TreeBASE — A database of phylogenetic knowledge.
General	
http://clue.genome.ad.jp/	Links for Genetic research.
http://www.ebi.ac.uk/biocat/	European Bioinformatics Institute — The Biocatalog.
http://iubio.bio.indiana.edu/	IUBio Archive for Biology data and software.
http://www.keil.ukans.edu/~fungi/	Mycological resources — from basics, to links, to genome sequencing projects.

References

1. Hillis, D. M. and Dixon, M. T. (1991). *Q. Rev. Biol.*, **66**, 411.
2. Elder, J. F. and Turner, B. J. (1995). *Q. Rev. Biol.*, **70**, 297.
3. Dover, G. A. (1982). *Nature*, **299**, 111.
4. Zimmerman, R. A. and Dahlberg, A. E. (1996). *Ribosomal RNA: structure, evolution, processing and function in protein biosynthesis.* CRC Press, Boca Raton.
5. Jeffreys, A. J., Wilson, V., and Thein, S. L. (1985). *Nature*, **314**, 67.
6. Dowling, T. E., Moritz, C., Palmer, J. D., and Riesenberg, L. H. (1996). In *Molecular systematics* (ed. D. M. Hillis, C. Moritz, and B. K. Mable). Sinauer Associates, Sunderland.
7. Tautz, D. and Renz, M. (1984). *Nucleic Acids Res.*, **12**, 4127.
8. Tautz, D. (1989). *Nucleic Acids Res.*, **17**, 6463.
9. Charlesworth, B., Sniegowski, P., and Stephan, W. (1994). *Nature*, **371**, 215.
10. Taylor, J. W. (1986). *Exp. Mycol.*, **10**, 259.
11. Mullis, K. B. and Faloona, F. A. (1987). In *Methods in enzymology* (ed. R. Wu), Vol. 155, p. 335. Academic Press, New York.
12. Palumbi, S. R. (1996). In *Molecular systematics* (ed. D. M. Hillis, C. Moritz, and B. K. Mable). Sinauer Associates, Sunderland.
13. White, T. J., Bruns, T. D., Lee, S. B., and Taylor, J. W. (1990). In *PCR protocols: a guide to methods and applications* (ed. M. A. Innis, D. H. Gelfand, J. J. Sninsky, and T. J. White). Academic Press, New York.
14. Southern, E. M. (1975). *J. Mol. Biol.*, **98**, 503.
15. Egger, K. N. and Fortin, J. A. (1990). *Can. J. Bot.*, **68**, 1482.
16. Lessa, E. P. and Applebaum, G. (1993). *Mol. Ecol.*, **2**, 119.
17. Hillis, D. M., Mable, B. K., Larson, A., Davis, S. K., and Zimmer, E. A. (1996). In *Molecular systematics* (ed. D. M. Hillis, C. Moritz, and B. K. Mable). Sinauer Associates, Sunderland.
18. Mindell, D. P. (1991). In *Phylogenetic analysis of DNA sequences* (ed. M. M. Miyamoto and J. Cracraft). Oxford University Press, Oxford.
19. Wheeler, W. C. (1995). *Syst. Biol.*, **44**, 321.
20. Hillis, D. M., Mortiz, C., and Mable, B. K. (ed.) (1996). *Molecular systematics.* Sinauer Associates, Sunderland.
21. Williams, J. G., Kubelik, A. R., Livak, K. J., Rafalski, J. A., and Tingey, S. V. (1990). *Nucleic Acids Res.*, **18**, 6531.
22. Welsh, J. and McClelland, M. (1990). *Nucleic Acids Res.*, **18**, 7213.
23. Burt, A., Carter, D. A., White, T. J., and Taylor, J. W. (1994). *Mol. Ecol.*, **3**, 523.
24. Piercey-Normore, M. D., Egger, K. N., and Bérubé, J. A. (1998). *Mol. Phylogenet. Evol.*, **10**, 49.
25. Olson, M., Hood, L., Cantor, C., and Botstein, D. (1989). *Science*, **245**, 1434.
26. Weising, K., Nyboom, H., Wolff, K., and Meyer, W. (1995). *DNA fingerprinting in plants and fungi.* CRC Press, Boca Raton.
27. Caetano-Anollés, G., Bassam, B. J., and Gresshoff, P. M. (1991). *Bio/Technology*, **9**, 553.
28. Vos, P., Hogers, R., Bleeker, M., Reijans, M., van de Lee, T., Hornes, M., *et al.* (1995). *Nucleic Acids Res.*, **23**, 4407.
29. Caetano-Anollés, G. (1998). In *Molecular approaches to ecology and evolution* (ed. R. DeSalle and B. Schierwater). Birkhäuser Verlag, Basel.
30. Wells, R. A. (1988). In *Genome analysis: a practical approach* (ed. K. E. Davies). IRL Press, Oxford.
31. Meyer, W., Mitchell, T. G., Freedman, E. Z., and Vilgalys, R. (1993). *J. Clin. Microbiol.*, **31**, 2274.
32. Meyer, W., Lieckfeldt, E., Kuhls, K., Freedman, Z., Börner, T., and Mitchell, T. G. (1993). In *DNA fingerprinting: state of the science* (ed. S. D. J. Pena, R. Chakraborty, J. T. Epplen, and A. J. Jeffreys). Birkhäuser Verlag, Basel.

33. Zietkiewicz, E., Rafalski, A., and Labuda, D. (1994). *Genomics*, **20**, 176.
34. He, S., Ohm, H., and Mackenzie, S. (1992). *Theor. Appl. Genet.*, **84**, 573.
35. McClelland, M., Arensdorf, H., Cheng, R., and Welsh, J. (1994). *Nucleic Acids Res.*, **22**, 1770.
36. Bachem, C. W., van der Hoeven, R. S., de Bruijn, S. M., Vreugdenhil, D., Zabeau, D., and Visser, R. G. (1996). *Plant J.*, **9**, 745.
37. Jones, J. T. and Harrower, B. E. (1998). *Fundam. Appl. Nematol.*, **21**, 81.
38. DeSalle, R. and Schierwater, B. (ed.) (1998). *Molecular approaches to ecology and evolution.* Burkhouser Press, Boston.
39. Brand, F. (1992). In *Mycorrhizas in ecosystems* (ed. D. J. Read, D. H. Lewis, A. H. Fitter, and I. J. Alexander). C. A. B. International, Kew.
40. Gardes, M. and Bruns, T. D. (1993). *Mol. Ecol.*, **2**, 113.
41. Egger, K. N. (1995). *Can. J. Bot.*, **73** (Supplement 1), S1415.
42. Kowalchuk, G. A., Gerards, S., and Woldendorp, J. W. (1997). *Appl. Environ. Microbiol.*, **32**, 3858.
43. Osborn, A. M., Moore, E. R. B., and Timmis, K. N. (2000). *Environ. Microbiol.*, **2**, 39.
44. Welsh, J., Chada, K., Dalal, S., Cheng, R., Ralph, D., and McClelland, M. (1992). *Nucleic Acids Res.*, **20**, 4965.
45. Rieseberg, L. H. (1996). *Mol. Ecol.*, **5**, 99.
46. Moritz, C. and Hillis, D. M. (1996). In *Molecular systematics* (ed. D. M. Hillis, C. Moritz, and B. K. Mable). Sinauer Associates, Sunderland.
47. Syvanen, M. (1994). *Annu. Rev. Genet.*, **28**, 237.
48. Pääbo, S., Higuchi, R., and Wilson, A. C. (1989). *J. Biol. Chem.*, **264**, 9709.
49. Anderson, J. B. and Kohn, L. M. (1995). *Annu. Rev. Phytopathol.*, **33**, 369.
50. Rayner, A. D. M. (1991). *Mycologia*, **83**, 48.
51. Fahselt, D. (1996). In *Lichen biology* (ed. T. H. Nash III). Cambridge University Press, Cambridge.
52. Aoki, T. and O'Donnell, K. (1999). *Mycologia*, **91**, 597.
53. Gargas, A. and DePriest, P. T. (1996). *Mycologia*, **88**, 745.
54. de la Bastide, P. Y., Kropp, B. R., and Piché, Y. (1994). *New Phytol.*, **127**, 547.
55. Goodwin, P. H. and Annis, S. L. (1991). *Appl. Environ. Microbiol.*, **57**, 2482.
56. Hadrys, H., Balick, M., and Schierwater, B. (1992). *Mol. Ecol.*, **1**, 55.
57. Jacobson, K. M., Miller, O. K. Jr., and Turner, B. J. (1993). *Proc. Natl. Acad. Sci. USA*, **90**, 9159.
58. Duncan, S., Barton, J. E., and O'Brien, P. A. (1993). *Mycol. Res.*, **97**, 1075.
59. Fabritius, A.-L. and Karjalainen, R. (1993). *Eur. J. For. Pathol.*, **23**, 193.
60. Garbelotto, M., Bruns, T. D., and Cobb, F. W. (1993). *Can. J. Bot.*, **71**, 565.
61. Nicholson, P. and Rezanoor, H. N. (1994). *Mycol. Res.*, **98**, 13.
62. Hamelin, R. C., Ouellette, G. B., and Bernier, L. (1993). *Appl. Environ. Microbiol.*, **59**, 1752.
63. Yoon, C.-S. and Glawe, D. A. (1993). *Mycologia*, **85**, 369.
64. Smith, J. J., Scott-Craig, J. S., Leadbetter, J. R., Bush, G. L., Roberts, D. L., and Fulbright, D. W. (1994). *Mol. Phylogenet. Evol.*, **3**, 135.
65. Trigiano, R. N., Caetano-Anollés, G., Bassam, B. J., and Windham, M. T. (1995). *Mycologia*, **87**, 490.
66. Pongam, P., Osborn, T. C., and Williams, P. H. (1998). *Phytopathology*, **88**, 1068.
67. Pongam, P., Osborn, T. C., and Williams, P. H. (1999). *Plant Dis.*, **83**, 149.
68. Julian, M. C., Acero, J., Salazar, O., Keijer, J., and Rubio, V. (1999). *J. Biotechnol.*, **67**, 49.
69. O'Neill, N. R., van Berkum, P., Lin, J.-J., Kuo, J., Ude, G. N., Kenworthy, W., *et al.* (1997). *Phytopathology*, **87**, 745.
70. Milgroom, M. G., Lipari, S. E., and Powell, W. A. (1992). *Genetics*, **131**, 297.

71. Kohli, Y., Morrall, R. A., Anderson, J. B., and Kohn, L. M. (1992). *Phytopathology*, **82**, 875.
72. Braithwaite, K. S. and Manners, J. M. (1989). *Curr. Genet.*, **16**, 473.
73. Meyer, W., Koch, A., Neimann, C., Beyermann, B., Epplen, J. T., and Börner, T. (1991). *Curr. Genet.*, **19**, 239.
74. Stenlid, J., Karlsson, J.-O., and Högberg, N. (1994). *Mycol. Res.*, **98**, 57.
75. Fisher, M. C., White, T. J., and Taylor, J. W. (1999). *Mol. Ecol.*, **8**, 1075.
76. Albee, S. R., Muller, G. M., and Kropp, B. R. (1996). *Mycologia*, **88**, 970.
77. Anderson, J. B. and Stasovski, E. (1992). *Mycologia*, **84**, 505.
78. Hibbett, D. S. (1992). *Trans. Mycol. Soc. Jpn.*, **33**, 533.
79. Siefert, K. A., Wingfield, B. D., and Wingfield, M. J. (1995). *Can. J. Bot.*, **73** (Supplement 1), S760.
80. Drehmel, D., Montcalvo, J.-M., and Vilgalys, R. (1999). *Mycologia*, **91**, 610.
81. Harrington, T. C. and Wingfield, B. D. (1995). *Mycologia*, **87**, 280.
82. Anderson, J. B., Petsche, D. M., and Smith, M. L. (1987). *Mycologia*, **79**, 69.
83. Karlsson, J.-O. (1994). *Exp. Mycol.*, **18**, 48.
84. Dahlberg, A., Jonsson, L., and Nylund, J.-E. (1997). *Can. J. Bot.*, **75**, 1323.
85. Erland, S. (1995). *Mycol. Res.*, **99**, 1425.
86. Kårén, O., Högberg, N., Dahlberg, A., and Jonsson, L. (1997). *New Phytol.*, **136**, 313.
87. Pritsch, K., Boyle, H., Munch, J. C., and Buscot, F. (1997). *New Phytol.*, **137**, 357.
88. Egger, K. N. and Sigler, L. (1993). *Mycologia*, **85**, 219.
89. Egger, K. N. (1996). *Can. J. Bot.*, **74**, 773.
90. Goffinet, B. and Bayer, R. J. (1997). *Fungal Genet. Biol.*, **21**, 228.
91. Hambleton, S., Egger, K. N., and Currah, R. S. (1998). *Mycologia*, **90**, 854.
92. Kretzer, A., Li, Y., Szaro, T., and Bruns, T. D. (1996). *Mycologia*, **88**, 776.
93. Taylor, D. L. and Bruns, T. D. (1997). *Proc. Natl. Acad. Sci. USA*, **94**, 4510.
94. Carbone, I., Anderson, J. B., and Kohn, L. M. (1999). *Evolution*, **53**, 11.
95. Gargas, A. and Taylor, J. W. (1995). *Exp. Mycol.*, **19**, 7.
96. Landvik, S., Egger, K. N., and Schumacher, T. (1997). *Nord. J. Bot.*, **17**, 403.
97. Bruns, T. D., Szaro, T. M., Gardes, M., Cullings, K. W., Pan, J. J., Taylor, D. L., *et al.* (1998). *Mol. Ecol.*, **7**, 257.
98. Swann, E. C. and Taylor, J. W. (1995). *Can. J. Bot.*, **73** (Supplement 1), S862.
99. Bruns, T. D., Vilgalys, R., Barns, S. M., Gonzalez, D., Hibbett, D. S., Lane, D. J., *et al.* (1992). *Mol. Phylogenet. Evol.*, **1**, 231.
100. Berbee, M. L. and Taylor, J. W. (1993). *Can. J. Bot.*, **71**, 1114.
101. Hibbett, D. S., Pine, E. M., Langer, E., Langer, G., and Donoghue, M. J. (1997). *Proc. Natl. Acad. Sci. USA*, **94**, 12002.
102. Radford, A. (1993). *J. Mol. Evol.*, **36**, 389.

Chapter 2
Nucleic acid isolation and analysis

Nicholas J. Talbot
University of Exeter, School of Biological Sciences, Washington Singer Laboratories, Perry Road, Exeter EX4 4QG, UK.

1 Introduction

The ability to carry out molecular genetic analysis of filamentous fungi is very heavily dependent on being able to extract and purify large amounts of DNA and RNA from fungal cells. Many procedures have been devised in order to purify nucleic acids from eukaryotic cells and a large number of commercially available kits have been formulated for this purpose. Most procedures devised, however, have been optimized for use in cells of model organisms such as yeast, *Arabidopsis*, *Drosophila*, and human cells. Adapting these procedures for use in filamentous fungi produces highly variable results and some species have proved recalcitrant to molecular analysis largely due to workers being unable to purify high-quality nucleic acid samples.

This chapter provides a small number of protocols that have proved effective for isolation of DNA and RNA from filamentous fungi. It is by no means exhaustive but instead is aimed at the fungal biologist with little, or no experience in isolating nucleic acids. By following the protocols listed, they should be able to produce DNA for use in restriction digests, Southern blot analysis, and for cloning procedures, and RNA for Northern blots, RT-PCR, and preparation of cDNA libraries. Of course the procedures may require some optimization for a given species, but are fairly robust and well-tested in a large number of research laboratories.

2 Isolation of genomic DNA

2.1 Genomic DNA extraction using cetyltrimethyl ammonium bromide (CTAB)

Isolating high molecular weight DNA from filamentous fungi has often proved problematic due to the large amounts of polysaccharide found in many species. Many of the storage polysaccharides and glycoprotein components of the cell wall co-purify with nucleic acids and can reduce the yield of both DNA and RNA, and inhibit restriction enzyme digestion of DNA. A widely used method to overcome these problems is the use of the cationic detergent cetyltrimethyl ammonium bromide (CTAB) and the reducing agent 2-mercaptoethanol. Nucleic acids complex with CTAB and the combination of CTAB and 2-mercaptoethanol protects them from degradation by endonucleases found in the fungal cell. Nucleic acids/CTAB complexes are then precipitated away from denatured protein, polysaccharide, and cell wall debris by extraction with chloroform, and isoamyl alcohol (or pentanol). CTAB is removed after precipitation by washing in 70% ethanol under low salt concentrations, leaving purified nucleic acids. The procedures for nucleic acid extraction using CTAB are well-established (1, 2) and were developed for isolation of DNA from plant species (3). They have since been applied to a number of fungal species and in my laboratory the protocol below (*Protocol 1*) has been used to extract DNA successfully from *Aspergillus nidulans*, *Neurospora crassa*, *Magnaporthe grisea*, *Cladosporium fulvum*, *Pennicillium chrysogenum*, *Stagonospora nodorum*, *Stagonospora pteridiicola*, *Mycosphaerella graminicola*, *Fusarium compactum*, *Cylindrocarpon destructans*, *Heliscus lugdunensis*, and *Rhizoctonia solani*. The procedure also works well for oomycete fungi such as *Phytophthora infestans* and *P. cactorum*. An example of the appearance of DNA produced using *Protocol 1*, digested with restriction enzymes, and fractionated by gel electrophoresis is shown in *Figure 1*.

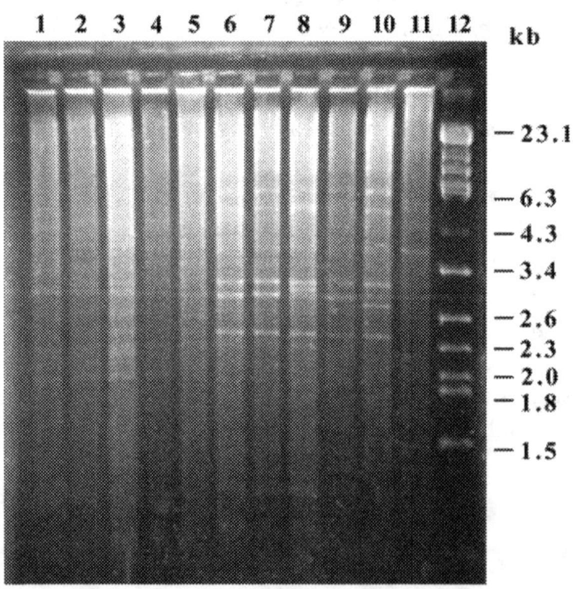

Figure 1 High molecular weight genomic DNA of *Magnaporthe grisea* after digestion with restriction enzymes and fractionation though 0.8% agarose gel. The gel is stained with ethidium bromide. Lanes 1–5 contain genomic DNA from five isolates of *M. grisea* digested with *Bam*HI. Lanes 6–10 contain genomic DNA from five isolates of *M. grisea* digested with *Eco*RI. Lane 11 contains genomic DNA from rice (*Oryza sativa*). Lane 12 contains the size marker, λ DNA digested with *Hin*dIII/*Sac*I.

Protocol 1

CTAB purification of DNA from fungi

Reagents

- 500 ml CTAB buffer: 2% hexadecyltrimethyl ammonium bromide (CTAB), 10 g (Sigma H-5882); 100 mM Tris, 6.06 g; 10 mM EDTA, 1.46 g; 0.7 M NaCl, 20.5 g—make up to 500 ml with ddH_2O and store at room temperature
- CIA: chloroform/isoamyl alcohol (24:1, v/v)

Method

1. Grow a fungal culture in 50–100 ml of rich growth medium for two to five days, depending on species. For best results the mycelium should be young. For most species this will be indicated by it being lightly coloured, prior to starvation when pigment production and autolysis can take place.
2. Remove the mycelium (around 5–10 g wet weight) from the flask either by filtration through Miracloth (Calbiochem), or as a mycelial mat using forceps. Blot the mycelium dry. Samples can be snap-frozen in liquid nitrogen and stored at -70 °C for future use.
3. Take a pestle and mortar and chill it by adding a small amount of liquid nitrogen and allowing it to evaporate. Place in a mortar with 10–20 ml of liquid nitrogen, allow the liquid nitrogen to evaporate, and then quickly grind the sample to a fine powder.
4. Transfer the powdered mycelium to an Oakridge tube (Nalgene) and add 4 ml CTAB lysis buffer which has been pre-heated to 65 °C. Gently disperse the fungal material using a pipette.
5. Immediately add 40 μl 2-mercaptoethanol and incubate for 30 min at 65 °C, inverting the tube every 10 min to ensure adequate mixing.
6. Add 3–4 ml CIA very carefully to the tubes and let them stand uncapped in a fume cupboard for a 1 min (the hot mixture will spill out of a sealed tube).
7. Incubate tubes on a shaking platform for 20 min at 60 r.p.m.
8. Extract by centrifugation at 9650 *g* for 20 min at 4 °C (in a fixed angle rotor).
9. Remove the aqueous (top) phase being careful not to remove denatured proteins and debris at the interface.
10. Add an equal volume of CIA, shake the tube gently, and repeat centrifugation.
11. Remove the aqueous phase and add to 0.54 vol. of isopropanol in a fresh Oakridge tube to precipitate DNA.
12. Leave the tubes on ice for 5 min and then centrifuge at 9650 *g* for 20 min at 4 °C in a swinging bucket rotor.
13. Wash pellet in 10 ml of 70% ethanol and repeat centrifugation.

Protocol 1 continued

14 Dry the pellet on the bench by inverting the tubes on paper towels. Leave them for 30 min to 1 h. Resuspend pellet in 100 μl TE buffer pH 8.0 and store at 4 °C.[a]

[a] Most DNA will digest perfectly with restriction enzymes after this protocol. If this does not occur, then re-extract with phenol/chloroform and re-precipitate using 0.1 vol. of 7.5 M ammonium acetate and 2 vol. of 95% (v/v) ethanol. DNA samples generated with this protocol may appear to be cloudy when resuspended, but this does not interfere with restriction digestion. This can be removed by centrifugation and removal of the clear supernatant if desired.

2.2 Isolation of high molecular weight DNA

Routine extraction of genomic DNA can be carried out using the CTAB procedure for most fungi. The protocol can also be amended to a small scale for quicker mini-preparations and carried out entirely in microcentrifuge tubes. In order to produce very high molecular weight DNA for construction of DNA libraries, however, it is often useful to use a procedure which ensures that fragile high molecular weight DNA does not become sheared and damaged during extraction. A very effective way of achieving this is to purify nuclei and lyse these gently. *Protocol 2* details a method developed for use in *Aspergillus nidulans* (4) and *Magnaporthe grisea* (5). DNA from this procedure can be used to construct genomic DNA libraries in phage vectors or cosmid vectors (cosmid vectors designed with selectable markers for filamentous fungi are available in ref. 6). The DNA is probably large enough to be cloned into BAC vectors, but chromosomal-sized DNA can be effectively isolated using procedures described in Chapter 6 of this volume. Further manipulations of DNA, including restriction digestion, gel electrophoresis, Southern blotting, and cloning procedures are very well described in ref. 7 and in other volumes in this series (8).

Protocol 2

Isolation of nuclear DNA from filamentous fungi

Equipment and reagents

- Dounce homogenizer
- Waring blender—chill blender cups at −70 °C
- Bead beater
- Washed Miracloth: rinse squares of Miracloth with ddH_2O and fold. Boil three times in 10 mM EDTA, boil four times in ddH_2O, rinse with ddH_2O, and squeeze dry. Store at 4 °C in 70% EtOH.
- Acid washed glass beads (appropriate size = 300 μm)
- 10 × SSE (recipe for 500 ml): 10 mM spermine-HCl (Sigma/Calbiochem), 10 mM spermidine-HCl, 1.0 M KCl, 100 mM EDTA, 100 mM Tris-HCl pH 7.0; store at −20 °C
- PMSF: 100 mM phenylmethylsulfonyl fluoride (Sigma), 0.871 g/10 ml 95% EtOH; store at −20 °C

Protocol 2 continued

- 0.5 × SSE (1 litre): 100 ml of 10 × SSE salts, 1 ml of 2-mercaptoethanol, 10 ml of 100 mM PMSF (0.871 g in 50 ml 95% EtOH), 171.2 g sucrose; make fresh
- 4 litres ddH_2O chilled (4 °C)
- TNE: 10 mM Tris–HCl pH 7.6, 15 mM NaCl, 1 mM EDTA
- 20% (w/v) *N*-lauryl sarcosine (sodium salt): 20 g/100 ml H_2O

Method

1 Grow a 50 ml fungal culture by removing a 4 cm square of mycelium from a plate culture and macerating it in a blender with 10 ml of rich growth media (for *Aspergillus nidulans* use 2YEG, ref. 5). Pour into a 250 ml flask and add medium to approx. 50 ml. Grow for three to four days with shaking. Blend the culture again and inoculate into four 1 litre flasks containing 500 ml rich medium. Grow with shaking at 24 °C for four to five days. Monitor growth carefully because young hyphae are always best for extraction of DNA (see *Protocol 1*).

2 Harvest mycelium by filtration through Miracloth and wash extensively with cold ddH_2O. Using two layers of Miracloth, gently squeeze the mycelium to remove any polysaccharide mucilage which surrounds it. Blot the mycelium, still wrapped in Miracloth, with paper towels until it is as dry as you can get it.

3 Place mycelium in a beaker on wet ice. Slowly, add 0.5 × SSE buffer, stirring constantly (with a spatula or spoon) until the cells will absorb no more buffer. The hyphae will resemble a wet paste.

4 Freeze the hyphae by spooning small amounts into a mortar (previously cooled to −70 °C) containing liquid nitrogen. Grind quickly, but gently to make small nuggets of mycelium. Store at −70 °C until used.

5 Pre-cool a stainless steel Waring blender cup to −70 °C and transfer nuggets to the blender. Grind at high speed until the mycelium is powdered.

6 Transfer the powder to a stainless steel bowl and allow to thaw, keeping the temperature at 0–4 °C. Stir constantly, monitoring temperature with a thermometer.

7 Pre-cool the bead beater with glass beads, in a wet ice-bath. Add the thawed mycelium and chilled glass beads (–20 °C) until the bead beater is full. Pack ice around the container and grind in three or four bursts of 30 sec. Monitor the breakage before and after each bead beater treatment. The preparation should not contain any intact hyphal fragments.

8 Filter the preparation through two layers of Miracloth and rinse the beads with ice-cold 0.5 × SSE. Increase the volume at this step to 250 ml. Rinse with approx. 100 ml of 0.5 × SSE.

9 Retain the filtrate and process by centrifugation at 6000 g for 20 min. Pour off the supernatant and gently transfer the pellet to a loose-fitting Dounce homogenizer. Gently resuspend pellet in 40 ml of fresh 0.5 × SSE by gentle use of the Dounce.

10 Decant suspension to an Oakridge tube (Nalgene) and process by centrifugation at 6000 g for 20 min in a swinging bucket rotor. Resuspend the pellet by adding 30 ml

Protocol 2 continued

of 0.5 × SSE and adding to the Dounce homogenizer once more. Repeat one more time.

11 Wash, spin, and resuspend using the Dounce homogenizer two more times with 0.5 × SSE, this time also containing 0.2% (v/v) Nonidet P-40 (Sigma), to lyse mitochondria.

12 Resuspend the final pellet in 10 ml of 0.5 × SSE (do not use a pipette). Add 20% N-lauryl sarcosine to a final concentration of 1%, with gentle agitation. Warm to room temperature. The solution should be very viscous.

13 Add proteinase K (about 0.5 g) to the suspension and incubate at 37 °C for 1 h.

14 Dialyse at room temperature against 1 litre of 1 × TNE. Exercise great care when transferring the DNA preparation to avoid shearing. Change the TNE dialysis buffer every 1–1.5 h. Cell debris will begin to settle to the bottom of the dialysis tubing.

15 Process the dialysed DNA sample to remove the cell debris by centrifugation in an Oakridge tube at 8000 g for 5 min at 4 °C in a swinging bucket rotor. Discard the pellet.

16 Prepare a standard caesium chloride gradient at 49% (w/w) (ref. 7). Do not expose to bright light to avoid DNA damage. Centrifuge overnight at 47 000 g.

17 Extract the DNA band using an 18 gauge needle with indirect UV. Do not increase the volume unnecessarily. Extract ethidium bromide with water-saturated butanol by gentle mixing (this may take all day). Wrap the tube in foil during this process and leave on a gently rocking platform.

18 Dialyse the DNA in three changes of 1 × TNE. Quantify the DNA by spectrophotometry. A concentration of 300–500 ng/μl is normal. The yield will be 5–10 ml from this procedure.

19 The DNA can be concentrated by isopropanol precipitation or using a Speed-Vac if required.

3 Isolation of total RNA from filamentous fungi

Isolation of total RNA from filamentous fungi can be achieved using a number of methods including the use of chaotropic agents such as guanidinium thiocyanate, together with reducing agents such as 2-mercaptoethanol. These agents denature cellular proteins and inactivate ribonucleases. Similarly CTAB or hot-phenol extraction can be used as an efficient means of extracting RNA. Most of these procedures depend on rapidly denaturing proteins and maintaining ribonuclease inhibition before removal and purification of RNA. Ribonucleases are stable enzymes and can still show residual activity even if partially denatured. For this reason, it has become standard laboratory practice to take precautions against ribonucleases (RNases). In this regard it is worth mentioning that although RNases will undoubtedly be present on many pieces of equipment, human skin, and in many reagents, they cannot act upon RNA if it is protected by detergents. Therefore if an RNA preparation will only be used for Northern blot analysis and stored routinely in a buffer containing SDS, then less rigorous precautions need be taken to ensure RNases are inhibited. If on the other hand, the RNA is to be poly(A)$^+$ enriched and manipulated enzymatically, then it will need to be stored in detergent-free conditions (water or Tris/EDTA) and all equipment should be rigorously treated. The protocols given in this chapter are used routinely to extract total RNA from fungal cells and have been tested on a range of species including true fungi from the ascomycetes and basidiomycetes, and oomycetes such as *Phytophthora infestans*. The first protocol describes a method that has been used for many years to isolate high-quality fungal RNA (4), while the second method describes the use of Trizol (Stratagene), a proprietary phenolic reagent which provides a fast alternative means of preparing RNA.

Protocol 3

Isolation of total RNA from filamentous fungi

Reagents

- Extraction buffer (recipe for 100 ml): 4.8 g PAS (*p*-aminosalicylic acid, sodium salt; Sigma), 0.8 g TNS (tri-isopropylnaphthalene sulfonic acid, sodium salt; Eastman-Kodak). Dissolve each detergent separately in 40 ml of ddH$_2$O. Add the PAS to the TNS with constant stirring.
- TELS: 10 mM Tris–HCl pH 7.6, 0.1 mM EDTA, 0.2% SDS
- 5 × RNB: 1.0 M Tris base, 1.25 M NaCl, 0.25 M EGTA. Titrate to pH 8.5 with NaOH and store at −20 °C. Add 20 ml 5 × RNB to the PAS-TNS mix (PAS = 4.8%, TNS = 0.8%). A light brown precipitate will form, but does not interfere with the buffer. Store the solution on ice and always use fresh.

Method

1. Harvest a fresh liquid culture by filtration (see *Protocol 1*). Blot the mycelium dry and grind in a mortar containing liquid N$_2$ until a fine powder.
2. Transfer the powder to an Oakridge tube containing 4 ml of extraction buffer and 4 ml phenol/chloroform. Vortex for 5 min in 30 sec bursts, followed by 30 sec on ice.
3. Process by centrifugation at 9650 g for 10 min at 4 °C. Transfer the aqueous phase to a new tube and keep on ice.
4. Add 5 ml of extraction buffer to the phenol/chloroform interface. Mix and heat at 68 °C for 5 min. Cool on ice and centrifuge as above. Pool the aqueous phases.
5. Re-extract the pooled aqueous phase twice (or until there is no interface) with phenol/chloroform.
6. Slowly add 0.25 vol. of 10 M LiCl (final concentration 2 M). Store overnight on ice. Pellet the RNA by centrifuging at 9650 g for 15 min at 4 °C.
7. Resuspend the pellet by gentle pipetting in 5 ml of 0.1 M sodium acetate pH 6/70% ethanol. Centrifuge as before, but for 20 min. Repeat once.

Protocol 3 continued

8 Resuspend the pellet in 50 μl TELS. If the RNA is to be used for poly(A)$^+$ purification and cDNA library construction, then store in ddH_2O which has been DEPC-treated (ref. 7).

Protocol 4

Isolation of total RNA from filamentous fungi using Trizol

Reagents

- Trizol (Gibco BRL)

Method

1 Harvest a fungal culture by filtration as described previously (*Protocol 1*). Grind mycelium in liquid N_2 in a pestle and mortar to a fine powder.

2 Transfer the powdered mycelium into an Oakridge tube containing 20 ml of Trizol. Vortex the mixture for 15 sec. Incubate the mixture at room temperature for 20 min.

3 Add a 4 ml aliquot of chloroform and shake the tubes vigorously for 15–30 sec. Incubate them for a further 3 min on the bench.

4 Process the RNA preparations by centrifugation at 12 000 g for 15 min at 4 °C in a fixed angle rotor.

5 Remove the aqueous phase to a fresh Oakridge tube being careful not to remove interface material.

6 Add 10 ml of isopropanol (1 vol.) and leave on the bench for 10–15 min.

7 Precipitate RNA by centrifugation at 12 000 g for 20 min at 4 °C.

8 Pour away supernatant and resuspend the pellet in 70% ethanol. Centrifuge again at 6000 g for 10 min at 4 °C.

9 Dry the pellet on the bench and resuspend in 200 μl TELS. Resuspend in ddH_2O for further manipulations other than Northern blotting.

3.1 Manipulation of RNA

Once isolated, RNA should be quantified using a spectrophotometer to determine the absorbance of a dilute solution at 260 nm (7). It is recommended that full spectrum of absorbances between 220 nm and 320 nm is recorded because RNA has a characteristic spectrum (7) which can be used to detect impurities. The ratio of the A_{260}/A_{280} for pure RNA should be 1.8–2.0 and a lower value is normally due to the presence of protein which raises the A_{280}. Fractionation of RNA through denaturing gels for Northern blot analysis is a standard procedure which is described very well elsewhere (7).

Enrichment of the messenger RNA fraction within total RNA can be carried out by taking advantage of the presence of poly(A) tails at the 3′ end of many mRNAs. Affinity chromatography using oligo(dT)-labelled gel matrix is very effective and a number of rapid commercial enrichment procedures now exist from companies such as Promega, Qiagen, and Stratagene. Having generated poly(A)$^+$ RNA, synthesis of cDNA for library construction can be readily carried out. Again, for these procedures it is probably now advisable to follow commercial procedures because these have been optimized greatly in the last few years and use polymerases, which have been modified by site-directed mutagenesis to increase their robustness and processivity. A number of companies provide these cDNA library construction kits, including Promega, Stratagene, Clontech, and many others.

Acknowledgements

I am grateful to John Hamer (Paradigm Genetics), Steve Harris (Connecticut), Kathy Dobinson (London, Ontario), and JinRong Xu (Purdue) for their insights into DNA and RNA extraction procedures.

References

1. Jones, A. S. (1953). *Biochim. Biophys. Acta*, **10**, 607.
2. Jones, A. S. (1963). *Nature*, **199**, 280.
3. Murray, M. G. and Thompson, W. F. (1980). *Nucleic Acids Res.*, **8**, 4321.
4. Timberlake, W. E. (1980). *Dev. Biol.*, **78**, 497.
5. Hamer, J. E., Farrall, L., Orbach, M. J., Valent, F. G., and Chumley, F. G. (1989). *Proc. Natl. Acad. Sci. USA*, **86**, 9981.
6. Fungal Genetics Stock Center, Dept. of Microbiology, University of Kansas Medical Center, Kansas City, KS 66160-7420, USA. *http://www.fgsc.net/methods*
7. Sambrook, J., Fritsch, E. F., and Maniatis, T. (ed.) (1989). *Molecular cloning: a laboratory manual.* Cold Spring Harbor Laboratory Press, Cold Spring Harbor, NY.
8. Frischauf, A.-M. (1991). In *Essential molecular biology: a practical approach* (ed. T. A. Brown), Vol. 2, pp. 15–38. Oxford University Press, Oxford, UK.

Chapter 3
DNA-mediated transformation of fungi

Paul Bowyer
IACR-Long Ashton Research Station, Department of Agricultural Sciences, Long Ashton, Bristol BS41 9AF, UK.

1 Introduction

The advent of techniques for manipulating and isolating genes from plant pathogenic fungi has lead to a rapid increase in our understanding of how these organisms cause disease (1). An important factor in these advances has been the ability to re-introduce homologous or heterologous genes into these organisms. Transformation systems for model fungi such as yeast, *Neurospora crassa*, and *Aspergillus nidulans*, have existed since the 1970s and these techniques are being adapted for use in a growing array of pathogens. This chapter concentrates on techniques developed for pathogenic fungal species, but the techniques have often been found to be broadly applicable among fungal species. The vast majority of transformation protocols for filamentous fungi are based on permeabilizing cell membranes with polyethylene glycol (PEG), or electroporation. Both methods typically require preparing protoplasts, or osmotically-sensitive cells, prior to transformation (2) (a detailed review of fungal transformation systems is given in ref. 3). Many phytopathogenic fungi have been transformed (see *Table 1*) using these methods, however the obligately parasitic nature of some of these organisms has rendered them difficult to transform. Methods used for transformation of such 'difficult' pathogens, such as particle bombardment and *Agrobacterium*-mediated DNA transfer (4) are therefore also included.

2 Selectable markers and vectors for transformation

In general, fungi can be transformed with either nutritional markers which complement an auxotrophic requirement or dominant, selectable antibiotic resistance markers. Most plant pathogens are not well-characterized genetically and have few defined auxotrophic markers, or complementary genes that can be used in the former approach. This has lead to the development of a variety of antibiotic resistance cassettes that can be used in most pathogens. The antibiotic resistance genes described below have been placed under the control of a variety

Table 1 Transformed filamentous ascomycete and basidiomycete fungi

Organism	**Method of transformation**[c]	**Type of transformation**
Agaricus bisporus[b]	PEG/*Agrobacterium*	Integrating
Alternaria brassicola	PEG/biolistic	Integrating
Botrytis cinerea	PEG/*Agrobacterium*	Integrating
Cercospora nicotianae, kikuchii	PEG	Integrating
Cladosporium fulvum	PEG	Integrating
Claviceps purpurea	PEG	Integrating
Cochliobolus heterostrophus, carbonum	PEG	Integrating
Colletetrichum magna, gloeosporoides	PEG/*Agrobacterium*	Integrating
Coprinus cinerea	PEG	Integrating
Erysiphe graminis	Biolistic	Integrating
Fusarium oxysporum, moniliforme	PEG	Integrating/replicating
Gibberella fujikora, zeae	PEG	Integrating
Gloeocercospora sorghi	PEG	Integrating
Glomerella cingulata	PEG	Integrating
Leptosphaeria maculans	PEG	Integrating
Magnaporthe grisea	PEG	Integrating
Mycosphaerella graminicola	PEG	Integrating
Nectria hematococca	PEG	Integrating
Ophiostoma novo-ulmi	PEG	Integrating
Phytophthora infestans, capsici, parasitica[a]	PEG/biolistic	Integrating
Pyrenopeziza brassica	PEG	Integrating
Pyrenophora graminea, teres	PEG	Integrating
Ryncosporium secalis	PEG	Integrating
Septoria lycopersici, avenae	PEG	Integrating
Stagonospora nodorum	PEG	Integrating
Tapesia yallundae, acuformis	PEG/*Agrobacterium*	Integrating
Ustilago maydis[b]	PEG	Integrating/replicating
Venturia inaequalis	PEG/biolistic	Integrating

[a] Oomycete pathogen.

[b] Basidiomycete phytopathogen.

[c] PEG: transformation by means of incubating fungal protoplasts with polyethylene glycol, Ca^{2+} ions, and DNA.

of promoters and cloned into plasmid vectors that can be conveniently replicated in *Escherichia coli*. The most commonly used fungal promoter for transformation is the *A. nidulans gpdA* (glyceraldehyde-3-phosphate dehydrogenase) promoter (5), although it is now widely accepted that use of a promoter from the organism to be transformed will aid both expression of the selectable marker gene and transformation frequency.

2.1 Hygromycin B resistance

Hygromycin B is an aminoglycoside antibiotic which inhibits growth of prokaryotic micro-organisms (bacteria), eukaryotic micro-organisms (yeasts), and mammalian cells, by inhibiting protein synthesis at the translocation step causing misreading of mRNA. Hygromycin B is active against most fungi in the range 50–250 μg/ml, although higher amounts are needed to inhibit growth of some species. The *hph* gene from *E. coli* has been widely used for transformation of pathogens. The most common vector used in the transformation of pathogens is pAN7-1, containing the *E. coli hph* gene under the control of the *A. nidulans gpdA* promoter and flanked at the 3′ end by the *A. nidulans trpC* terminator region (5). Recently, more streamlined hygromycin resistance vectors (such as pCB1004 and pCB1535) have been produced using the *trpC* promoter from *A. nidulans*, controlling variants of the *hph* gene from which common restriction sites have been removed (6).

2.2 Phleomycin resistance

Phleomycin inhibits DNA synthesis without interfering directly with the production of proteins or RNA. Phleomycin (the related compounds Zeocin or Bleomycin can also be substituted), like actinomycin appears to owe its specificity of action to selectivity in its binding with DNA. It appears to have an affinity for polynucleotides rich in A-T base pairs and inhibits DNA polymerase activity at concentrations that do not interfere with RNA polymerase. Phleomycin is active against most fungi in the range 100–250 μg/ml although higher amounts are needed to inhibit growth of some species. The most common phleomycin resistance vector in use is pAN8-1 which is identical to pAN7-1 except that the *hph* gene is replaced with the phleomycin resistance gene from *Streptomyces hindustanus* (5).

2.3 Bialophos resistance

Bialophos is the active ingredient of the herbicide formulation Basta or Ignite. Bialophos is a naturally occurring tripeptide produced by *Streptomyces hygroscopicus* which is broken down to phosphinothricin—an inhibitor of glutamine synthesis. A gene (*bar*) coding for phosphinothricin acetyltransferase has been isolated from *S. hygroscopicus* and is widely used as a selective marker for the transformation of higher plants and several filamentous fungi, including *M. grisea*, *N. crassa*, *Cochliobolus sativa*, and *Colletotrichum* species (7, 8). A number of useful vectors are available including pCB1534 and the pBARGEM series (6, 7).

2.4 Sulfonylurea resistance

A sulfonylurea resistant allele of the *Magnaporthe grisea ILV1* gene has been cloned. This gene encodes acetolactate synthase, an enzyme involved in isoleucine and valine synthesis. Sulfonylurea inhibits acetolactate synthase at 100 μg/ml. Sulfonylurea (Chlorimuron ethyl) can be purchased from Chem Service, PO Box 3108,

West Chester, PA 19381-3108, USA. Currently pCB1536 is the most widely used vector for production of sulfonylurea resistance in fungi (6).

2.5 Benomyl resistance

Benomyl is a fungicide which binds to β-tubulin causing dissociation of microtubules. This compound is active against many ascomycete and basidiomycete fungi in the 1–10 μg/ml range. The β-tubulin gene of *N. crassa* is often used to confer resistance to this compound but naturally resistant β-tubulin genes from many pathogenic species are now available (9).

2.6 Acetamidase (*AmdS*)

The *amdS* gene is not an antibiotic resistance marker but encodes acetamidase, an enzyme conferring the ability to grow using acetamide as sole carbon source (11, 12).

3 Quality of DNA required for transformation

A general rule here is that vector DNA used for fungal transformation should be of the best quality available. We routinely use DNA that has been purified on a caesium chloride gradient, or using a commercially available kit such as Qiagen midi-preparation (Qiagen, Crawley, West Sussex). DNA from standard alkaline lysis type mini-preparations is unsuitable for fungal transformation.

4 Entry of DNA into fungi

DNA can be introduced into fungi in a variety of ways. Fungi possess tough cell walls which block direct uptake of DNA, necessitating either removal of the cell wall (protoplast formation) or forcible injection of DNA into the cell by particle bombardment (12–14), electroporation (15, 16), or *Agrobacterium*-mediated transformation (4). Traditional methods have relied on PEG-stimulated uptake of DNA by protoplasts (5, 10, 11) but more recent developments are detailed below.

4.1 Protoplast formation

4.1.1 Enzymatic digestion of fungal cell walls

Enzymatic digestion of the complex polysaccharide, chitin cell wall of fungi results in blebbing of cytoplasm and release of spherical protoplasts. Protoplasts contain fungal cytoplasm and are heterogeneous, with some containing single nuclei, others multiple nuclei, and some anucleate. The commercially available enzyme mixtures which are most suitable for fungal protoplast formation are Novozyme 234 and Glucanex, which are both derived from mycolytic *Trichoderma* species. Where the action of these enzymatic preparations mixtures is not sufficient to produce sufficient numbers of protoplasts, small amounts of helicase, driselase, or other lytic enzymes may need to be added. In general 5–10 mg/ml of Novozyme or Glucanex is sufficient to produce usable numbers of protoplasts.

4.1.2 Osmotic stabilization of protoplasts

Protoplasts will immediately lyse unless they are released into osmotically-stabilized medium. Typical media for osmotic stabilization consist of roughly 1 M salt or sugar solutions. Common osmotic stabilizers include 0.6–0.8 M KCl, 1–1.2 M $MgSO_4$, 0.8–1.2 M NaCl, and 1 M sucrose. These are added to conventional media for fungal propagation.

4.1.3 Starting material

Not all ages or types of mycelium will yield protoplasts with equal efficiency. A rule-of-thumb is to use hyphae that are as young as possible—ideally very young hyphae that have just germinated from spores should be prepared. An overnight culture, or two-day-old culture of most ascomycete, or badsiomycete fungi is sufficient for effective protoplast preparation.

5 REMI transformation

Restriction enzyme-mediated integration (REMI) has been used as a means of introducing DNA into fungi for mutagenesis (17–22). Addition of restriction enzymes has also been used to facilitate integration of transforming DNA (17, 19). Restriction enzymes are included with transforming DNA and are thought to facilitate integration by nicking chromosomal DNA at their cognate recognition site (17). This technique has been widely used for production of insertional mutants as the transforming DNA integrates in the genome in a predictable manner and can be easily rescued from mutants of interest. Many enzymes have been found that increase transformation from several-fold to a hundred-fold. These include four base pair recognizing enzymes and six base pair recognizing enzymes. The increase (or decrease) in transformation frequency is dependent on the amount of enzyme included with the transforming DNA which must be determined empirically.

Protocol 1

Transformation of *Magnaporthe grisea*

Reagents

- ST buffer: 0.6 M sucrose, 0.1 M Tris-HCl pH 7.0
- STC: 1.2 M sucrose, 10 mM Tris-HCl pH 7.5, 10 mM $CaCl_2$
- TB3 medium (per litre): 3 g yeast extract, 3 g casamino acids, 20% sucrose
- Restriction PEG: 50 mM Tris, 40% PEG_{8000}, 20% sucrose, 50 mM $CaCl_2.2H_2O$, 50 mM NaCl, 50 mM KCl, 10 mM $MgCl_2.6H_2O$
- OM: 1.2 M $MgSO_4$, 10 mM $NaPO_4$ buffer pH 5.8

Method

1 Using a sterile spatula, cut out about 25 cm^2 of mycelium from a freshly prepared plate culture. Transfer to a sterile blender cup and blend (five pulses, 5 sec each) in 100 ml of supplemented minimal medium.

Protocol 1 continued

2 Transfer the macerate to a sterile 500 ml flask. Place the liquid culture in a swirling incubator for three days at room temperature (about 23 °C).

3 Pour the culture into a sterile blender cup and pulse blend as before. Transfer the macerate to a sterile 1 litre flask, add 100 ml fresh supplemented minimal medium, and incubate for two days at room temperature with swirling.

4 Collect the mycelium on sterile Whatman No. 4 filter paper by vacuum filtration. Wash with sterile distilled water.

5 Weigh mycelial pad. Resuspend approx. 5 g wet weight of mycelium in 40 ml OM buffer supplemented with 100 mg Novozyme 234/Glucanex. Place in an Oakridge tube. Incubate with gentle swirling (100 r.p.m.) at 25 °C for 60–90 min.

6 Transfer protoplasts to two sterile Oakridge tubes and overlay with an equal volume cold ST buffer. Spin at 5000 g in a swinging bucket rotor at 4 °C. Protoplasts are recovered from the OM/ST interface using a Pasteur pipette.

7 Transfer protoplasts to two or three sterile Oakridge tubes, and fill the tubes with STC. This will adequately dilute out the OM and ST. Pellet protoplasts at 3000 g for 10 min at 4 °C in a swinging bucket rotor. Wash protoplasts twice more with 10 ml STC. Completely resuspend the pellet each time.

8 Mix 200 μl protoplasts (at a concentration of 5×10^7 protoplasts/ml) with plasmid DNA. Incubate at room temperature for 15 min. For REMI, add restriction enzymes (generally 5–50 U) at this point, immediately before addition of restriction PEG.

9 Add 1–1.25 ml of restriction PEG, pre-heated to 37 °C and incubate for 20 min at room temperature.

10 Add 3 ml of TB3 medium and incubate with shaking for 1 h to allow protoplast recovery.

11 Pellet protoplasts as above and resuspend the protoplasts in 100 μl STC. Pour 7–10 ml of molten (42–50 °C) top agarose (TB3 with 20 g LMP agarose/litre), containing the appropriate selective antibiotic (for hygromycin B add 200 μg/ml to top agarose), into each tube and pour onto appropriate TB3 plates. Incubate plates at 25–28 °C.

Protocol 2

REMI transformation of *Septoria nodorum*

Reagents

- CzV8 Czapek-Dox liquid medium (Oxoid)
- 1 M $MgSO_4$
- Novozyme 234
- Sorbitol
- PEG solution: 66% PEG_{4000}, 10 mM Tris pH 7.4, 100 mM $CaCl_2$
- Regeneration medium: 1 M sucrose, 0.5% myco-peptone, 0.1% yeast extract

Protocol 2 continued

Method

1. Transfer 100 ml CzV8 Czapek-Dox liquid medium supplemented 10% V8 juice (Campbells), aseptically to 2 × 50 ml Falcon tubes. Spin to remove V8 debris. Pour into 250 ml Erlenmeyer flask.
2. Inoculate with 5 × 2–3 mm^2 mycelial plugs from a fresh plate culture of *S. nodorum*, prepared on potato dextrose agar (Difco).
3. Shake for three days at 25 °C to give a dispersed mycelial culture of young hyphae.
4. Harvest mycelium by centrifugation at 3000 g for 5 min. Resuspend in 50 ml of 1 M $MgSO_4$.
5. Centrifuge again and resuspend mycelium in 10 ml of filter sterilized Novozyme 234 (5 mg/ml) in 1 M $MgSO_4$ in a Falcon tube. Place on rocking platform for 60–90 min at room temperature.
6. Protoplasts appear after an hour and microscopic examination should show them budding (blebbing) away from hyphae, considerably before this.
7. Overlay protoplast suspension carefully with an equal volume of 0.6 M sorbitol. Spin at 4000 g for 30–45 min. Carefully transfer protoplasts at the interface to a new tube.
8. Wash protoplasts with 10 ml of 1 M sorbitol and centrifuge at 2000 g for 5 min. Resuspend in 10 ml of 1 M sorbitol. Repeat and resuspend protoplasts at a final concentration of 5×10^8/ml in 1 M sorbitol, 100 mM $CaCl_2$.
9. Gently mix 100 µl protoplasts (5×10^7) with 25 µl PEG solution and 5 µg DNA. For REMI, carry out treatment of protoplasts with 20–40 U restriction enzyme (e.g. *Bam*HI or *Eco*RI) for 30–60 min at room temperature before addition of PEG.
10. Incubate protoplasts on ice for 20 min. Add 1 ml PEG drop-wise over 5 min, with gentle mixing. Immediately add 10 ml of 1 M sorbitol, 100 mM $CaCl_2$. Centrifuge at 200 g for 5 min. Resuspend in 10 ml regeneration medium.
11. Optionally, the protoplasts can be incubated overnight at room temperature with gentle rocking at this stage.
12. Mix aliquots of protoplasts with 50 ml molten bottom agar (regeneration medium plus 1.5% agarose) and pour into five Petri dishes. Overlay with 10 ml top agar (as bottom agar but with the inclusion of 200 µg/ml hygromycin). Incubate at 17 °C under black light, until colonies appear (usually about eight to ten days).

6 *Agrobacterium*-mediated transformation of fungi

Agrobacterium tumefaciens is a soil bacterium, which has the ability to transfer sections of plasmid borne DNA to plants. This ability has long been exploited for DNA-mediated transformation of higher plants. *A. tumefaciens* has recently been shown to transfer DNA to fungi in a similar manner (4). This method of transformation is potentially simpler and more reliable than standard methods, although it has not been proven in many species to date. DNA for transformation, including a selectable marker and any genes of interest, must first be cloned between left and right T-DNA borders in a binary transformation vector such as pBIN19. This vector must then be transformed into *A. tumefaciens*. Bacteria containing the DNA of interest are then incubated with fungus to allow DNA transfer, before being killed by antibiotic treatment.

Protocol 3
Preparation of competent *Agrobacterium* cells

Reagents

- 2YT media
- TE pH 7.5
- Liquid nitrogen

Method

1 Inoculate 10 ml of 2YT with a single colony of *Agrobacterium* from a fresh plate and grow with shaking at 28–30 °C overnight.

2 Add the overnight culture to 200 ml 2YT and grow with shaking for 3–4 h.

3 Harvest cells by centrifugation at 3000 g for 10 min. Decant the supernatant and wash the surface of the pellet with 10 ml ice-cold TE pH 7.5.

4 Resuspend the pellet in 20 ml ice-cold 2YT and immediately flash-freeze 0.5 ml aliquots in liquid nitrogen.

5 Competent *Agrobacterium* cells can be stored at −70 °C for 6–12 months.

Protocol 4
Transformation of competent *Agrobacterium* cells

Reagents

- See *Protocol 3*
- Competent *Agrobacterium* cells
- Plasmid DNA

Method

1 Thaw the frozen competent *Agrobacterium* cells on ice and add 1 μg plasmid DNA with gentle mixing.

2 Incubate on ice for 5 min and then quickly freeze in liquid nitrogen for 5 min.

3 Incubate cells at 37 °C for 5 min and add 1 ml of 2YT and incubate at 28–30 °C for 2–4 h with shaking.

4 Plate 1–200 μl aliquots on selective medium (LB + 50 μg/ml kanamycin for pBIN19) and incubate at 28 °C for two days until colonies appear.

Protocol 5

DNA-mediated transformation of fungi using *Agrobacterium*

Reagents

- *Agrobacterium* minimal medium: 3 g K_2HPO_4, 1 g NaH_2PO_4, 1 g NH_4Cl, 0.3 g $MgSO_4.7H_2O$, 0.15 g KCl, 0.01 g $CaCl_2$, 0.0025 g $FeSO_4.7H_2O$, 5 g sucrose, H_2O to 1 litre, pH to 7.0 with HCl, and autoclave
- Induction medium (IM) for 100 ml use: 1 g NH_4Cl, 0.3 g $MgSO_4.7H_2O$, 0.15 g KCl, 0.01 g $CaCl_2$, 0.0025 g $FeSO_4.7H_2O$, 40 mM MES pH 5.6, 10 mM glucose, 5 ml glycerol, 4 mM acetosyringone

Method

1. Inoculate a single *Agrobacterium* colony into 2 ml minimal medium. Grow for 48 h with shaking at 28 °C.
2. Dilute the culture to OD_{660} of 0.15 in induction medium and grow for 6 h at 28 °C with agitation at 200 r.p.m.
3. Mix 200 μl *Agrobacterium* with 200 μl fungal spores at 10^3, 10^4, and 10^5 spores/ml. Pipette the mixtures onto nitrocellulose filters placed on IM plates. Incubate at 20 °C for 48 h.
4. Transfer the filters to fungal minimal medium agar plates containing the selective antibiotic (e.g. hygromycin B) and 500 μg/ml Claferin (Cefotaxime). Incubate plates at 20 °C for 48 h.
5. Remove filters and place in 10 ml fungal minimal medium containing selective antibiotic and Claferin in a 50 ml Oakridge tube. Vortex briefly to resuspend the fungal spores. Plate out the suspension on selective medium—0.5 ml per 9 cm Petri dish is usually adequate to give well-separated colonies.

7 Particle bombardment

Microprojectile bombardment involves projecting DNA coated gold or tungsten particles into intact cells. Over the last few years, microprojectile bombardment has become a powerful tool for transformation of intractable plant species, such as the commercial cereals. Several filamentous fungi have been transformed by means of microprojectile bombardment (*Table 1*). This technique is particularly useful for transformation of fungi which cannot be cultured outside their host, such as obligate biotrophs (23).

Protocol 6

Preparation of tungsten particles and transformation of filamentous fungi by particle bombardment

Reagents

- M10 particles (Bio-Rad Laboratories, Hercules, California) were prepared and coated with plasmid DNA according to manufacturer's instructions

Protocol 6 continued

Method

1 To a 1.5 ml microcentrifuge tube add 25 μl tungsten particle suspension (1.5 mg in 50% glycerol), 5 μl plasmid DNA (0.5 μg/μl), 25 μl of 2.5 M $CaCl_2$, and 5 μl of 1 M spermidine free base. After each addition the suspension is vigorously vortexed and incubated on ice for 10 min.

2 Spin down tungsten particles for 5–10 sec at full speed in a microcentrifuge. Decant the supernatant and wash the particles twice in 100 μl of 95% ethanol.

3 Resuspend particles in 50 μl ethanol, and pipette 10 μl particles onto a macrocarrier disc (Bio-Rad).

4 Spread conidia from fungus to be transformed onto 10 ml of selective agar medium in a 9 cm Petri dish, briefly air dried under sterile conditions. Use for micro-projectile bombardment within 1–3 h.

5 In general a helium pressure of 1000 psi, and a target distance of 6 cm can be used for bombardment. Transformants are visible three to six days after bombardment.

8 Transformation of fungi by electroporation

Electroporation describes the use of a transmembrane electric field pulse to induce microscopic pores to be generated in a plasma membrane. These pores allow ingress of DNA or other molecules into a cell. If the electric field pulse has the proper parameters, then the pores reseal spontaneously and cells will continue to grow. The induction of pores is affected by three major factors:

(a) Variability in target cell populations causes some cells to be more sensitive to electroporation pulses than others.

(b) The amplitude and duration of the pulse must exceed a certain minimum level, which is thought to depend on the reciprocal of cell size.

(c) As the product of amplitude and duration increases, the number and size of pores increases. If amplitude and duration of the pulse exceed a certain threshold, pores become too large for repair and the cells will die.

As very little is understood about the mechanism of electroporation, most pulse parameters (amplitude, duration, number, interpulse interval) are developed by trial and error for each organism and tissue. The first electroporation systems described for filamentous fungi used germinated conidia treated with lytic enzymes to open the cell wall. Subsequent researchers have found that protoplasts result in more reliable transformation frequency (15, 16).

Protocol 7
Transformation of *Tapesia yallundae* by electroporation

Reagents

- Czapek Dox medium (Oxoid)
- Sorbitol
- Novozyme 234
- Miracloth (Calbiochem)
- STC: 1.2 M sorbitol, 10 mM Tris-HCl pH 7.4, 10 mM $CaCl_2$
- Regeneration medium: 1 M sucrose, 0.5% yeast extract, 0.1% bacto peptone

Method

1. Harvest 10^6 conidia into 10 ml of Czapek Dox medium. Allow the conidia to germinate overnight at 18 °C in a shaking incubator (100 r.p.m.).
2. Collect the germinated conidia by centrifugation at 2500 g for 5 min. Wash each pellet once with 5 ml of ice-cold sterile water and once with 5 ml of ice-cold 1 M sorbitol. Centrifuge at 2500 g for 10 min to collect the cells between the wash steps.
3. Resuspend the four germling pellets in a total of 10 ml of 1 M sorbitol, containing 5 mg/ml Novozyme 234. Incubate in a shaking 25 °C incubator at 50 r.p.m. for 2 h.
4. Filter the resulting protoplast suspension through four layers of sterile Miracloth.
5. Transfer the protoplasts into a sterile 30 ml Corex tubes and collect the cells by centrifugation at 100 g for 5 min. Wash the protoplast pellets three times by gently resuspending the cells in 25 ml of 1 M sorbitol.
6. After the third wash, resuspend the protoplasts in 10 ml ice-cold sterile STC. Count the protoplasts using a haemocytometer then centrifuge at 100 g for 5 min and remove the STC with an aspirator.
7. Gently resuspend the protoplast pellet to a final concentration of 10^8/ml. Add 50 μl of protoplasts to 3–6 μl of DNA (at 1 μg/μl). Flick the tube to mix, then transfer to a pre-chilled electroporation cuvette with a 0.2 cm gap (Bio-Rad Cuvette No. 165–2086).
8. Electroporate at 200 Ohms, 25 μF, and 0.7 kV (field strength of 3.5 kV/cm, time constant of 0.4 msec).
9. Following the electroporation, resuspend protoplasts immediately in 1 ml of sterile ice-cold regeneration medium. We allow the protoplasts to regenerate their cell wall by incubation at room temperature overnight and then plate the cells on selective medium.

9 Fate of transforming DNA

9.1 Integration/replication of transforming DNA

DNA transformed into fungi can either integrate into the chromosome or remain separate as a replicating plasmid. Few replicating plasmids are available for filamentous fungi, the most notable being in *Ustilago maydis*, *Fusarium oxysporum*, and *Aspergillus nidulans*. Replicating plasmids require special elements such as autonomously replicating sequences or telomeres for maintenance in the nucleus or cytoplasm, whereas any plasmid seemingly can integrate into fungal chromosomes. The factors governing site and frequency of plasmid integration are, however, unknown but a few features of this process are described below. During integrative transformation most markers give a background of small abortive transformants which may result from transient expression of the marker without stable integration. Transforming DNA can integrate into the chromosome in a number of different ways, which can be exploited for a variety of gene manipulations.

9.1.1 Types of plasmid integration event

Three main types of integration event can occur in transformation. These are homologous integration, non-homologous or random integration, and double homologous integration. Homologous integration occurs when DNA in the transforming vectors is homologous to DNA in the genome. It is assumed that recognition and recombination machinery from the transformed organism recognizes the similarity between the two DNAs and integrates the vector into the chromosome. The frequency of homologous integration varies greatly between organisms and may vary between different pieces of homologous DNA from the same organism. In general, longer stretches of homology promote homologous recombination, therefore when such events are desired it is advisable to include as much homologous DNA as possible in the transforming vector. Non-homologous, or random, integration occurs when the vector integrates at a non-homologous site in the genome. This can occur even when the vector carries sequences homologous to the host genome and is probably mediated by the same cellular machinery as homologous integration. Double integration events occur when two separate regions in a homologous stretch of DNA undergo homologous recombination with the respective regions of the host genome. Single and double homologous recombination events can be detected using Southern blot and PCR-based approaches. Single homologous integration can be used to replace gene promoters or for simple gene disruption, and double homologous integration can be used to replace regions of genomic DNA with a region of choice. For example, a gene coding region can be replaced with a selectable marker for gene replacement, or with a reporter gene for locus-specific expression studies.

9.1.2 Marker rescue

When a plasmid integrates by homologous recombination the bacterial plasmid backbone is also usually incorporated into the genome. This means that it is possible to rescue the plasmid from genomic DNA after integration, by transformation of digested and re-ligated genomic DNA into *E. coli*. If the fungal genomic DNA is digested with an enzyme which does not cut within the transforming plasmid it is also possible to rescue regions of genomic DNA which flank the integration site. This is important for determining the site of integration and can be of great significance, especially if the integration has created a mutant phenotype.

Protocol 8

Plasmid rescue from filamentous fungi

Reagents

- Restriction enzyme
- Phenol/chloroform/isoamyl alcohol
- Chloroform
- Ethanol
- 3 M sodium acetate pH 5.6
- Ligase
- Competent *E. coli*

Method

1 Prepare genomic DNA from the fungal transformant. The genomic DNA used must be of high quality.

2 Set up a series of restriction enzyme digestions of genomic DNA containing from 1 μg–10 ng genomic DNA and > 20 U restriction enzyme. The enzyme chosen should not cut the plasmid DNA.

3 Digest the DNA overnight. Add an equal volume of phenol/chloroform/isoamyl alcohol (25:24:1). Vortex for 10 sec, then centrifuge at 10 000 g in a microcentrifuge to separate the aqueous (upper) and organic (lower) phases.

4 Extract the upper aqueous phase with an equal volume of chloroform.

5 Add 2.5 vol. of ethanol and 0.1 vol. of 3 M sodium acetate pH 5.6. Mix gently and leave on ice for 30 min.

6 Centrifuge at 10 000 g for 20 min. Wash the pellet in 70% (v/v) ethanol and air dry.

7 Resuspend the DNA pellet in 20 μl water and add ligase and buffer. Incubate overnight at 16 °C.

8 Transform the ligation into competent *E. coli* using standard procedures.

In our experience the plasmids rescued from genomic fungal DNA often contain rearrangements or deletions. However, it is usually possible to gain some information about the site of insertion by sequencing out from the universal primers site within the plasmid DNA.

References

1. Oliver, R. P. and Osbourn, A. E. (1995). *Microbiology*, **141**, 1.
2. Goldman, G. H., Geremia, R., van Montagu, M., and Herrera-Estrella, A. (1990). Bulletin 1352, Bio-Rad Laboratories, Hercules, CA, USA.
3. Lemke, P. A. and Peng, M. (1995). In *The mycota* Volume II, *Genetics and biotechnology* (ed. U. Kuck), pp. 109–39. Springer, Berlin.
4. DeGroot, M. J. A., Bundock, P., Hooykaas, P. J. J., and Beijersbergen, A. G. M. (1998). *Nat. Biotechnol.*, **16**, 839.
5. Punt, P. J., Oliver, R. P., Dingemanse, M. A., Powels, P. H., and van den hondel, C. A. M. J. J. (1987). *Gene*, **56**, 117.

6. Carroll, A. M., Sweigard, J. A., and Valent, B. (1994). *Fungal Genet. Newsl.*, **41**, 22.
7. Pall, M. L. and Brunelli, J. P. (1994). *Fungal Genet. Newsl.*, **41**, 63.
8. Avalos, J., Geever, R. F., and Case, M. E. (1989). *Curr. Genet.*, **16**, 369.
9. Orbach, M. J., Porro, E. B., and Yanofsky, C. (1986). *Mol. Cell. Biol.*, **6**, 2452.
10. Tilburn, J., Scazzocchio, C., Taylor, G. G., Zabicky-Zissman, J. H., Lockington, R. A., and Davies, R. W. (1983). *Gene*, **26**, 205.
11. Kelly, J. M. and Hynes, M. J. (1985). *EMBO J.*, **4**, 475.
12. Armaleo, D., Ye, G. N., Klein, T. M., Shark, K. B., Sanford, J. C., and Johnston, S. A. (1990). *Curr. Genet.*, **17**, 97.
13. Bailey, A. M., Mena, G. L., and Herrera-Estrella, L. (1993). *Curr. Genet.*, **23**, 42.
14. Hilber, U. W., Bodmer, M., Smith, F. D., and Koller, W. (1994). *Curr. Genet.*, **25**, 124.
15. Fungaro, M. H. P., Rech, E., Muhlen, G. S., Vainstein, M. H., Pascon, R. C., deQueiroz, M. V., *et al.* (1995). *FEMS Microbiol. Lett.*, **125**, 293.
16. Robinson, M. and Sharon, A. (1999). *Curr. Genet.*, **36**, 98.
17. Schiestl, R. H. and Petes, T. D. (1991). *Proc. Natl. Acad. Sci. USA*, **88**, 7585.
18. Bölker, M., Böhnert, H. U., Braun, K. H., Görl, J., and Kahmann, R. (1995). *Mol. Gen. Genet.*, **248**, 547.
19. Granado, J. D., Kersetz-Chaloupková, K., Aebi, M., and Kües, U. (1997). *Mol. Gen. Genet.*, **256**, 28.
20. Shi, Z., Christian, D., and Leung, H. (1995). *Phytopathology*, **85**, 329.
21. Balhadère, P. V., Foster, A. J., and Talbot, N. J. (1999). *Mol. Plant-Microbe Interact.*, **12**, 129.
22. Sweigard, J. A., Carroll, A. M., Farrall, L., Chumley, F. G., and Valent, B. (1998). *Mol. Plant-Microbe Interact.*, **11**, 404.
23. Chaure, P., Gurr, S. J., and Spanu, P. (2000). *Nat. Biotechnol.*, **18**, 205.

Chapter 4
Genetic analysis of ascomycete fungi

Steven D. Harris
Department of Microbiology, University of Connecticut Health Center, Farmington, CT 06030–3205, USA.

1 Introduction

Filamentous fungi, in particular *Aspergillus nidulans* and *Neurospora crassa*, have historically played an important, yet under-appreciated, role as model eukaryotic organisms (1, 2). In this context, the primary attribute of these fungi is the ease with which they can be manipulated using both classical and molecular genetic techniques. Because they are haploid, the generation and analysis of mutants is a relatively straightforward process in *A. nidulans* and *N. crassa*. A variety of approaches can be used to generate mutations; ranging from traditional chemical mutagenesis to more recent transformation-based strategies. Similarly, an unlimited array of screening protocols can be used to select the desired mutant. Once obtained, a mutant must be subject to rigorous characterization to determine if the relevant phenotype is caused by one or more mutations, and if the mutation is recessive or dominant. The protocols employed to answer these questions in the filamentous fungi have remained essentially unchanged over the past forty years. The purpose of this review is to summarize the approaches used for genetic analysis in the ascomycete fungi, with particular emphasis on *A. nidulans*, *N. crassa*, and the plant pathogen *Magnaporthe grisea*. The first section will cover strategies used to identify and isolate mutants, whereas the second section will summarize the protocols employed for the analysis of a mutant.

2 Identification and isolation of mutants

2.1 Important considerations

Relatively straightforward protocols are available for the mutagenesis of ascomycete fungi. However, two important considerations must be taken into account before designing a mutagenesis strategy. First, since filamentous fungi are multicellular, care must be taken to choose the appropriate cell type for mutagenesis. Critical factors that influence the choice of cell type include:

(a) Nuclear content and ploidy (uninucleate cells possessing a haploid genome, such as *A. nidulans* conidia, are ideal).

(b) The ability to obtain large numbers of pure cells (typically at least 1×10^7/ml).

These factors are particularly important if a chemical or physical mutagenesis approach is being employed. In contrast, if a transformation-based mutagenesis strategy (i.e. REMI; Chapter 3, *Protocol 2*) is being used, these problems are circumvented by clonally purifying the transformants prior to selecting for the desired mutant.

The second important consideration is the nature of the desired mutation. Approaches such as REMI typically involve the insertion of a selectable marker at quasi-random sites throughout the genome. Although an advantage of REMI is that it allows rapid cloning of the mutated gene, this approach typically yields only null mutations. Often, simple hypomorphic or conditional mutations are much more informative than simple null alleles. In this case, chemical or physical mutagens with a propensity for causing base substitutions (for example UV light, or mono-functional alkylating agents such as 4NQO, MNNG, or EMS) should be used. Alternatively, if deletion mutations are sought, physical mutagens such as ionizing radiation, or bifunctional alkylating agents such as DEB and DEO, are useful. Notably, DEB and DEO appear to be particularly effective at inducing mutation in *A. nidulans* (3). As a cautionary note, certain chemical mutagens (EMS, MNNG) generate a strongly biased spectrum of mutations (4). Since this would have the adverse effect of constraining the number of amino acid substitutions, it is potentially dangerous to rely excessively on the use of a single mutagen.

2.2 Chemical mutagenesis

Although the following protocol was developed for the mutagenesis of fungal spores with 4NQO (5), it is equally effective using other chemical mutagens in its place. Freshly harvested spores should always be used, because older stocks usually contain non-viable spores. Most chemical mutagens (i.e. EMS, MMS, DEO, DEB) can be purchased as ready-to-use liquid solutions. Those chemicals only available as powder (4NQO and MNNG) should be resuspended in acetone at 50 mg/ml (note that 4NQO exhibits poor solubility). Before initiating a mutagenesis program, this protocol should first be used to identify the concentration of the mutagen that yields the greatest recovery of mutants. In most ascomycete fungi, the level of mutagenesis can be readily assessed by visually scoring the appearance of spore colour or non-sporulating mutants amongst surviving colonies. Once the ideal mutagen dose has been identified, mutagenized spores are typically plated onto undefined rich media to allow optimal recovery of all mutants. Survivors are then tested under specific growth conditions to identify the mutants of interest (i.e. temperature-sensitive, auxotrophic, drug resistant). A useful resource of media recipes for the growth of *N. crassa* and *A. nidulans* is provided by the FGSC (http://www.fgsc.net/methods/genmthds.html).

Protocol 1

Chemical mutagenesis of fungal spores

Reagents

- 4NQO (Sigma), EMS, MMS, MNNG, DEB, or DEO (all from Aldrich Chemical Co.)
- 50% sodium thiosulfate (Sigma)

Method

1 Dilute 4NQO in sterile double-distilled water to a working stock of 200 μg/ml.[a]

2 Harvest conidia in sterile double-distilled water containing 0.01% Tween 80 (to minimize spore clumping). Determine the concentration of conidia by counting on a haemocytometer, and dilute in sterile double-distilled water to 1×10^7 conidia/ml.

3 Prepare the mutagenesis mix in a 1.5 ml Eppendorf tube as follows:
- 200 μl conidia
- 8 μl 4NQO
- 800 μl sterile double-distilled water

4 Incubate the mutagenesis mix at 30 °C for 30 min with constant shaking.

5 Inactivate the mutagen by adding 100 μl of 50% sodium thiosulfate to the mutagenesis mix, and continue the incubation for a further 5 min at 30 °C.

6 Centrifuge the mutagenized conidia for 1 min at full speed in a microcentrifuge. Pour off the supernatant and resuspend the conidia in 1 ml of sterile double-distilled water.

7 Dilute the mutagenized conidia and plate at an estimated density of 100–200 survivors per plate on rich media. Plates are usually incubated at 28 °C or 37 °C.[b]

[a] The concentration of working stocks of other mutagens should be empirically determined.

[b] The addition of detergents such as Triton X-100 (0.01%, v/v) or sodium deoxycholate (0.08%, v/v) to the growth medium promotes the formation of compact colonies that can easily be replica plated.

2.3 UV mutagenesis

UV light is an effective mutagen that generates a broad spectrum of mutagenic events (4). An advantage of UV mutagenesis is that spores can be mutagenized directly after plating. However, it is critical to predetermine the spore dilution and UV dose that maximizes the recovery of mutants, while retaining a density of 100–200 colonies per plate. A variety of sources of UV light can be used, including:

- a hand-held UV lamp
- a UV transilluminator
- a UV crosslinker

Protocol 2

UV mutagenesis of fungal spores

Equipment

- UV transilluminator

Method

1. Harvest conidia (*Protocol 1*, step 2).
2. Dilute the conidia and plate to an estimated density of 100–200 survivors per plate.
3. Irradiate the plated conidia by inverting the plates over a UV transilluminator. Be sure to remove the lids from the plates. Exposure to light should be minimized by irradiating in the dark and wrapping the irradiated plates in aluminium foil.[a]
4. Plates are usually incubated at 28 °C or 37 °C.

[a] When using a transilluminator, the dose can be changed by varying the time of exposure.

2.4 Recovery of mutants

Regardless of the type of mutant being sought, the most effective means of recovery is to replica plate surviving colonies from non-selective to selective conditions. This technique allows the pattern of colonies growing on the surface of one plate to be duplicated on another. The selective plates are subsequently scored to identify colonies that failed to grow under a specific condition. These colonies are then recovered from the corresponding position on the original master plate.

Replica plating blocks can be purchased from a supplier (VWR Scientific Products; Bio 101, Inc.) or can be custom-made. The ideal material for transferring colonies is velveteen, although damp paper towels also work well. When purchased, cut the velveteen into 15 cm^2 pieces and launder them. Double wrap the velveteen squares in aluminium foil and sterilize by autoclaving prior to each use. Used squares can be washed and re-sterilized. Conidiating colonies transfer remarkably well by replica plating, whereas predominantly hyphal colonies do not. The undesired spreading of colonies on replica plates can be minimized by the inclusion of detergents in the growth media (*Protocol 1*, step 7).

2.5 Types of mutants

Auxotrophic mutants can only grow on media supplemented with a component not normally required for the growth of wild-type strains. In most fungi, the ability to construct diploids and/or heterokaryons requires the availability of non-complementing, auxotrophic mutations that can be used to select against the growth of the parental strains. In addition, several selection systems for fungal transformation are based on auxotrophic markers (Chapter 3). Auxotrophs are typically detected by their failure to grow on minimal medium. The nature of the auxotrophy is determined by adding back pools of supplements and identifying those that restore growth. Typically, pools containing five different amino acids, vitamins, or purines are tested (the desired concentration of each supplement can be obtained from the FGSC web site; http://www.fgsc.net/methods/genmthds.html). Once a pool that restores growth is identified, each individual component is tested to identify the specific supplement required by the auxotroph.

Mutants that cannot grow at high (Ts) or low (Cs) temperature are referred to as conditional mutants. Other, less frequently used, types of conditional selections include sensitivity to solvents such as dimethyl sulfoxide (6). Conditional mutations are primarily used to characterize the function of essential gene products. In the ascomycete yeasts *Saccharomyces cerevisiae* and *Schizosaccharomyces pombe*, and to a lesser extent in *A. nidulans*, large collections of Ts mutants have proven to be valuable resources that greatly expedited the characterization of vital cellular functions (see refs 7, 8).

Other useful types of mutants include those that display altered growth in response to specific pharmacological agents. A compendium of such drugs, and the concentrations to which *A. nidulans* exhibits sensitivity, is provided in ref. 9.

Protocol 3

Characterization of auxotrophic mutants

Reagents

- Vitamins: *p*-aminobenzoic acid, riboflavin, thiamine, biotin, pantothenic acid, nicotinic acid, pyridoxine (Sigma)
- Amino acids (Sigma)

Method

1. Mutagenize conidia and recover survivors on rich media at a density of 100–200 colonies per plate (*Protocols 1* and 2).
2. Replica plate colonies onto plates containing unsupplemented minimal media.[a] Incubate the replica plates for a sufficient period of time to allow colony formation. Store the master plates at 4 °C.
3. Compare each replica plate to its master plate. Identify and recover colonies that fail to grow on minimal media.
4. Re-test putative auxotrophic mutants to verify that they cannot grow on unsupplemented media. Store confirmed mutants in 15% glycerol at −70 °C.
5. Determine the nature of the auxotrophy by testing the ability of each mutant to grow on minimal media supplemented with pools of different amino acids or vitamins. When a pool is identified that restores growth, repeat this step to identify the individual component required by the mutant.

[a] Draw a mark on the master plate and the replica plate so that they can be oriented with respect to each other.

Protocol 4

Generating a collection of Ts mutants

1. Mutagenize conidia and recover survivors on rich media at a density of 100–200 colonies per plate (*Protocols 1* and 2).
2. Replica plate colonies onto plates containing rich media.[a] Incubate the replica plates at restrictive temperature (usually 42 °C) for a sufficient period of time to allow colony formation. Store the master plates at 4 °C.
3. Compare each replica plate to its master plate. Identify and recover colonies that fail to grow at restrictive temperature.
4. Re-test putative Ts mutants to verify that they cannot grow at restrictive temperature. Store confirmed mutants in 15% glycerol at −70 °C.[b]

[a] Draw a mark on the master plate and the replica plate so that they can be oriented with respect to each other.

[b] Stored mutants can be examined by microscopy to obtain insight into the nature of the Ts defect. Alternatively, they can be subjected to secondary screens to obtain the desired class of Ts mutants.

3 Classical genetic analysis of mutants

3.1 Setting up crosses

The most important step in the genetic analysis of a new mutant is to cross it to a wild-type strain. This cross serves several purposes, including:

(a) Verification that the mutant phenotype is caused by a mutation in a single gene.

(b) Generation of appropriately marked strains for complementation analysis.

(c) In fungi such as *A. nidulans*, the construction of heterozygous diploids that can be used to determine if the mutation is dominant or recessive.

Two general approaches can be used to establish a cross. The first approach is to simply streak spores from the two parental strains in adjacent lines down the centre of a plate containing rich media. Mating will occur in a zone between the two lines where the parental mycelia converge. Alternatively, spores ($\sim 1 \times 10^5$) from the parental strains are co-inoculated into 2 ml of rich liquid media in a well of a 24-well microtitre plate. Mating occurs in the resulting mixed mycelium, which is then transferred to plates containing media that select for growth of heterokaryons. An advantage of the latter approach is that heterokaryotic mycelia can be readily isolated and screened for diploid spores. Regardless of the approach used, the cross plates are typically incubated for one to three weeks to allow the formation and maturation of fruiting bodies (cleistothecia or perithecia). The yield of cleistothecia in *A. nidulans* appears to be higher if the plates are sealed with tape and incubated in the dark. In addition, yields tend to be higher if crosses are performed on minimal media as opposed to richer complete media.

3.2 Random ascospore analysis

A. nidulans forms darkened fruiting bodies known as cleistothecia. Since *A. nidulans* is homothallic, both selfed and out-crossed cleistothecia form on cross plates. However, out-crossed cleistothecia are usually distinguished from selfs by their larger size. Immature *A. nidulans* cleistothecia typically contain thousands of unordered asci with eight ascospores (octads). As cleistothecia age, individual ascospores are released through degradation of the ascus wall (10). Accordingly, mature cleistothecia often contain $> 10^5$ free ascospores, which are recovered by simply crushing the cleistothecium with a pair of fine tweezers (*Protocol* 6).

N. crassa and *M. grisea* are heterothallic fungi which produce fruiting bodies known as perithecia. For *M. grisea*, perithecia are buried in the agar surface of the cross plate with only their 'beaks' protruding. They can be isolated by excavating them from the agar with a pair of fine tweezers. They typically contain 20–30 ordered octads that are recovered by carefully squeezing them out of the perithecium. Incubation of recovered asci on rich media permits degradation of the ascal wall and allows access to individual asci. Alternatively, monoconidial isolation (*Protocol* 5) is performed after the ascospores from a single ascus are allowed to grow into a heterogeneous colony. For *N. crassa*, perithecia are buried in the hyphal mass on the surface of the cross plate. The ascospores are 'shot' out of the perithecium and are usually recovered from the condensed moisture on the surface of the plate using a sterile loop (11).

Protocol 5

Isolation of single conidia

Equipment

- Stereomicroscope (Olympus SZH10)
- Scalpel

Method

1. Spread approx. 10^3 conidia onto the surface of a Petri dish containing 4% solidified agar.[a]
2. Under the stereomicroscope, identify a well-isolated conidium and carefully excise it and a small block of the underlying agar using a sharp scalpel.
3. Transfer the conidium (and underlying agar) to a Petri dish containing the appropriate growth media.

[a] The increased agar concentration provides a harder surface for the spreading and excavation of conidia.

Protocol 6

Octad analysis in *A. nidulans*

Equipment

- Stereomicroscope (Olympus SZH10)
- Fine tweezers (Ted Pella, Inc.)
- Dissecting microscope
- Micromanipulator (Singer Instruments Company, Ltd.; also see ref. 12 for a list of suppliers)

Method

1. Under the stereomicroscope, pick two week-old cleistothecia using a pair of fine tweezers. Transfer them to a Petri dish containing 4% solidified agar.[a]
2. Using tweezers, roll the cleistothecium on the surface of the agar plate to remove adhering Hulle cells, asexual conidia, and hyphal fragments. Do not apply too much pressure as it will prematurely crush the cleistothecium.
3. Transfer the cleaned cleistothecium to a 1.5 ml Eppendorf tube containing 200 µl of sterile water. Crush the cleistothecium against the side of the tube.[b]
4. Immediately examine the cleistothecial contents for the presence of octads.[c] If sufficient numbers are present (usually ~ 1% of the cleistothecial contents), streak a loopful in a straight line along the edge of a plate containing rich media.
5. Invert the plate on the stage of a dissecting microscope such that the streak line lies above the needle (*Figure 2A*).[d]
6. Identify a well-separated octad in the streak line, and, using the micromanipulator, raise the needle and gently pick it off the surface of the plate.
7. Using the stage micrometer as a guide, transfer the octad to a site approx. 10 mm from the streak line.
8. Liberate individual ascospores from the octad using the needle and, if necessary, gentle tapping on the microscope table (mature octads readily break apart).
9. Use the needle to array individual ascospores at 3 mm intervals along a line perpendicular to the streak line (*Figure 2B*).
10. Dissect an additional three octads by repeating steps 6–9. Array the octads in parallel lines that are separated by 3 mm (*Figure 2B*).
11. Incubate the dissection plate at 28 °C for two to three days to allow ascospores to grow into conidiating colonies.
12. Using toothpicks, grid the colonies onto a master plate that can subsequently be used to test for the presence or absence of segregating markers.

[a] The increased agar concentration provides a harder surface for rolling the cleistothecium.

[b] Since *A. nidulans* ascospores are coloured red, their release from the cleistothecium can be easily detected.

[c] Octads are shaped like three-dimensional diamonds. Avoid tetrads and other aberrant clumps of four ascospores, which usually look like one-dimensional diamonds.

[d] See ref. 12 for a thorough discussion of how to set up the micromanipulator and construct a dissecting needle.

3.3 Octad analysis

Octad analysis represents an alternative approach to random spore analysis for the characterization of sexual crosses. Among the advantages of octad analysis are:

(a) The unambiguous identification of double mutants.

(b) The ease of mapping mutations, particularly with respect to centromeres.

Octad analysis has been used extensively to analyse *N. crassa* crosses. Asci are typically dissected out of perithecia using fine tweezers, and individual ascospores are recovered following a period of incubation on rich media to allow degradation of the ascal wall (11). Since *N. crassa* asci are ordered, the position of a spore within the octad is informative.

Unlike *N. crassa*, octad analysis has typically not been used to characterize *A. nidulans* crosses. This is presumably because octads are only present during an intermediate stage of cleistothecial development (10). Furthermore, the modest rate of random ascospore inviability in *A. nidulans* crosses may hinder the analysis of octads. However, in certain situations, the potential benefits of octad analysis may outweigh these caveats. For example, in *A. nidulans*, the construction of double mutants by meiotic crosses is often a laborious process whereby each candidate has to be backcrossed to a wild-type strain to verify that the two parental mutations are segregating. In contrast, by dissecting octads, double mutants can be unequivocally obtained by identifying octads that display a non-parental ditype (NPD) segregation pattern (*Figure 1*).

3.4 Parasexual genetics in *A. nidulans*

In certain filamentous fungi, such as *A. nidulans*, the parasexual cycle provides an effective alternative to sexual crosses for the construction of double mutants. In addition, this approach is frequently used to assign a new mutation to a linkage group. In asexual fungi such as *A. fumigatus* and *A. niger*, the parasexual cycle is

yA* x *chaA

PD: **4 yellow segregants**
4 chartreuse segregants

NPD: **4 green segregants**
4 pale yellow segregants

TT: **2 yellow segregants**
2 chartreuse segregants
2 green segregants
2 pale yellow segregants

Figure 1 Expected octad classes from an *A. nidulans* cross between strains possessing complementing *yA* and *chaA* mutations. The *yA* mutation causes the formation of bright yellow-spored colonies, whereas the *chaA* mutation leads to the production of chartreuse spores. The *yA chaA* double mutant produces a readily distinguishable pale yellow-spored colony. Note that the NPD octad is composed solely of wild-type and double mutant segregants.

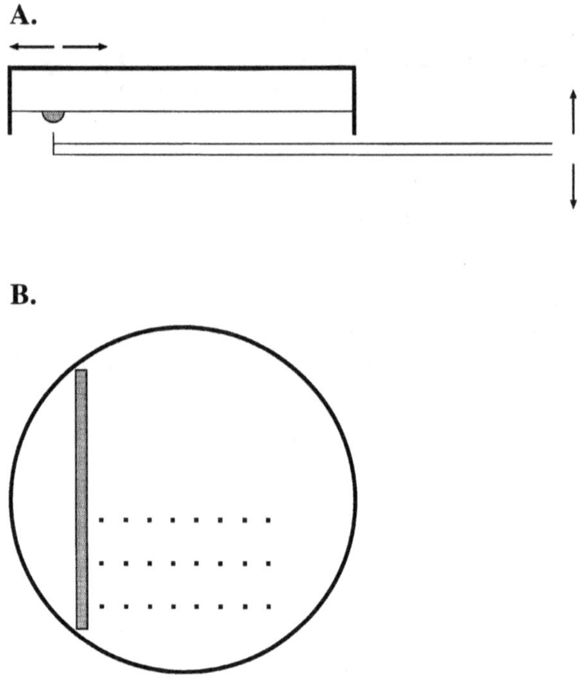

Figure 2 Octad dissection. (A) Arrangement of the plate. The plate is inverted on the microscope stage, and the dissecting needle is brought up underneath the streaked cleistothecial contents. Arrows indicate the range of needle movement using the micromanipulator. (B) Arrangement of dissected ascospores on the plate. The line of streaked cleistothecial contents is indicated by the hatched box. The ascospores are arrayed in rows that are perpendicular to the streak line.

usually the only means available for the rigorous genetic analysis of a new mutation (13). An essential prerequisite for parasexual genetic analysis is the ability to construct stable diploids (14). Nuclear fusion normally occurs at low frequency in heterokaryotic mycelia (i.e. $\sim 10^{-6}$). However, since heterokaryons only produce uninucleate conidia, diploid spores can be readily isolated if the parent strains possess complementing auxotrophic and/or spore colour markers. Once constructed, diploids can be propagated and stored as per haploids.

Protocol 7

Linkage determination by parasexual reassortment

Reagents

- Benomyl (also known as 'Benlate')[a]

Method

1 Use the procedure outlined in Section 3.1 to set up a cross between a strain containing the desired mutation and the appropriately marked mapping strain.[b]

Protocol 7 continued

2 Transfer the myecial mat to a minimal agar plate to select for growth of heterokaryons, which can be identified as fans of wispy mycelia emerging from the mat. Excise a block of heterokaryotic material and transfer to a second minimal agar plate.

3 Harvest conidia from the heterokaryon in sterile double-distilled water. Add fivefold dilutions (i.e. 50 μl, 5 μl, 1 μl) of conidia to 20 ml of molten minimal agar and pour into a Petri dish.

4 Incubate the plates for two to three days to allow colony formation. Diploid colonies can usually be distinguished from heterokaryons or haploids by examining conidial colour and size.[c]

5 Patch conidia from an isolated diploid colony onto a minimal agar plate with a sterile toothpick. Following two to three days growth, harvest conidia from the diploid to generate a stock.

6 Streak a thick smear of diploid conidia down the centre of a fully supplemented CM + Benomyl plate and incubate at 28 °C for five to seven days. During the incubation, haploid sectors will emerge from the dense streak.[d]

7 Prepare a master plate by patching conidia from haploid sectors in a gridded pattern on a supplemented CM plate (i.e. one patch/sector; 26 patches/plate).

8 Score haploids for the presence or absence of segregating markers by toothpick transfer to appropriately supplemented media.

[a] Brand name fungicides such as Bonomyl© (Bonide) contain a benzimidazole precursor as the active ingredient and are usually cheaper and more readily available (i.e. at garden centres or nurseries) than Benomyl. The chemical should be resuspended in 50% ethanol (20 mg/ml) and added to autoclaved CM media (at a concentration range of 5–50 μg/ml).

[b] Mapping strains possessing at least one marker on each chromosome that can be readily scored are available from the Fungal Genetics Stock Center (Dept. of Microbiology, University of Kansas Medical Center, Kansas City, KS 66160-7420; http://www.fgsc.net).

[c] Diploids typically produce larger conidia than haploids. Moreover, when establishing a parasexual cross, it is advisable to use parents with complementing spore colour mutations (i.e. *wA* = white spore, and, *yA* = yellow spore), such that diploid colonies will be green, whereas heterokaryotic colonies will exhibit a mottled white-yellow appearance. The Benlate sensitivity test (16) can also be used to distinguish diploids from haploids.

[d] Haploid sectors expressing recessive spore colour markers are the easiest to identify.

The key aspect of the parasexual cycle is the ability to reduce the diploid genome back to a haploid state. Normally, haploidization occurs at a low frequency (i.e. $\sim 10^{-3}$). However, exposure of diploids to microtubule-depolymerizing chemicals enhances this rate and triggers reassortment by chromosome loss (15). As a result, visible haploid sectors are produced, which can then be tested for the segregation of the relevant markers.

References

1. Perkins, D. D. (1992). *Genetics*, **130**, 687.
2. Martinelli, S. D. and Kinghorn, J. R. (ed.) (1994). *Aspergillus: 50 years on*. Elsevier, Amsterdam.
3. Hynes, M. J. (1979). *Genetics*, **91**, 381.
4. Miller, J. H. (1983). *Annu. Rev. Genet.*, **17**, 215.
5. Harris, S. D., Morrell, J. M., and Hamer, J. E. (1994). *Genetics*, **136**, 517.
6. Goldman, G. H. and Morris, N. R. (1995). *Genetics*, **139**, 1223.
7. Pringle, J. R. and Hartwell, L. H. (1981). In *The molecular biology of the yeast Saccharomyces: life cycle and inheritance* (ed. J. N. Strathern, E. W. Jones, and J. R. Broach), p. 97. Cold Spring Harbor Laboratory Press, NY.
8. MacNeill, S. A. and Fantes, P. A. (1995). In *Cell cycle control* (ed. C. Hutchison and D. M. Glover), p. 63. IRL Press, Oxford.
9. de Souza, C. C., Pellizzon, C. H., Hiraishi, M., Goldman, M. H. S., and Goldman, G. H. (1998). *Curr. Genet.*, **33**, 60.
10. Kirk, K. E. and Morris, N. R. (1991). *Genes Dev.*, **5**, 2014.
11. Fungal Genetics Stock Center, Dept. of Microbiology, University of Kansas Medical Center, Kansas City, KS 66160-7420, USA. http://www.fgsc.net/methods/stanford.html
12. Sherman, F. and Hicks, J. (1991). In *Methods in enzymology* (ed. C. Guthrie and G. R. Fink), Vol. 194, p. 21. Academic Press, NY.
13. Pontecorvo, G., Roper, J. A., and Forbes, E. (1953). *J. Gen. Microbiol.*, **8**, 198.
14. Pontecorvo, G. (1956). *Annu. Rev. Microbiol.*, **1**, 393.
15. Hastie, A. C. (1970). *Nature*, **226**, 771.
16. Upshall, A., Giddings, B., and Mortimore, I. D. (1977). *J. Gen. Microbiol.*, **100**, 413.

Chapter 5
Genetic analysis of basidiomycete fungi

Piers J. Walser, Marcel Hollenstein, Michaela J. Klaus, and Ursula Kües
ETH Zürich, Institut für Mikrobiologie, CH-8092 Zürich, Switzerland.

1 Introduction

The basidiomycetous fungi are divided into two main groups, the heterobasidiomycetes and the homobasidiomycetes. For the most part, genetic studies of heterobasidiomycetes have been restricted to *Ustilago maydis*, *Microbotryum violaceae*, and other smut fungi, while *Coprinus cinereus* and *Schizophyllum commune* have served as model organisms for the homobasidiomycetes. Other homobasidiomycetes are studied for industrial purposes, for example the white rot fungus *Phanerochaete chrysosporium* which produces exoenzymes such as laccases, and various edible mushrooms including *Agaricus* spp. These species can easily be cultured on relatively simple media, and they complete their entire life cycle within a few weeks under laboratory conditions. Since we are working with *C. cinereus*, here we present genetic techniques that are applied to this fungus. The approaches, however, are broadly applicable to the full range of basidiomycetes.

1.1 Life cycles

The characteristic life cycle of the basidiomycetes starts with the germination of the basidiospore (meiospore) to give rise either to yeast-like cells called sporidia, which proliferate by budding (heterobasidiomycetes), or to a primary homokaryotic (monokaryotic) mycelium (homobasidiomycetes). By fusion of two cells, which are compatible in mating type, a secondary mycelium called the dikaryon is formed (*Figure 1*). The dikaryon typically contains two, genetically distinct, haploid nuclei per hyphal compartment. Nuclear division is performed in a highly synchronized fashion. One daughter nucleus of the pair of dividing nuclei migrates into a clamp cell formed laterally on the hyphal tip cell. A septum is then formed in the spindle plane between the hypha and clamp cell, and another septum emerges between the other pair of dividing nuclei, more distantly located from the hyphal tip. As a result of this, two genetically distinct nuclei are situated in the apical hyphal cell; the two remaining nuclei being trapped in the

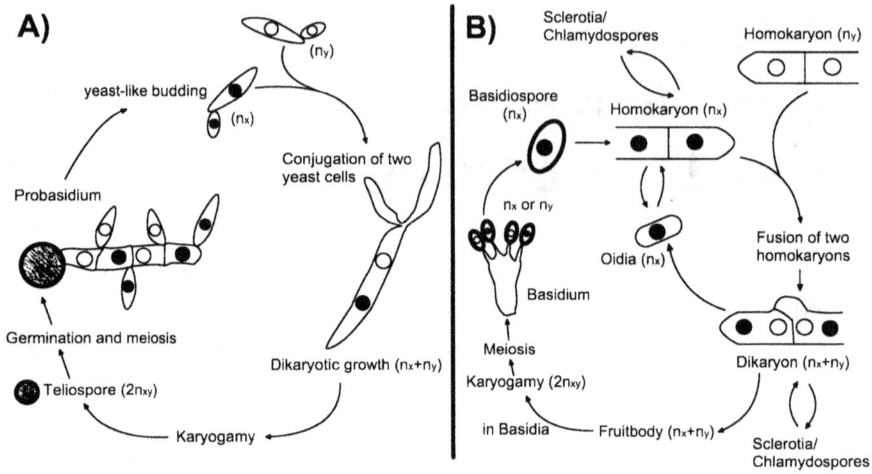

Figure 1 Schematic presentation of a life cycle of the heterobasidiomycete *Ustilago maydis* (A) and the homobasidiomycete *Coprinus cinereus* (B). For simplification, subsidiary cycles in *C. cinereus* are shown for only one homokaryon.

clamp and sub-apical cell. By fusion of the clamp to the sub-apical cell, the dikaryotic nuclear distribution is restored (1).

Dikaryotic mycelium is capable of undergoing sexual development. Karyogamy, subsequent meiosis, and basidiospore formation occur within specialized cells, called basidia which are borne by fruit bodies in the case of homobasidiomycetes (1). Karyogamy in smut fungi occurs *in planta* and results in thick-walled diploid teliospores (brand spores) which, as a consequence of meiosis, form the promycelium (phragmobasidium), budding off four types of haploid sporidia (2). The diploid phase in basidiomycetes is therefore restricted to a very short period immediately prior to meiosis, while the haploid state is maintained throughout the majority of the life cycle. Only in a few exceptional cases (*Armillaria* species), is the ploidy state in the vegetative mycelium also diploid (3, 4).

In addition to the sexual cycle, basidiomycetes may exhibit subsidiary developmental cycles of asexual reproduction. Asexual aerial spores, either conidia or oidia, are formed either by the primary or the secondary mycelium, or by both, and those produced on dikaryons may be homokaryotic or dikaryotic. Large thick-walled chlamydospores and multicellular sclerotia can also be produced as part of the generating mycelium and reflect either the homo- or heterokaryotic state (1).

1.2 Breeding systems

In the basidiomycetes, we distinguish homothallic (self-fertile) and heterothallic (self-sterile) species. Homothallic mycelia undergo sexual development without fusion to another mycelium of the same species (*P. chrysosporium*), restricting genetic studies. Sexual development in heterothallic species requires fusion of two homokaryotic strains that differ in their mating type specificities—they must contain different alleles at their mating type locus/loci. In this way outbreeding is promoted. Heterothallic species are further divided into two subgroups, bipolar (unifactorial) and tetrapolar (bifactorial). In bipolar species (for example *M. violoceae*, *Ustilago hordei*, *Coprinus congregatus*), self-incompatibility is regulated by a single multiallelic locus whose different forms segregate 1:1 in meiosis. In tetrapolar species, such as *C. cinereus*, *S. commune*, and *U. maydis*, two unlinked mating type loci are present, resulting in four types of mating type segregants from meiosis (5).

Secondarily homothallic (pseudohomothallic) species such as *Agaricus bisporus* can either be bipolar or tetrapolar. In these species two of the four products of meiosis migrate into a basidiospore, and when of compatible mating type, circumvent the need for fusion with another mycelium. However, secondarily homothallic breeding systems are usually not as tightly controlled and because only a proportion of spores contain one or two nuclei with the same mating type, classical genetic analysis can be carried out, albeit with more effort (6).

1.3 Mating types

The mating type loci regulate sexual development in a highly concerted way, each controlling discrete steps in dikaryon formation and fruit body development (1, 5), and, in the smut fungi, also pathogenicity (2). The two mating type loci in *U. maydis*, *C. cinereus*, and *S. commune* have been shown to encode homeodomain transcription factors and pheromones and pheromone receptors, respectively. Unfortunately, the classical mating type nomenclature *a* and *b* in *U. maydis*, *A* and *B* in *C. cinereus* and *S. commune*, respectively, does not correspond to loci of similar molecular function and more confusion in nomenclature has to be expected as more species are analysed. It has therefore been proposed to call the transcription factor encoding loci *matT*, and the pheromone/pheromone receptor loci *matP*. Analysis of *U. hordei* suggest that in bipolar species *matT* and *matP* are so closely linked that they fail to recombine in meiosis. In many species one or both loci are multiallelic. Different specificities at the mating type loci are indicated by different numbers (e.g. *A1*, *A2*, ... *Ax*, and *B1*, *B2*, ... *Bx*). The molecular basis of the high variability in mating type specificity has been established and can be found in recent reviews (1, 5).

2 Strain isolation

2.1 Isolation of haploid strains from dikaryons through meiosis

To obtain haploid yeast-like cells from smut fungi, diploid teliospores (which are obtained by grinding infected plant tissues, filtering through glass wool, and incubating overnight in 1.5% $CuSO_4$ solution to remove contaminants and vegetative cells) are spread on nutritive medium (see Appendix 1 to this chapter). Two days later, after meiosis, pro-basidium formation and budding at 27–32 °C, sporidia can be washed off the plates with sterile water. Genetically homogeneous colonies grow out from single cells after plating of appropriately diluted sporidial suspensions (50–100 cells per plate) on fresh agar medium (7, 8).

Haploid strains are obtained from filamentous heterothallic, and with a low frequency from secondary homothallic, basidiomycetes by plating and germinating basidiospores on rich medium such as YMG (see Appendix 1). Basidiospores may be collected from fruit bodies in the wild, either by harvesting spores shed overnight from the fruit body cap, placed in a sterile Petri dish or, in the case of a *Coprinus*, by collecting the spore liquid of an autolysing cap. To avoid bacterial contamination, antibiotics can be added to the germination medium (a solution of 25 000 units penicillin/ml, 25 mg streptomycin/ml, 50 mg neomycin/ml or a pre-prepared SIGMA P-3364 solution). Production of fruit bodies from dikaryotic mycelium in the laboratory is limited to a few species and usually requires very special conditions (for *C. cinereus* see *Protocol 1*). Fruiting of *S. commune* occurs at a somewhat lower temperature at 22 °C after a brief period of white (blue) light illumination on CYM medium (see Appendix 1) (9).

Protocol 1

Production of fruit bodies and basidiospores on *C. cinereus* dikaryons and *Amut Bmut* homokaryons on artificial medium

Equipment and reagents

- A sharp tungsten needle: tungsten wire, *ca.* 2–3 cm in length, 0.3 mm in diameter, fixed in a needle holder[a]
- 25 °C (–28 °C) incubation chamber/room equipped with light sources (light intensity 50–75 μE m^{-2} s^{-2}, light source, e.g. Osram L36W) and a humidifier[b]
- Ventilated, light-proof boxes (home-made) with a layer of wet paper towels[b]
- YMG agar plates (see Appendix 1) or YMG/T agar plates (YMG supplemented with 100 mg tryptophan/litre)

Method

1. Cut a small plug of mycelium (2 mm^2) from a grown plate culture with the flamed tungsten needle and place it onto the middle of a fresh agar plate.
2. Incubate plate at 37 °C in the dark (in boxes) until the first hyphal tips reach the edge of the Petri dish.[b]
3. Transfer plate into a 12 h light/12 h dark regime at 25 °C, 80–90% humidity; fruit bodies grow and maturate in about a week.[b]
4. Harvest the spores into a sterile Eppendorf tube with 1 ml of sterile H_2O when spores have blackened and the fruit body cap autolyses.
5. Plate spores in suitable dilutions on fresh medium or store spores at 4°C until needed.[c]

Protocol 1 continued

6 Incubate plates with spores for one to two days at 37 °C.

7 Cut out small agar pieces with individual colonies with the tungsten needle[d] and transfer them onto fresh medium for further growth.

[a] Dikaryotic mycelium is often very tough and tungsten needles have been found to cut the mycelial matting best.

[b] The dark–light schedule, a high humidity both at growth and during the fruiting period, and the time point of transfer from 37 °C to 25 °C are of importance for reliable fruiting.

[c] Spores keep for several months without loss of germination ability.

[d] Although more arduous, we usually do this under a dissecting microscope to avoid picking up closely neighbouring colonies not detectable as such by eye.

2.2 Isolation of haploid strains through dedikaryotization of dikaryotic mycelium

When fruit bodies are not available or basidiospores do not germinate, haploid cultures might be obtained from the dikaryon through 'dedikaryotization'. This process gives rise to both or, more often, to one type of homokaryotic mycelium containing the unrecombined component nuclei of the dikaryon, in contrast to meiotic basidiospores which usually represent recombinants of the two different genotypes. Techniques for dedikaryotization, which are also useful for secondary homothallic species, are as follows.

(a) Isolation and germination of haploid asexual spores (*Protocol* 4) which some species (*C. cinereus*, *Pholiota nameko*) produce on the dikaryotic mycelium (10, 11). Sporulation on the *C. cinereus* dikaryon needs to be induced by light (*Protocol* 4).

(b) Production and regeneration of protoplasts (*Protocol* 7) which is especially interesting for species such as *S. commune*, having no haploid mitotic spores on the dikaryon (12).

(c) Maceration of mycelium in a sterile Waring blender and plating of macerates —a less efficient method that might be used when protoplasts can not be obtained (13).

(d) Microsurgery.

   (i) The tip cell of a growing hyphae, its clamp cell, and the cells downstream of the sub-apical cell are killed by piercing them with a sharp micromanipulator needle directly after septum formation but before clamp cell fusion, leaving only the uninucleate sub-apical cell alive for growth (14). Since the two nuclei of a dikaryon change the leading position at each division, determining which nucleus migrates into the clamp and which into the sub-apical cell (1), it should theoretically be possible to obtain homokaryons of both component nuclei.

(ii) On solid medium isolated dikaryotic chlamydospores and the large-sized outer veil cells of *C. cinereus* fruit bodies often germinate at opposite ends into homokaryotic hyphae. A little less skill and only a dissecting microscope are needed to separate these homokaryotic hyphae, using a sharp scalpel, before they form dikaryons once again (15, 16).

(e) Growth on media with sub-lethal concentrations of toxic substances (bile salts/oxgall, sodium cholate, Cu^{2+}, Zn^{2+}, Cr^{6+}, arsenic) selects for survival of homokaryotic hyphae containing the more resistant nucleus (17, 18).

(f) Starvation of dikaryotic mycelium on low phosphate, low Mg^{2+} medium containing 3–8% glycine or peptone and 2% glucose leads to outgrowth of homokaryotic mycelia (19).

2.3 Mating for dikaryon formation

Haploid sporidia of smut fungi of different mating type conjugate on special charcoal medium (*U. maydis*; see Appendix 1) or 2% water agar (*M. violaceae*) and short dikaryotic filaments develop at the permissive temperature (32 °C for *U. maydis*, 15–20 °C for *M. violaceae*). Further dikaryotic development occurs only in host plants however. To obtain teliospores from these species, dense suspensions of mixed haploid cells (10^7 to 10^8 cells/ml) are injected into plant nodes or leaves (*U. maydis* into four- to ten-day-old maize seedlings, *M. violaceae* into young *Silene alba* plants). About ten days later, diploid teliospores appear within infected tumorous tissue and can be harvested and stored in sterile desiccated silica gel for several years (2, 7, 8).

Mating of filamentous basidiomycetes is performed by placing two small plugs of mycelium from a plate culture (one from each parent strain) close together (1–2 cm apart) onto YMG agar or any other suitable medium (*Figure 2*). Plate cultures have to be grown at a suitable temperature (for *C. cinereus* usually at 37 °C, although sometimes mating is better at lower temperatures such as 24–30 °C, as for *S. commune*). Following hyphal fusion, outgrowth of a dikaryon can occur at three different places; at the zone in between the two homokaryons, and at the outer border of each homokaryon (*Figure 2*). During mating, nuclei will be exchanged and migrate through the homokaryotic mycelium of the mating partner until they reach a hyphal tip cell where they pair and divide synchronously with the resident nucleus, to form the dikaryotic hyphae (1). Since mitochondria are not exchanged, the dikaryons at the distal borders of the homokaryons will be genetically distinct. Those in the middle are initially mitochondrial chimeras but the mitochondrial load is quickly sorted in favour of only one type (1, 20). Recombination of mtDNA can occur in the interim zones of a cross (20), and examples are known where the mitochondrion influences the phenotypes of the dikaryons (21). Thus, dikaryons should preferentially be taken from the outer borders of a homokaryon to be sure about the origin of mitochondria. However, as seen in *C. cinereus*, not all homokaryotic strains do accept nuclei —in some crosses dikaryons will only form in between the mating partners (22) (see *Figure 2*).

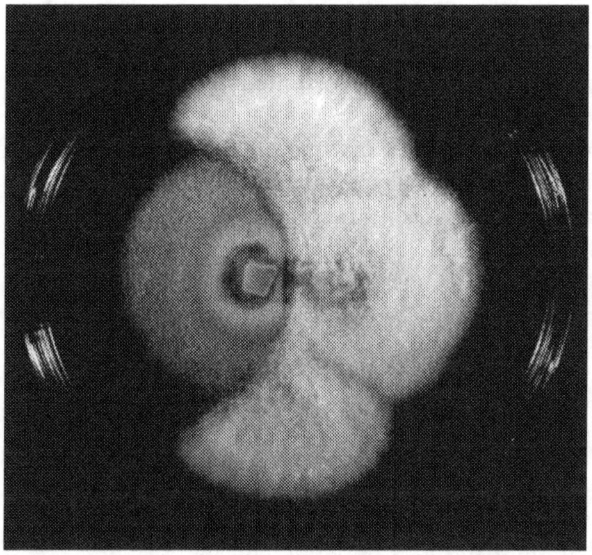

Figure 2 Mating between the compatible *Coprinus cinereus* monokaryons PG78 (*A6 B42 pab-1 trp-1.1,1.6*) at the left and AT8 (*A43 B43 trp-3 ade-8*) at the right. Note the vigorously growing dikaryotic mycelium between the two monokaryotic colonies and at the edge of the colony of monokaryon AT8. At the edge of the colony of monokaryon PG78, no dikaryon is formed.

2.4 Isolation of strains of special genetic condition: forced heterokaryons, diploid strains, solo-pathogenic strains, and *Amut Bmut* homokaryons

When analysing functions of the mating type loci in filamentous fungi, it is often of interest to form heterokaryons containing the same genetic information at one or both mating type loci (for example, common *A*, common *B*, and common *AB* heterokaryons). Usually, such heterokaryons are very unstable and a strong selective pressure is needed for their selection and maintenance. Mycelial plugs of two strains carrying different auxotrophic markers are placed directly together on minimal medium. Upon hyphal fusion, heterokaryotic mycelium will grow due to complementation of the different nuclear defects (1). For homothallic species such as *P. chrysogenium*, formation of forced heterokaryons between different auxotrophic mycelia (obtained from mutagenesis of uninucleate conidia), on minimal medium is the trick used to obtain balanced heterokaryotic mycelia that will give rise to recombined progeny following meiosis (23).

In *C. cinereus*, somatic diploids can be isolated from forced common *A* and common *AB* heterokaryons (although so far this has not been shown for common *B* heterokaryons), by harvesting oidia (see *Protocol 4*) and plating them on minimal medium (see Appendix 1). At frequencies of 10^{-3} to 10^{-4}, oidia can contain a single diploid nucleus formed by fusion of the two different haploid nuclei and only these oidia will germinate on minimal media (1). Diploid mycelia of *S. commune* arise when common *AB* heterokaryons, carrying different auxotrophic markers in their nuclei, are transferred onto minimal medium (24). Diploid prototrophic cells from smut fungi are selected by incubating mating-compatible strains with different auxotrophies for 24 hours on mating medium at room temperature or 15 °C (Section 2.3), followed by spreading the cells on minimal medium (7, 25). The $a{\neq}b{\neq}$ diploids obtained of *U. maydis* (heterozygotes for both mating type loci) are filamentous on charcoal medium. If irradiated with low UV doses (15–30% survival rate), diploid $a{\neq}b{=}$ and diploid $a{=}b{\neq}$ yeast-like cells (heterozygous for *a* and *b*, respectively) appear on the plates due to mitotic recombination. $a{\neq}b{\neq}$ and $a{=}b{\neq}$ diploids are 'solo-pathogenic' and can infect plants and produce teliospores in the host without prior mating to another strain (25). Haploid solo-pathogenic strains, which have proved useful for the isolation of pathogenicity mutants (26), are generated through incorporation of compatible *b* mating type genes at the endogenous *b* locus by homologous recombination in transformation (27).

In the homobasidiomycetes, self-compatible homokaryons with defects in the mating type loci (*Amut Bmut* homokaryons) are of special interest for studying homokaryotic functions of the dikaryon (fruit body development, meiosis) in a haploid condition. These homokaryons may arise spontaneously or through chemical mutagen treatment (Section 6.2). The respective mutations are rare but selection and isolation is easy when using a positive selection, the formation of a fruit body containing basidiospores with the mutated loci. Mutations in the two mating type loci are best selected independently from each other by co-culturing two strains, differing only in one of the two loci (either *A* or *B*). Fruit bodies will appear upon a mutation in the common mating type information. Mutated *A* and *B* loci in different isolates can be combined within the same nucleus through crosses. *Amut Bmut* homokaryons from *C. cinereus* and *S. commune* are available (28–30). Since both *A* and *B* genes are now available for transformation (1, 5), it is also possible to obtain dikaryon-like homokaryotic mycelia through ectopic chromosomal integration of the genes (U. K. and M. J. K., unpublished).

2.5 Preservation and maintenance of cultures

The example in *Figure 3* shows that long-term culturing of the same strain might lead to the accumulation of mutations causing divergence of phenotypes. Thus, in order to keep a genetic condition unchanged it is advisable to prepare frozen stock cultures as soon as an interesting strain is isolated or obtained from another laboratory. Yeast-like cells, such as those of *U. maydis* grown in liquid culture, can be diluted with sterile 15–35% glycerol for storage at −80 °C. These frozen stocks last for a significant period and only tiny amounts need to be used to inoculate fresh medium for re-culturing. For mycelial basidiomycetes, plate cultures are prepared and sliced into small squares. Around 15–20 of these are transferred into 1 ml of sterilized 15% glycerol in 2.5 ml cryo-tubes (Nunc™) for freezing at −80 °C. For re-culturing, the whole vial has to be defrosted and the contents emptied onto a fresh plate. Since glycerol can have negative effects on growth, the liquid should be poured off separately. According to Pukkila (31) it is

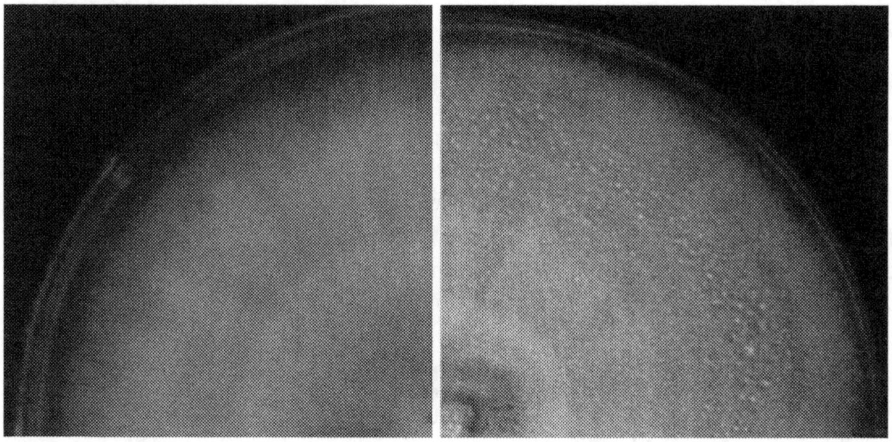

Figure 3 *Coprinus cinereus* monokaryon 218 as obtained from P. J. Pukkila (left) and L. A. Casselton (right). Both strains have the mating type specificities *A3 B1* and are auxotroph for tryptophan (*trp-1.1,1.6*) as originally described for the strain by Binninger *et al.* in 1986 (84). The strains were grown at 37 °C on YMG/T.

possible when storing *C. cinereus* to freeze and thaw a vial several times with no loss of viability, so that each time only one or a few of the agar plugs may be removed. However, we routinely keep ten or more original frozen stocks of each important *C. cinereus* strain. An original stock will then be used to produce a subset of frozen cultures. Working cultures are produced from these latter stocks and are kept for some months in continuous culture before restarting from another frozen stock. The length of the culturing period can depend on the strain and the stability of characters of importance. For example, homokaryon LT2 (*A6 B6 trp1.1,1.6*), a common transformation strain, easily loses its ability to produce aerial mycelium and oidia. In contrast, in five years of continuous research, homokaryon *Amut Bmut* (*A43mut B43mut pab1*) has proven to be completely stable and we have only once noticed a loss in fruiting ability.

3 Genetic complementation

Complementation tests in smut fungi are carried out by synthesizing appropriate diploid strains (Section 2.4) on mating medium and streaking out the diploids on selection medium. Auxotrophic markers in the tetrapolar *U. maydis* must be in strains of different *a* mating type (controlling cellular fusion), whereas the specificity at the *b* mating type locus (controlling pathogenicity and later stages of sexual development) can be the same or variable (7, 24). In homobasidiomycetes, where hyphal fusions occur independently of the mating type (1, 5), complementation of metabolic defects can be tested either by dikaryon or forced heterokaryon formation (Section 2.3/2.4). Whether or not developmental defects in self-compatible *Amut Bmut* homokaryons can be complemented, is often established by crossing the respective *Amut Bmut* mutants (32).

4 Genetic mapping

Extensive classical genetic maps obtained from sexual and parasexual recombination are available for *C. cinereus* (33), *Schizophyllum commune* (27), and *M. violaceae* (34), assigning so far 108, 66, and 43 markers, respectively, to specific chromosomes. Although the former importance of classical genetics has been overtaken to an extent by molecular genetics, it is still of value in mutant analysis—a well performed genetic analysis can greatly reduce the molecular efforts to obtain mutated genes (see Chapter 6). Moreover, new molecular techniques also create new tasks for classical genetics (Section 7.2.3).

4.1 Genetic mapping by sexual recombination

Classical gene mapping employs crosses of mutants and wild-type strains. In most basidiomycetes, recombination frequencies between two genetic markers, reflecting linkage distances, are followed using random progeny from a sexual cross. Usually, basidiospores will be germinated on non-selective medium before analysing the distribution of the markers among the progeny. In the case of morphological markers, colonies will be directly analysed. Progeny are transferred onto a specific selection medium in the case of auxotrophic or resistance markers, or into specific selection conditions in the case of conditionally lethal markers, or when studying defects in pathways such as fruit body development. When handling auxotrophic and resistance markers, parallel plating on selective and non-selective medium is the fastest way to determine the proportion of the different possible progeny. Linkage between two genetic markers is given when the frequency of parental phenotypes is significantly greater than the frequency of non-parental phenotypes. In the case of unlinked genes, parental and non-parental phenotypes will appear to equal proportions (33) (see also Chapter 4 of this book).

The genetic backgrounds of strains have often been neglected in genetic crosses of basidiomycetes. As a consequence, skewed Mendelian segregations have regularly been recorded, caused for example by translocations or by the presence of alleles within one or both parents which hinder spore germination (7, 31, 32, 36–38). To avoid any unnecessary conflicts in later analysis, classical genetic mapping is therefore ideally performed using co-isogenic strains. Co-isogenic strains are generated by repetitive backcrossing to a fixed parent. As indicated by the algorithm $C(n) = 2c(\frac{1}{2})^n = c(\frac{1}{2})^{n-1}$ of Leslie (39)—where $C(n)$ is the average number of chromosome tips remaining allogenic, c the haploid chromosome number, and n the number of backcrosses. The absolute number of different chromosomes of a given species determines the number of backcrosses that have to be performed. Due to the work of Raper's laboratory, co-isogenic strains distinguished just by their mating type (obtained by ten backcrosses) are available for *S. commune* (9, 30).

4.2 Electrophoretic karyotyping and RFLP analysis

Electrophoretic karyotyping can now be used to assist genetic mapping. Classical genetics defines only those chromosomes where marker genes are available, while karyotyping identifies the absolute number of chromosomes. Usually, isolates of the same species differ in length of some, if not all, homologous chromosomes (40, 41). These length polymorphisms can be followed in progeny by CHEF gel electrophoresis (*Protocol 2*). Phenotypic markers can be consecutively assigned to individual chromosomes present in all clones exhibiting the phenotype of interest. Instead of analysing a high number of isolates from a first cross, it is more efficient to reduce the number of variable chromosomes by repeated backcrosses of mutant progeny to the wild-type parent. Each backcross will eliminate some chromosomes of the original mutant leaving, in the optimal case, only one chromosome difference between the last mutant progeny and the wild-type parent, this being descended from the original mutant, or being a recombinant chromosome of both strains (42, 43). The number of backcrosses to be performed will dependent on the absolute chromosome number of the fungus. Three to four backcrosses is normally enough for *C. cinereus* and *S. commune* which have 13 and 11 different chromosomes, respectively (40, 44).

Protocol 2

Separation of *C. cinereus* chromosomes by CHEF gel electrophoresis[a]

Equipment and reagents

- CHEF-DRII™ system (Bio-Rad) including a Bio-Rad CHEF 5 × 5.5 inch gel mould and plug mould; set up in a cold room
- Cooling water-bath and pump linked with a tubing system to the electrophoresis chamber
- Hybond N^+ membranes (Amersham)
- MM buffer: freshly prepared from 2 vol. sterile 1 M mannitol and each 1 vol. of sterile 0.2 M maleate pH 5.5 and sterile H_2O
- Protoplasts (well digested) in MM buffer at a density of 6×10^8/ml (for protoplasting see *Protocol 7*)
- 2% low melting point agarose solution in MM buffer held at 50 °C
- Tris/EDTA: 0.01 M Tris–HCl pH 9.5, 0.3 M EDTA, filtered and autoclaved
- NDS: dissolve 2% *N*-lauroyl sarcosine, 1 mg/ml proteinase K in Tris/EDTA immediately before use
- TBE: 45 mM Tris base, 45 mM boric acid, 1.25 mM EDTA pH 8.3
- Bio-Rad chromosomal grade agarose (for 3–4 Mb sized chromosomes) or Beckman LE agarose (for 1–2 Mb sized chromosomes)

A Plug preparation

1. Mix aliquots of protoplasts 1:1 with the agarose solution, immediately fill into the plug mould, and chill at 4 °C for 20 min.
2. Remove plugs with metal spatula into a Petri dish, cover the plugs completely with NDS, and incubate for 24 h at 50 °C.
3. Remove NDS with 20 ml Tris–HCl and store plugs at 4 °C until use.

Protocol 2 continued

B Gel electrophoresis

1 Prepare a 140 ml 0.5 × TBE agarose gel (1%, w/v Beckman or 0.9% Bio-Rad agarose) in the gel mould.

2 Load agarose-protoplast plugs (6 × 9 mm) into the wells of the gel and seal with 1% low melting point agarose.

3 Run gels in the CHEF-DRII chamber in a cold room at 60 V for about six days with a 22 min pulse in 0.5 × TBE.

4 Maintain TBE buffer temperature at 14°C during the run by buffer circulation through a 4–9 °C water-bath.

5 Change the 0.5 × TBE buffer daily.

6 Stain gels in 0.5 μg/ml ethidium bromide for 1 h when run is finished and photograph.

7 If Southern blot analysis has to be performed, transfer chromosomes to Hybond N^+ membranes and hybridize with suitable gene probe following the procedures as described by the manufacturer.

[a] Protocol was originally developed by Zolan and co-workers (48) and can be adapted to other fungi according to the length of their chromosomes.

Electrophoretically separated chromosomes, in combination with Southern blot analysis (*Protocol 2*), can give information regarding the chromosomal location of an available gene sequence (38, 41, 42). However, this analysis cannot show genetic distances between markers present on the same chromosome. To define the exact gene localization on a chromosome, crosses can be performed using markers whose positions have already been assigned on the chromosome. Traditionally, genetic mapping requires different gene alleles leading to distinct phenotypes. Many such markers, originating mostly from the 1960s and 70s, are still obtainable from type culture collections such as ATCC and from the researchers in the field, but others might have been lost over the years. However, molecular markers are easily obtainable using RFLPs (restriction fragment length polymorphisms), detected in Southern blot analysis of unrelated digested genomic DNAs, with either known gene sequences or random chromosomal DNA fragments. Molecular markers can be treated in linkage determination in the same way as physical markers (38). When close linkage is encountered, this information can be used in chromosome walking in order to clone a gene of interest (45) (see also Chapter 6). Contig maps, currently only available for the smallest chromosome of *C. cinereus* (46), and genomic sequencing projects, approaching completion for *U. maydis* (47) will greatly assist molecular mapping approaches in the future.

4.3 Tetrad analysis

Tetrad analysis has great advantages over random spore analysis. Defects in spore viability are easily detected and the number of progeny to be analysed is reduced. Furthermore, in ordered tetrads, centromeres act as additional markers for linkage determination indicating whether genetic markers are separated in the first or in the second division of meiosis (see Chapter 4). Basidiomycetes, however, generally have unordered tetrads so that the spore arrangement does not allow any conclusion regarding the mode of meiotic reduction, and two genetic markers for genetic analysis are usually required (35).

In smut fungi, unordered tetrad analysis is easily performed because teliospores (and thus each individual meiotic event) are simply separated by plating suitably diluted spore solutions onto solid medium (Section 2.1). Sporidia (the asexual progeny of the four products of meiosis), budding from the germinated promycelium can also be isolated as a whole colony from the agar (using a sharp needle and possibly a dissecting microscope). Cells are suspended in a small volume of sterile water and re-plated on fresh medium to produce monosporidial colonies. To minimize the loss of particular genotypes with a poor growth rate, it is advisable to take very small teliospore colonies of 50–100 cells (8, 9, 50). Growing colonies can be analysed for morphological markers (colony shape, surface texture, colour) or replica plated on selection media for analysis of marker distribution (Section 6.2.2). Once a marker has been identified in closest proximity of a centromere, it can then be used as an indicator of the specific centromere. In *M. violaceae*, a gene cluster for colony colour marks a centromere and allows half-tetrad analysis: teliospores are directly plated onto selection medium enabling only two of the four basidiospores to bud off sporidia. Uniformly-coloured colonies indicate first meiotic division segregation, sectoring colonies represent second meiotic division events, and an even rate between colonies of parental (parental ditype, *P*) and of non-parental colour (recombinant ditype, *R*), represents independent segregation of the second marker. Any significant deviation from an equal *P*:*R* distribution represents linkage between the colour marker (centromere) and second marker. Linkage of a marker to its centromere is determined from the frequency of teliospore colonies of mixed colour (tetratype, *T*) using the equation $(T/2)/N \times 100$ where $T =$ absolute number of tetratypes and $N =$ absolute number of colonies (50).

In most homobasidiomycetes, the distribution of genetic traits during meiosis can only be studied by random spore analysis after plating spore suspensions on germination media. However, because of a relatively large spore size (9–13 $\times$ 6–7 μm), synchronized karyogamy and meiosis within a given fruit body and black spore pigmentation indicating maturation (1), protocols are available for dissecting tetrads of *C. cinereus* using a micromanipulator (31, 51). The rapidly proceeding cap autolysis of *C. cinereus* at the point of maturation can be stopped by transfer of the fruit body to 4 °C and single gills can be removed as required.

Protocol 3
Tetrad dissection in *C. cinereus*

Equipment and reagents

- Microscope with a × 10 objective and a × 16 ocular equipped with a micromanipulator (see ref. 52 for model descriptions)
- 0.5 cm^2 agar slabs cut with a sterile spatula from 10 ml YMG (4% agar), solidified in a Petri dish
- Sterile slides fitting to the respective dissecting chamber of the microscope
- Glass needle made from a 3 mm soft glass tubing, clamped into the micromanipulator
- Petri dish with a support for the microslide and a layer of dried silica gel

Method

1. Split a double-sided gill with sterile forceps into two layers, lay the 'half-gill' segment spore-side up at one end of a microscope slide and allow to dry.
2. (Optional: Disperse the spores onto the surrounding glass by sweeping the micromanipulator needle across the dried gill.)[a]
3. Quickly dry the sample by placing the slide into a Petri dish containing a layer of dry silica gel to aggregate the four spores of a tetrad.
4. Store slides at room temperature over silica gel.[b]
5. When needed, take tetrads directly from the gill slice or from the glass slide with a micromanipulator needle and transfer them for dissection onto agar slabs placed onto the free end of the microscope slide.
6. Separate the four spores of a tetrad on the agar slab under the micromanipulator, or by hand with a suitable glass needle under a dissecting microscope.
7. Germinate the separated spores overnight at 37 °C and transfer germlings using a tungsten needle to plates for further growth.

[a] To variable degrees, proportions of spores spread onto the glass will stay together in whole tetrads.

[b] Slides keep for several months without loss of spore germination.

4.4 Gene mapping by parasexual recombination

The parasexual cycle in fungi includes the production of a diploid cell (Section 2.4), mitotic recombination to yield diploid segregants, and haploidization to give haploid or aneuploid segregants. In species performing meiosis under laboratory conditions, a parasexual cycle can be used for linkage determination but is usually not required (9, 10). In contrast, parasexual recombination is the most effective means of genetic analysis in the vegetative diploid *Armillaria mellea* and of organisms that do not easily undergo meiosis (3). In some species (*M. violaceae*), mitotic crossing-over in diploid nuclei occurs at frequencies of up to 5% (50). Mitotic recombination in other species can be increased by low doses of UV irradiation (5–8 mJ/cm^2) (53). To select for recombination events in diploid nuclei, haploidization is often induced by low doses of UV irradiation (7, 50) or treatment with 0.3 g/litre *p*-fluorophenylalanine (50), 9–15 mg/litre thiabendazole (54), 25 μg/litre Benomyl (3), 20 mg/litre griseofulvin (55), and other chemicals (7). Linkage analysis of marker genes is carried out as in ascomycetes (53, 56, 57) (Chapter 4 of this book).

Dikaryons of homobasidiomycetes are capable of contributing fertilizing nuclei to a haploid homokaryon resulting in a new dikaryon. This process, called the 'Buller phenomenon', can also be exploited for parasexual genetic analysis in 'illegitimate' dikaryon–homokaryon matings, where the haploid nucleus of the homokaryon is only half-compatible with each nucleus of the dikaryon. The haploid nucleus shares with one nucleus of the dikaryon, the *A* specificity, and with the other the *B* mating type specificity. In such illegitimate matings, dikaryons arise from the homokaryon invaded by a nucleus with a new combination of mating type specificities, indicating the transient existence of a diploid nucleus and mitotic recombination within the original dikaryon. The recombinant nuclei of new mating type specificity usually contain additional markers from both parental nuclei (1, 58). Analogously, in matings with diploid and haploid strains of *Armillaria gallica*, recombinant nuclei appear carrying, equally or unequally, genes from both the haploid and the diploid parental nucleus (4).

5 Natural mutants

When studying developmental processes such as fruit body formation, it can be useful to isolate various strains from the wild. In *C. cinereus*, four different genes acting in fruit body initiation and sclerotia formation have been identified from a very small collection of 11 wild-type isolates. Recessive alleles of these genes which have defects in the performance of fruit body development, are obviously maintained within natural populations because of the high chance of allele complementation in the dikaryon (37). A developmental mutant isolated from the progeny of a fruit body collected in the field, caused the isolation of *ich1*, a gene essential for pileus formation (45). Another fruit body from the wild gave rise to homokaryon CopD5-12 (*A12 B12*) which has a trait of homokaryotic fruiting, due to a recessive mutation in gene *pcc1* encoding an HMG box transcription factor (59, 60).

6 Classical mutagenesis

The general principles of classical mutagenesis in basidiomycetes (methods, type of mutations obtained) do not differ substantially from ascomycetes (Chapter 4) nor any other organisms. Mutagenesis of any form is best performed with cells containing only one haploid nucleus. These are easily available from heterobasidiomycetes having a haploid yeast form. In the filamentous *C. cinereus*, haploid oidia can be employed (*Protocol 4*). For those homobasidiomycetes lacking any mitotic spores but producing fruit bodies in culture (for example *S. commune*), the use of basidiospores can partly overcome the problem. When taking basidiospores from an *Amut Bmut* homokaryon the spores are even genetically homogeneous. However, basidiospores of heterothallic species are often binucleate due to a post-meiotic mitosis (1) and in cases of strong pigmentation or impermeable thick cell walls, basidiospores have to be germinated before use in mutagenesis.

Two mycelial mutagenesis approaches are possible: mycelial pellets obtained from liquid cultures can be heavily macerated in a sterile Waring blender, or the mycelium can be protoplasted. Macerated mycelium is usually multicellular and thus only of limited use. There is, however, a good chance that protoplasts contain only one nucleus. Whenever possible, protoplasts are therefore the best choice when suitable spores are missing.

6.1 UV mutagenesis

UV mutagenesis can be performed with open UV lamps switched on and off by the investigator, but to avoid any accidental irradiation of the researcher and for the accuracy of the UV dose given, a UV Stratalinker (Stratagene) should be used if available (*Protocol 4*). UV radiation is kept within the chamber and the machines are equipped with an internal photodetector designed for compensating the natural shift in power output of ageing ultraviolet bulbs. Before performing large UV mutagenesis experiments, it is essential to determine a sensitivity profile of a strain. The strain is subjected to various doses of UV and survival rates in c.f.u. (colony-forming units) are determined. *C. cinereus* oidia usually have a germination rate of 10–20% and we apply energy doses giving a 90% killing rate and a final germination rate of about 1%.

Protocol 4

UV mutagenesis of *C. cinereus* oidia using a Stratalinker

Equipment and reagents

- Fresh cultures grown at 37 °C on YMG or on YMG/T (YMG plus 100 mg tryptophan/litre)[a]
- Sterile twistle funnels containing a layer of glass wool, the outlet (5–10 cm) of which is fixed with cotton wool in a glass test-tube
- Haemocytometer
- Fresh YMG or YMG/T plates
- UV Stratalinker 2400 (Stratagene), positioned in a dark room
- Ventilated, light-proof boxes (home-made) with a layer of wet paper towels

Method

1 Pour 10–15 ml sterile H_2O onto a grown culture and harvest the oidia from the aerial mycelium by scraping with a sterile scalpel.

Protocol 4 continued

2 Briefly flame-sterilize the border of the Petri dish and pour the spore solution into the twistle funnel for separation of mycelial debris from oidia through the glass wool.

3 Determine cell density of the spore solution with a haemocytometer and adjust cell density to 10^5 cells/ml.

4 Plate 3×10^3 to10^4 cells onto fresh medium.[b,c]

5 Place plates into the chamber of the UV Stratalinker without their lids.

6 Irradiate the plates with an energy setting of 100–200 (= 10–20 mJoule/cm^2).[d]

7 Take plates out of the UV Stratalinker in complete darkness, cover with lids, and transfer them directly into light-proof boxes.[e]

8 Incubate at 37 °C for two days, isolate grown colonies with a tungsten needle onto fresh medium for further growth.

[a] In case of *Amut Bmut* strains (as for dikaryons!), oidia production has to be induced by light (11): transfer fully grown plates for two to four days at 37 °C into constant light (light intensity 20–25 $\mu E\ m^{-2}\ s^{-1}$, light source, e.g. Osram L40/25).

[b] Do not plate cells on solid medium in cases where a selective mutant enrichment technique (Section 6.2.3) can be applied. Perform UV mutagenesis with a thin layer of a cell suspension in a sterile Petri dish and transfer cell suspension into a suitable liquid medium after mutagenesis.

[c] If selecting for specific resistances, respective substances might be added in concentrations toxic for the wild-type.

[d] These energy doses cause a killing rate of 90% in our *Coprinus* strains, giving rise to ca 50–100 surviving colonies per plate (J. D. Granado and A. P. F. Bottoli, unpublished observations).

[e] It is important to perform these steps in the dark in order not to induce the organisms DNA repair system.

6.2 Chemical mutagenesis

DNA alkylating chemicals, such as *N*-methyl-*N*′-nitro-*N*-nitrosoguanidine (MNNG) and ethyl methane sulfonate (EMS) which predominantly promote base pair substitutions and DNA deletions, as well as agents causing oxidative deamination and crosslinking of DNA (e.g. HNO_2), have been used for mutagenesis in basidiomycetes. As in the case of UV mutagenesis, it is imperative to determine the dose-dependent viability of fungal spores, protoplasts, or mycelial macerates in response to the compound and incubation time, before setting the cell density in the mutagenesis to obtain a manageable number of survivors. Due to the lipophilic nature of the chemicals in use, it can prove beneficial to prepare protoplasts for mutagenesis. Note that mutagens must be considered potential carcinogens and therefore have to be handled with appropriate precautions.

Protocol 5

NTG (MNNG)/EMS mutagenesis of basidiomycetes

Equipment and reagents

- Spores or mycelial macerates at desired densities in sterile H_2O or protoplasts at desired densities in sterile MM buffer (*Protocol 7*) in transparent conical Sterilin™ tubes (2 cm diameter), fitting bench centrifuge
- Sterile H_2O or MM buffer (*Protocol 2*)
- MNNG (Sigma 12994-1; 1 mg/ml freshly dissolved in 0.2 M citrate pH 5.0 or MM buffer) or liquid EMS (Sigma M-0880)[a]
- Fresh solid medium plates and sterile liquid selection medium[b]

Method

1. Add either 1 ml of MNNG solution or 1–4% (v/v) (final concentration) EMS to cell suspension.[a]
2. Incubate at RT while gently shaking.[c]
3. Centrifuge cell suspension for 5 min,[d] collect supernatant in container.[e]
4. Wash cells with sterile H_2O or MM buffer, centrifuge again, and discard supernatant.[e]
5. Resuspend cell pellet in 200–500 μl of H_2O or MM buffer and plate aliquots on solid medium.[f]
6. Incubate at appropriate temperature until small colonies (1–2 mm diameter) appear.
7. Isolate grown colonies with a tungsten needle onto fresh medium for further growth.[b]

[a] When handling reagents, work in a fume hood and wear disposable rubber gloves.

[b] If selecting for resistance, the substance of interest might be added at concentrations toxic to the wild-type.

[c] Individual times of incubation vary between 1–3 h for MNNG and 4–12 h for EMS; to stop the EMS reaction, 5% (w/v) $Na_2S_2O_3$ is added at the end of the incubation period.

[d] Centrifuge spores or macerates at 2600 g for 5 min, protoplasts at lower speed, i.e. 640 g for 10 min.

[e] Collect all contaminated material (utensils, solutions) in a safe container and dispose of by incineration.

[f] When enrichment methods, Section 6.2.3, can be applied, inoculate liquid selection medium instead.

6.3 Mutant selection

Procedures for mutant selection do not differ regardless of whether UV or chemical mutagenesis have been employed for their generation. In general, mutants that are 'leaky' in phenotype and easily reverting mutants are better avoided for further analysis. Clear-cut phenotypes should be selected for.

6.3.1 Morphological and developmental mutants

Mutants with defects in cell and mycelial morphology and/or in growth are easily detected by their altered colony appearance and/or growth behaviour on solid medium. Screening for mutants with defects in non-essential developmental pathways or in pathogenicity is in most cases work-intensive. For example, solo-pathogenic haploid mutants of smut fungi need to be individually tested on the plant to identify defects in pathogenicity (26). Similarly, thousands of *C. cinereus* mutants from *Amut Bmut* homokaryons and other homokaryotic fruiters have individually been subjected to a day/night regime at 25–28 °C to detect defects in initiation and performance of fruit body development (60) (U. K. in preparation). Defects in asexual spore formation have often discovered by an altered mycelial morphology. However, especially when spores are colourless, like the oidia of *C. cinereus*, replica plating of spores can assist in discovering mutations affecting sporulation (61).

6.3.2 Replica plating

For screening large numbers of mutants for growth on specific media, and for detection of temperature-sensitivity, replica plating is a very useful technique when dealing with colonies of yeast-like cells (52) (Chapter 4). A commonly used protocol is given below (*Protocol 6*). The same technique is applicable with mycelial cultures of *C. cinereus* and other filamentous basidiomycetes producing sticky mitotic spores within the aerial mycelium. However, *Coprinus* spores also stick to bare metal surfaces, enabling replica plating of microtitre dishes with the help of a commercially available multi-pin replicator with stout blunt pins (61). Kaplan and Walls (62) exploited strips of Velcro® to tear off mycelial pieces from colonies with the hooks of the material for transfer onto fresh plates. This latter technique might be useful for filamentous basidiomycetes, such as *S. commune*, producing mycelial mats without spores. Replica plating of mycelial colonies is only efficient when a sufficient number of colonies is present on a plate (50–500). Individual colonies might be reduced in growth by adding sucrose or other substances to the medium.

Protocol 6

Replica plating

Equipment and reagents

- Squares of velvet (*ca.* 15 × 15 cm, preferentially with hemmed edges), packed in tin foil, autoclaved
- A wooden, plastic, or metallic round block fitting into Petri dishes
- A rubber band or an O-ring slightly larger in diameter than the block
- Plates with grown colonies (master plates) and fresh plates of suitable selection media

Protocol 6 continued

Method

1. Label the side or the bottom of the master plate and the fresh Petri dish with an index mark.
2. Stretch the sterile velvet over the block and fix it with the rubber or the O-ring.
3. Press the agar surface of the grown culture onto the velvet (or, alternatively, the block with the velvet onto the agar surface).
4. Press the agar surface of the fresh culture onto the velvet (or, alternatively, vice versa) with its index mark at exactly the same position as the master plate.
5. Incubate the replicated plates under suitable growth conditions and examine presence and absence of colonies in comparison with the master plate.

6.3.3 Filtration enrichment and enrichment by selective killing

The technique of filter enrichment is an efficient way for a selective removal of prototrophs after mutagenesis of *C. cinereus* oidia (31), and basidiospores of *S. commune* (63). Mutagenized spores are allowed to germinate in liquid minimal medium on specific carbon sources. Prototrophic oidia or basidiospores grow, while those with genetic defects leading to auxotrophies do not. Germinated mycelium is separated from ungerminated spores by filtration of the cultured suspension through sterile glass wool (see *Protocol 4*). This step can be repeated in regular intervals before rescuing ungerminated mutants by plating on fully supplemented media.

Cycloheximide, an inhibitor of protein biosynthesis, strongly affects the vitality of dividing cells, but affects non-dividing cells only weakly. Thus, an 800-fold enrichment for UV-sensitive and auxotrophic mutants has been achieved by adding 100 μg/ml cycloheximide to mutagenized *S. commune* mycelial macerates. This technique has also been shown to be applicable to mutagenized cell suspensions of the ascomycetous yeast *Saccharomyces cerevisiae*, suggesting it to be valid for selection of any type of conditional mutant in a broader range of fungal organisms (64). Selective killing of prototrophs through cell wall degrading enzymes attacking growing cells (final concentration 0.5% β-glucuronidase) enriched for auxotrophic mutants in *U. hordei* (49). To our knowledge, other selective killing strategies using fungal antibiotics like nystatin and *N*-glycosyl polifungin, and starvation of specific double mutants, well established in ascomycetes (65), have not been applied to basidiomycetous fungi. There is clearly potential to develop further elegant ways for effective selection.

6.3.4 Selection by inhibition

Many substances that block particular metabolic and cellular functions are known, and have been applied in screenings for resistant mutants. In *C. cinereus*, these include fluoroacetate (66), 5-fluoroindole (67), cycloheximide (68), special sugars (69), Benomyl (70), calcoflour-white (71), cytochalasin E (72), and caffeine

(A. P. F. Bottoli, unpublished). Many other substances might be applicable for selection, of which metabolic and cellular targets are known from studies with *S. cerevisiae* and from medical drug research. The basic principle of inhibition selection is to increase the concentration of the respective substance in the growth medium up to a level that is toxic or inhibiting for the unmutagenized wild-type, allowing only resistant mutants to grow. Thus, the tolerance level of the wild-type strain has to be determined beforehand with a series of different inhibitor concentrations. If the tolerance level is known, it is also possible to screen for over-sensitive mutants that will not grow at concentrations accepted by the wild-type (72).

Mutagen treatment can lead to a temporary inhibition of germination or growth, and for some mutations it can take time before full phenotypic expression is realized. Thus, introduction of a propagation step after mutagenesis and before inhibition selection can be crucial (73). This can either be done by growing mutagenized cells for a short time in liquid unselective media before plating on solid selective media. Alternatively, cells might be plated before or after mutagenesis on unselective media and, after a short period of unselective growth, plates are overlaid with soft agar containing the selective inhibitor. Per plate, 4–5 ml sterile, liquified soft agar (0.9%) will be needed, kept in a test-tube at a temperature close to the point of gelation (40–42 °C) until use.

7 Insertional mutagenesis

7.1 Transposon mutagenesis

Tagging genes by DNA insertion can be a powerful approach to obtain mutants and subsequently disrupted genes. A transposon-like element named 'Scooter' has recently been detected in *S. commune*, enabling the cloning of the *thn* gene from a spontaneous mutant with an altered hyphal morphology (74). At least in *S. commune*, we can therefore expect a well-developed transposon tagging system in the near future.

7.2 REMI mutagenesis

7.2.1 The principles of the technique

In REMI (restriction enzyme-mediated integration) mutagenesis, restriction enzymes are added to plasmid DNA and protoplasts in transformation (Chapter 3). The enzymes appear to enter cells and nuclei and probably introduce double-strand breaks into the chromosomes (75, 76). Since REMI is not yet a perfectly established technique we discuss several aspects of REMI to which attention needs to be paid to when designing experiments.

REMI in basidiomycetes is most advanced in *U. maydis* (26) and *C. cinereus* (77, 78), but first attempts of REMI have also been reported for *C. congregatus* (79) and *Lentinus edodes* (80). Parameters (vector construct and its conformation, the kind of restriction enzyme and its concentration, etc.) varied between these studies (*Table 1*) currently making it difficult to define the best conditions for REMI.

Table 1 Parameters of REMI transformation experiments with basidiomycetes

Species	Plasmid, selection marker	DNA conformation	Restriction enzyme(s)	Comments	Reference(s)
Ustilago maydis	pSMUT, bacterial hygromycin resistance gene (*hph*) under control of *Saccharomyces cerevisiae* sequences	Circular	50 U *Bam*HI	5–25 transformants/μg DNA, limited influence of enzyme concentration on number of transformants Most REMI transformants contain one insert Restoration of the *Bam*HI restriction sites in about 50% of the cases Defects in pathogenicity arose at a frequency of 1.4% Tagged genes have been isolated by plasmid rescue in *Escherichia coli* Linkage between mutation and tagging is shown by gene knock-outs in wild-type strains	26
Coprinus cinereus	pPAB1-2, homologous *pab1*$^{+}$ gene	Circular	20–100 U *Bam*HI 20–100 U *Eco*RI 20–225 U *Pst*I	Up to 400–800 transformants/μg DNA, enzyme concentrations strongly influence number of transformants REMI reduces numbers of insert sites per transformant to one (in 30–60% of the cases), less often to two to three Tandem insertions and plasmid rearrangements occur, which hinders plasmid rescue in *E. coli* Frequency of restoration of restriction sites varies (8–56%) Instabilities of inserts were observed Transformation *per se* is mutagenic independent of enzyme application Enzymes have mutagenic effects independent of plasmid integration Mutation rates in sexual and asexual development are high (*ca.* 20–40%) Linkage between insertion and mutation needs to be shown by genetic crosses; one out of three analysed mutants showed linkage between insert and mutation	32, 77, 85, J. D. Granado and U. K., unpublished

Coprinus cinereus	pPHT1, bacterial *hph* gene between homologous promoter and terminator	Linear	20–100 U *Kpn*I	400–700 transformants/μg DNA, addition of enzyme increases number of transformants REMI reduces number of insert sites per transformant to one (in 75% of the cases), less often to two Tandem insertions occur, which hinders plasmid rescue in *E. coli* Restoration of the *Kpn*I restriction sites in about 50% of the cases Sporulation-defective mutants appeared at a frequency of 1.2% Linkage between insert and mutation has been shown for some REMI mutants in crosses Inserts can be lost in meiosis Tagged sequences have been isolated by semi-random PCR	78
Coprinus congregatus	pBARGEM 7-1, bacterial phosphinothricin resistance gene (*bar*) under control of an *Aspergillus nidulans* promoter	Linear	30 U *Eco*RI	500–600 transformants/μg DNA No increase of transformation efficiency by addition of *Eco*RI One or two inserts per transformants, no perfect REMI integration Three laccase mutants have been detected (mutation frequency 6%), linkage data are missing	79
Lentinus edodes	pLC1-hph, bacterial *hph* gene under homologous promoter	Circular	50 U *Sal*I 2.5 U *Sph*I 50 U *Dra*I 50 U *Kpn*I	10–15 transformants/2.5 μg DNA, enzyme concentrations strongly influence number of transformants Restoration of restriction sites in up to 50% of plasmid integrations 10–50% of transformants have an insert at a single site	80

Systematic studies on the technical side of REMI are so far not extensive enough and results of different groups appear to contradict each other. For example, Granado *et al.* (77) reported, in *C. cinereus*, an increase in the number of transformants by increasing enzyme concentrations and a decrease in the number of transformants once an optimal concentration had been exceeded. In contrast, Cummings *et al.* (78) found no reduction of transformation rates within the same fungus with high enzyme concentrations. It is not clear whether this relates to the type and source of the enzyme, the plasmid, the amount of DNA or the form of the plasmid, to the strains and/or to the specific batches of protoplasts used.

It is possible that certain parameters are important for a specific organism but not for all. For example, in *Ustilago* which has an efficient recombination system, homologous vector sequences are best avoided (26), whereas in *C. cinereus*, where vector integrations are usually ectopic, homologous sequences can be used (77, 78). Background knowledge about transformation of the organism of interest is therefore important for designing the most promising REMI experiments. Generally, it seems best to perform a series of preliminary experiments, determining effects of different enzymes and enzyme concentrations in obtaining transformants. In *C. cinereus*, an increase in mutation frequency has been observed with higher enzyme concentrations, but without a simultaneous increase of plasmid integration. This result indicates the mutagenic effects of restriction enzymes, probably due to incorrect repair of induced DNA breaks (77). In ascomycetous fungi, deletions have indeed been detected after REMI mutagenesis (81). Thus, the lowest possible enzyme concentration showing an optimal effect in transformation should preferentially be applied. In many filamentous fungi, insertion at multiple sites, as well as tandem vector integrations are common in standard transformation procedures (82). In REMI transformation, the number of integrated plasmids is usually reduced (26, 77, 78, 80). In the optimal case, a plasmid linearized by a restriction enzyme fuses to a related restriction site at just one chromosomal place. However, small scale degradation of DNA ends of the plasmid and/or the opened chromosome site can occur before fusion has been observed, and compatible ends are therefore not always necessary for fusion (75, 76). The conditions of the transformation protocol appear to play a role on frequency of single vector integrations and base pair exact reconstruction of the integration sites (83). How REMI transformation of *C. cinereus* is performed in our laboratory is given in *Protocol 7*. Our experience shows that it is not necessary to linearize the vector before transformation, as restriction enzymes obviously act *in vivo* on both plasmid and chromosomal DNA (26, 77). Nevertheless, there may be advantages in vector linearization before transformation. DNA and protoplast concentration have been adapted from standard transformation protocols for *C. cinereus* (84). It is not known to what extent changing these parameters will influence the outcome of REMI.

As not all genes will contain restriction sites for every possible restriction enzyme, we recommend the use of several different enzymes when generating large collections of REMI mutants. Rare cutting enzymes are obviously a bad choice for the technique. Standard enzymes, obtainable in high enzyme concentrations and at a reasonable price such as *Bam*HI, *Eco*RI, *Hin*dIII, *Pst*I, and *Xba*I are commonly applied in REMI (76). All these enzymes recognize defined six base pair long palindromic sequences and produce 3′ or 5′ overhanging DNA ends, but blunt-end producing restriction enzymes may equally be valid for REMI mutagenesis. Enzymes with four base pair recognition sites will more frequently recognize potential digestion sites within the chromosomal DNA, and might therefore lead to a more randomized distribution of plasmid insertions within the genome. Such enzymes have rarely been employed for REMI, probably due to fear the enzymes might cut too often. However, titrating for the optimal enzyme concentration should help to avoid this.

Protocol 7

REMI mutagenesis in *Coprinus cinereus*

Equipment and reagents

- Fresh cultures grown at 37 °C on YMG or YMG/T[a]
- Transparent conical Sterilin tubes (2 cm in diameter, sterile) and fitting bench centrifuge
- Enzyme, per 20 ml MM buffer: 0.8 g Onozuka R10 cellulase, 0.02 g Sigma C-6137 chitinase. Sterilize by centrifugation at 30 000 g for 30 min at 4 °C. Carefully dispense 2 ml aliquots of the enzyme solution into sterile 2 ml Eppendorf tubes. Store at −20 °C until use.
- MMC buffer freshly prepared, as MM buffer but with 25 mM $CaCl_2$
- Filter sterilized PEG solution, per 100 ml: 25 g PEG 4000 (BDH), 1 ml of 1 M Tris–HCl pH 7.5, 2.5 ml of 1 M $CaCl_2$
- MM buffer (*Protocol 2*)
- 1 μg linearized or circular DNA of pPAB1-2 or pPHT1 (*Table 1*) depending on the strain to be transformed
- Restriction enzymes appropriately diluted in 10 μl sterile TE pH 8.0
- STC buffer, per 100 ml: 18.2 g sorbitol, 1 ml of 1 M Tris–HCl pH 7.5, 2.5 ml of 1 M $CaCl_2$; autoclave
- Fresh regeneration agar: 100 mg adenine (optional), 5 g soluble starch, 172 g sucrose, 12 g agar/litre *Coprinus* minimal medium (containing only 5 g glucose/litre)[b]
- Standard *Coprinus* minimal medium (see Appendix 1), supplement with 100 mg amino acids/litre, if necessary

Method

1. Harvest spores and filter spore solution as described in *Protocol 4*.[c]
2. Pour the spore solution into a Sterilin tube, centrifuge at 2600 g for 5 min, and discard the supernatant.
3. Wash the pellet with 10 ml MM buffer, centrifuge again, and discard the supernatant.
4. Resuspend in 2 ml of enzyme solution and incubate at 37 °C for 3–4 h.
5. Monitor protoplast formation under the microscope; when 50–70% of all oidia have formed protoplasts[d] stop the reaction by adding 5 ml of MMC.

Protocol 7 continued

6. Centrifuge the protoplasts at low speed 640 g for 10 min and cautiously discard supernatant.
7. Adjust protoplast concentration to $0.2–2 \times 10^8$/ml by adding 200–500 μl MMC to the pellet.
8. On ice: mix 1 μg of plasmid DNA, 12.5 μl PEG solution, and 10 μl restriction enzymes in TE pH 8.0 and add 50 μl protoplast suspension; leave out restriction enzymes for control transformation.
9. After 20 min on ice, add 0.5 ml PEG solution, mix carefully, incubate for 5 min at room temperature, dilute with 1 ml STC buffer, and plate aliquots (0.3–0.5 ml) onto regeneration agar.
10. Incubate plates at 37 °C until transformants appear (usually after three to four days), slow down growth by transferring plates to room temperature.
11. Cut out small agar pieces containing individual colonies (check under binocular!) with a tungsten needle and transfer clones onto MM plates for further growth.

[a] See *Protocol 4* for induction of oidia production in *Amut Bmut* strains.

[b] Be exact with the volume to keep the correct osmolarity! Take care that plates are not wet from condensation!

[c] Take two or more plates for strains producing less than $0.5–1 \times 10^9$ oidia per plate.

[d] We find that higher rates in protoplasting negatively interferes with the number of transformants obtained.

7.2.2 Mutant selection

Since REMI involves gene disruption, like transposon mutagenesis, it is restricted to isolation of mutants with defects in non-essential genes—those acting in non-essential developmental pathways, or in pathogenicity of non-obligate pathogenic fungi. Currently it is not clear whether all possible non-essential genes can be tagged by REMI or whether there is a preference for integration into specific chromosomal regions. Repeated isolation of tagged genes within different screens of the same fungus strongly suggests the latter. Nevertheless, tagging even only a subset of genes makes REMI worthwhile (75, 76). Screening for mutations in development and/or pathogenicity does not differ from screening after UV or chemical mutagenesis. However, because instability of plasmid integrations has been observed (77), it is advisable to store isolated REMI transformants under selection conditions.

7.2.3 Linkage studies

Several, but not all, interesting mutations in REMI transformants of basidiomycetes have been found to be linked to the insertion event (27, 33, 78, 79). In *Ustilago*, linkage between insertion and mutation can be readily shown by using a recovered tagged sequence to disrupt the intact sequence in a wild-type strain by homologous recombination in transformation. Linkage is shown when the gene disruptant strain has the same phenotype as the original mutant (26). In the homobasidiomycetes, where homologous plasmid integration is rare (82), genetic analysis is required to establish a link between the plasmid insertion and mutation. Current attempts to analyse developmental REMI mutants of *C. cinereus* homokaryon *Amut Bmut* has employed strains of different genetic backgrounds. *pab1*$^{+}$ insertions were followed by prototrophy using the *pab1* auxotrophic homokaryon PG78 (*A6 B42 pab1 trp1.1,1.6*) in crosses (32, 85), *hph* insertions by hygromycin resistance using wild-type homokaryon JV6 (*A42 B42*) as a mating partner (78). In some instances, plasmid integrations were lost in high frequencies during meiosis (78, 85). Because nothing is known about this type of plasmid eviction, the presence of the mutation in the absence of the insert does not disprove linkage. Clones of the progeny expressing the mutant phenotype but lacking the insert can be treated like UV and chemical-induced mutants and complemented by transformation (85). Similarly, mutations that have been clearly shown in crosses not to be linked to an insert might still be studied further using mutant progeny lacking the plasmid (77).

Due to the heterogeneity of genomes, other problems can be encountered such as loss of certain progeny or new gene combinations independent of the REMI mutation. This is especially true for the complex regulation of fruit body initiation (32). In some instances, for example when investigating defects in oidia formation, and defects in later stages of fruit body development, the analysis of only a proportion of the progeny can still be sufficient to obtain a clear-cut result about co-inheritance of plasmid selection marker and mutation (32, 78). Generating co-isogenic homokaryotic strains of compatible mating type, with or without a *pab1* defect (depending on the REMI strategy employed), through repeated backcrossing to homokaryon *Amut Bmut* should overcome the encountered problems in future (Section 4.1).

7.2.4 Rescuing tagged sequences

Single and multiple and/or tandem plasmid integrations, as well as deletion of plasmid sequences and complex rearrangements, have been observed following REMI mutagenesis of basidiomycetes (77) (J. D. Granado and U. K. unpublished). Before recovering an integrated plasmid together with chromosomal DNA of the host, it is therefore important to establish the pattern of DNA integration by Southern blot analysis. In *Ustilago*, full-length plasmids with flanking genomic DNA segments have successfully been recovered by isolation and digestion of chromosomal DNA of mutants, ligation, and *Escherichia coli* electroporation (26). Rescue attempts of plasmids including flanking chromosomal DNA of *C. cinereus* have been unsuccessful because of tandem integrations and the high probability of secondary plasmid contaminations in *E. coli* electroporation (J. D. Granado, unpublished). Zolan and co-workers (78) established semi-random PCR (*Protocol 8*) to obtain tagged gene sequences from hygromycin resistant REMI mutants carrying tandem copy insertions of plasmid pPHT1.

Protocol 8

Semi-random, two-step PCR to recover genomic DNA adjacent to plasmid inserts in *C. cinereus* transformants

Equipment and reagents

- Chromosomal DNA isolated from a REMI mutant (Chapter 2)
- dNTPs, *Taq* DNA polymerase and *Taq* buffer (final concentration: 50 mM KCl, 10 mM Tris-HCl pH 8.3, 200 μg/ml gelatin, 3 mM $MgCl_2$, or as supplied by the manufacturer)
- Primer pair 1: choose one primer sequence internal from your vector construct (e.g. for plasmid pHTP1 within the *hph* gene: 5′ATCCACGCCCTCCTACATCG3′); the second primer is random (e.g. 5′GGCCACGCGTCGACTAGTAC(N_{10})GATAT3′)
- Primer pair 2: choose one primer sequence from a region of your vector expected to be amplified in the 1st PCR reaction (e.g. for plasmid pHTP1 within upstream or downstream control regions of gene *hph*: 5′CAUCAUCAUCAUCAGCTTCATTTTCCGTGTGG3′); the second primer overlaps in sequence to the second primer of the 1st PCR (e.g. 5′CUACUACUACUAGGCCACGCGTCGACTAGTAC3′)
- Thermal cycler

Method

1. Prepare a 20 μl reaction mixture with 100 ng chromosomal DNA, each dNTP at 200 mM, 2.5 U *Taq* in *Taq* buffer.
2. 1st PCR cycle: 94 °C, 2 min; 6 × (94 °C, 30 sec; initial temp. = 42 °C, 30 sec, −1 °C for each subsequent cycle); 24 × (94 °C, 30 sec; 65 °C, 30 sec; 72 ° C, 3 min); 4 °C, end.
3. Prepare a 20 μl reaction mixture with 1 μl of a 1:5 dilution of the first step reaction, each dNTP at 200 mM, 2.5 U *Taq* in *Taq* buffer.
4. 2nd PCR cycle: 20 × (94 °C, 30 sec; 65 °C, 30 sec; 72 °C, 3 min); 4 °C, end.[b]

[a] This protocol has been developed by Cummings *et al.* (78) for REMI transformants harbouring plasmid pHTP1. For REMI transformants harbouring plasmids pPAB1-2, primers might be designed from the *C. cinereus pab1* sequence (Genbank accession number AF166096).

[b] The amplified product can be purified through agarose gels, and further amplified in another PCR reaction following the 2nd PCR cycle or be used for cloning by the CloneAmp^{R}pAmp1 System of Gibco BRL (therefore the CUA sequences within the second set of primers!), hybridization etc.

7.2.5 Complementation of mutant phenotypes with isolated DNA sequences

After isolation, tagged sequences can then be used to screen a suitable genomic library in order to obtain the uninterrupted sequence by standard DNA techniques (86). Unlike in *U. maydis*, where gene knock-outs can prove linkage between a tagged sequence and a mutant phenotype (Section 7.2.3), the final proof in *C. cinereus* can only be made via complementation of the mutant phenotype by the isolated sequence in transformation. A library containing the $pab1^+$ selection marker (86) can be used for complementation of hygromycin-resistant REMI mutants of the *pab1* auxotrophic strains. The *hygr* gene (78) can be used for selection of transformants of prototrophic *C. cinereus* REMI mutants. Alternatively, the *pab1*$^{+}$ gene encoding *para*-aminobenzoic acid synthase and mediating sulfonamide resistance can serve as selection marker in repetitive transformation (U. K. *et al.* unpublished).

Acknowledgements

We thank our colleagues Alan Bottoli and José Granado for permission to cite their unpublished work. Lorna Casselton and Pat Pukkila for *Coprinus* strains, Jörg Kämper and Cosima Hohenlohe for advice on *Ustilago,* Frank Schuren for advice on *Schizophyllum*, and Markus Aebi for critically reading the manuscript. Research leading to the development of *Coprinus* techniques in our laboratory has been supported by the Swiss National Science Foundation (Grants 31-46'940.96 and 31-46'940.96/2) and the ETH Zürich.

References

1. Kües, U. (1999). *Mol. Biol. Microbiol. Rev.*, **64**, 316.
2. Banuett, F. (1995). *Annu. Rev. Genet.*, **29**, 179.
3. Anderson, J. B., Petsche, D. M., and Franklin, A. L. (1984). *Can. J. Genet. Cytol.*, **27**, 47.
4. Carvalho, D. B., Smith, M. L., and Anderson, J. B. (1995). *Mycol. Res.*, **99**, 641.
5. Hiscock, S. J. and Kües, U. (1999). *Int. Rev. Cytol.*, **193**, 165.
6. Raper, C. A. (1976). *J. Gen. Microbiol.*, **95**, 54.
7. Holliday, R. (1974). In *Handbook of genetics* (ed. R. C. King), Vol. I, p. 575. Plenum Press, New York.
8. Day, A. W. and Garber, E. D. (1988). In *Advances in plant pathology* (ed. G. S. Sidhu), Vol. 6, p. 457. Academic Press, London.
9. Raper, J. R. and Hoffman, R. M. (1974). In *Handbook of genetics* (ed. R. C. King), Vol. I, p. 597. Plenum Press, New York.
10. Kertesz-Chaloupková, K., Walser, P. J., Granado, J. D., Aebi, M., and Kües, U. (1998). *Fungal Genet. Biol.*, **23**, 95.
11. Hui, C., Yamamoto, H., Ohta, T., Takeo, K., and Kitamoto, Y. (1999). *Mycoscience*, **40**, 199.
12. Raper, C. A. (1985). *Exp. Mycol.*, **9**, 149.
13. Kemp, R. O. F. (1974). *Trans. Br. Mycol. Soc.*, **62**, 547.
14. Harder, R. (1927). *Zeitschrift für wissenschaftliche Mikroskopie und mikroskopische Technik*, **44**, 173.
15. Zolan, M. E. and Pukkila, P. J. (1986). *Mol. Cell. Biol.*, **6**, 195.
16. Lewis, D. (1961). *Genet. Res.*, **2**, 141.
17. Takemaru, T. (1964). *Rep. Tottori Mycol. Inst.*, **4**, 35.
18. Kerruish, R. M. and Da Costa, E. W. B. (1963). *Ann. Bot.*, **27**, 653.
19. Leal-Lara, H. and Eger-Hummel, G. (1982). *Theor. Appl. Genet.*, **61**, 65.
20. Röhr, H., Kües, U., and Stahl, U. (1998). *Prog. Bot.*, **60**, 39.
21. de la Bastide, P. Y., Sonnenberg, A. S. M., van Griensven, L. J. L. D., Anderson, J. B., and Horgen, P. A. (1997). *Appl. Environ. Microbiol.*, **63**, 3426.
22. May, G. and Taylor, J. W. (1988). *Genetics*, **118**, 213.

23. Krejci, R. and Homolka, L. (1991). *Appl. Environ. Microbiol.*, **57**, 151.
24. Banuett, F. and Herskowitz, I. (1989). *Proc. Natl. Acad. Sci. USA*, **86**, 5878.
25. Mills, D. I. and Ellingboe, A. H. (1969). *Genetics*, **62**, 271.
26. Bölker, M., Böhnert, H. U., Braun, K. H., Görl, J., and Kahmann, R. (1995). *Mol. Gen. Genet.*, **248**, 547.
27. Gillissen, B., Bergemann, J., Sandmann, C., Schroer, B., Bölker, M., and Kahmann, R. (1992). *Cell*, **68**, 647.
28. Swamy, S., Uno, I., and Ishikawa, T. (1984). *J. Gen. Microbiol.*, **83**, 73.
29. Koltin, Y. (1970). *Arch. Mikrobiol.*, **74**, 123.
30. Raper, C. A. (1990). In *Genetic maps, 5th edition. Book 3, Lower eukaryotes* (ed. S. J. O'Brien), p. 3.90. Cold Spring Harbor Press, Cold Spring Harbor.
31. Pukkila, P. (1993). In *Genetics and breeding of edible mushrooms* (ed. S.-T. Chang, J. A. Buswell, and P. G. Miles), p. 249. Gordon and Breach Science Publishers, Y-Parc, Switzerland.
32. Liu, Y., Granado, J. D., Polak, E., and Kües, U. (1999). *Fungal Genet. Newsl.*, **46**, 16.
33. North, J. (1990). In *Genetic maps, 5th edition. Book 3, Lower eukaryotes* (ed. S. J. O'Brien), p. 3.76. Cold Spring Harbor Press, Cold Spring Harbor.
34. Day, A. W. and Gardner, E. D. (1990). In *Genetic maps, 5th edition. Book 3, Lower eukaryotes* (ed. S. J. O'Brien), p. 3.94. Cold Spring Harbor Press, Cold Spring Harbor.
35. Esser, K. (1996). In *Fungal genetics. Principles and practice* (ed. C. J. Bos), p. 43. Marcel Dekker, New York.
36. Raper, J. R. and Oettinger, M. T. (1962). *Rev. Biol. (Lisbon)*, **3**, 205.
37. Moore, D. (1981). *Curr. Genet.*, **3**, 145.
38. Kerrigan, R. W., Royer, J. C., Baller, L. M., Kohli, Y., Horgen, P. A., and Anderson, J. B. (1993). *Genetics*, **133**, 225.
39. Leslie, J. F. (1981). *Genet. Res.*, **37**, 239.
40. Zolan, M. E. (1995). *Microbiol. Rev.*, **59**, 686.
41. Sonnenberg, A. S. M., de Groot, P. W. J., Schaap, P. J., Baars, J. J. P., Visser, J., and van Griensven, L. J. L. D. (1996). *Appl. Environ. Microbiol.*, **62**, 4542.
42. Zolan, M. E., Heyler, N. K., and Yeager Stassen, N. (1994). *Genetics*, **137**, 87.
43. Arima, T., Okida, T., and Moringa, T. (1996). *Mycoscience*, **37**, 111.
44. Ásgeirsdóttir, S. A., Schuren, F., and Wessels, J. G. H. (1994). *Mycol. Res.*, **98**, 689.
45. Muraguchi, H. and Kamada, T. (1998). *Development*, **125**, 3123.
46. Arima, T., Tagashira, N., Kamada, T., Tanaka, N., and Moringa, T. (1997). *Mycoscience*, **38**, 127.
47. *Ustilago* sequencing project, Bayer, Germany.
48. Zolan, M. E., Crittenden, J. R., Heyler, N. K., and Seitz, L. C. (1992). *Nucleic Acids Res.*, **20**, 3993.
49. Henry, C. E., Bullock, B., Smith, V., and Setward-Clark, E. (1988). *Bot. Gaz.*, **149**, 101.
50. Garber, E. D. and Day, A. W. (1985). *Bot. Gaz.*, **146**, 449.
51. Moore, D. (1966). *Nature*, **209**, 1157.
52. Spencer, J. F. T. and Spencer, D. M. (1988). In *Yeast: a practical approach* (ed. I. Campbell and J. H. Duffus), p. 65. IRL Press, Oxford, UK.
53. Amirkhanian, J. D. and Cowan, J. W. (1985). *J. Hered.*, **76**, 348.
54. Garber, E. D., Gottlieb, M. L., and Cusack, E. M. (1987). *Bot. Gaz.*, **148**, 116.
55. North, J. (1977). *J. Gen. Microbiol.*, **98**, 529.
56. Frankel, C. and Ellingboe, A. H. (1977). *Genetics*, **85**, 417.
57. Bos, C. J. (1996). In *Fungal genetics. Principles and practice* (ed. C. J. Bos), p. 73. Marcel Dekker, New York.
58. Shalev, M., Stamberg, J., and Simchen, G. (1972). *Heredity*, **29**, 191.

59. Murata, Y., Fujii, M., Zolan, M. E., and Kamada, T. (1998). In *Proceedings of the Fourth Meeting on the Genetics and Cellular Biology of Basidiomycetes* (ed. L. J. L. D. Van Griensven and J. Visser), p. 143. Mushroom Experimental Station, Horst, The Netherlands.
60. Muraguchi, H., Takemaru, T., and Kamada, T. (1999). *Mycoscience*, **40**, 277.
61. Polak, E., Kües, U., and Aebi, M. (1997). *Fungal Genet. Newsl.*, **44**, 45.
62. Kaplan, P. J. and Walls, D. (1971). *Genet. Res.*, **17**, 279.
63. Anderson, M. R. and Deppe, C. S. (1977). *Genet. Res.*, **29**, 93.
64. Shneyour, Y., Stamberg, J., Hundert, P., Werczberger, R., and Koltin Y. (1978). *Mutat. Res.*, **49**, 195.
65. Bos, C. J. and Stadler, D. (1996). In *Fungal genetics. Principle and practice* (ed. C. J. Bos), p. 13. Marcel Dekker, New York.
66. Casselton, L. A. and Casselton, P. J. (1974). *Mol. Gen. Genet.*, **132**, 255.
67. Veal, D. and Casselton, L. A. (1985). *Arch. Microbiol.*, **142**, 157.
68. North, J. (1982). *J. Gen. Microbiol.*, **128**, 2747.
69. Moore, D. and Stewart, G. R. (1971). *Genet. Res.*, **18**, 341.
70. Kamada, T., Sumiyoshi, T., Shindo, Y., and Takemaru, T. (1989). *Curr. Microbiol.*, **18**, 215.
71. Bourne, A. N., Chiu, S.-W., and Moore, D. (1996) In *Patterns in fungal development* (ed. S.-W. Chiu and D. Moore), p. 126. Cambridge University Press, Cambridge, UK.
72. Tsukomoto, M., Yamamoto, M., and Kamada, T. (1996). *Fungal Genet. Biol.*, **20**, 52.
73. Bos, C. J., Debets, A. J. M., Nachtegaal, S. M., Slakhorst, S. M., and Swart, K. (1992). *Curr. Genet.*, **21**, 117.
74. Fowler, T. J. and Mitton, M. F. (2000). *Genetics*, **156**, 1585
75. Kahmann, R. and Basse, C. (1999). *Eur. J. Plant Pathol.*, **105**, 221.
76. Maier, F. J. and Schäfer, W. (1999). *Biol. Chem.*, **380**, 855.
77. Granado, J. D., Kertesz-Chaloupková, K., Aebi, M., and Kües, U. (1997). *Mol. Gen. Genet.*, **256**, 28.
78. Cummings, W. J., Celerin, M., Crodian, J., Brunick, L. K., and Zolan, M. E. (1999). *Curr. Genet.*, **XXX**, YYY.
79. Leem, Y., Kim, S., Ross, I. K., and Choi, H. T. (1999). *FEMS Microbiol. Lett.*, **172**, 35.
80. Sato, T., Yaegashi, K., Ishii, S., Hirano, T., Kajiwara, S., Shishido, K., *et al.* (1998). *Biosci. Biotechnol. Biochem.*, **62**, 2346.
81. Linnemannstöns, P., Voss, T., Hedden, P., Gaskin, P., and Tudzynski, B. (1999). *Appl. Microbiol. Biotechnol.*, **65**, 2558.
82. Kües, U. and Stahl, U. (1990). *Prog. Bot.*, **52**, 201.
83. Balhadère, P. V., Foster, A. J., and Talbot, N. J. (1999). *Mol. Plant-Microb. Interact.*, **12**, 129.
84. Binninger, D. M., Skrzynia, C., Pukkila, P. J., and Casselton, L. A. (1987). *EMBO J.*, **6**, 835.
85. Polak, E. (1999). PhD thesis, ETH Zürich, Switzerland.
86. Bottoli, A. P. F., Kertesz-Chaloupková, K., Boulianne, R. P., Granado, J. D., Aebi, M., and Kües, U. (1999). *J. Microbiol. Methods*, **35**, 129.

Appendix 1

Standard basidiomycete media

(a) **Growth medium for smut fungi**: (per litre) 10 g yeast extract, 20 g peptone, 20 g glucose, 10–20 g agar.

(b) ***Ustilago* complete medium**: (per litre) 10 g glucose, 1.5 g NH_4NO_3, 2.5 g hydrolysed casein, 0.5 g DNA, 1 g yeast extract, 10 ml vitamin solution, 62.5 ml salt solution, adjust to pH 7.0 with NaOH, 2% agar.

(i) Vitamin solution: (per litre) 100 mg thiamine, 50 mg riboflavin, 50 mg pyridoxin, 200 mg calcium pantothenate, 500 mg *p*-aminobenzoic acid, 200 mg nicotinic acid, 200 mg choline chloride, 100 mg myo-inositol, filter sterilize.

(ii) Salt solution: (per litre) 16 g KH_2PO_4, 4 g Na_2SO_4, 8 g KCl, 4 g $MgSO_4.7H_2O$, 1.32 g $CaCl_2.2H_2O$, 8 ml trace element solution, filter sterilize.

(iii) Trace element solution: (per litre) 60 mg H_3BO_3, 140 mg $MnCl_2.4H_2O$, 400 mg $ZnCl_2$, 40 mg $NaMoO_4.2H_2O$, 100 mg $FeCl_3.6H_2O$, 40 mg $CuSO_4.5H_2O$, filter sterilize.

(c) ***Ustilago* minimal medium**: (per litre) 10 g glucose, 3 g KNO_3, 62.5 ml salt solution, adjust to pH 7.0 with NaOH.

(d) **Conjugation medium for smut fungi**: 2% water agar or *Ustilago* complete medium plus 1% charcoal.

(e) **YMG**: (per litre) 4 g yeast extract, 10 g malt extract, 4 g glucose, 10 g agar.

(f) **CYM**: (per litre) 2 g peptone, 2 g yeast extract, 20 g glucose, 0.5 g $MgSO_4.7H_2O$, 0.46 g KH_2PO_4, 1 g K_2HPO_4, 12 g agar.

(g) **Standard *Coprinus* minimal medium**: (per litre) 25 ml stock A, 1 ml stock B, 10 ml stock C, 2 g asparagine, 10 g glucose, 10 g agar. Supplement with 100 mg amino acids/litre, if necessary.

(i) Stock A: (per litre) 40 g KH_2PO_4, 90 g Na_2HPO_4 (anhydrate), 11.6 g Na_2SO_4, 20 g di-ammonium tartrate (stored over 5 ml chloroform).

(ii) Stock B: (per litre) 40 mg thiamine (store at 4 °C).

(iii) Stock C: (per litre) 25 g $MgSO_4.7H_2O$ (stored over 5 ml chloroform).

(h) **Standard *Schizophyllum* minimal medium**: (per litre) 0.5 g $MgSO_4.7H_2O$, 0.46 g KH_2PO_4, 1.0 g K_2HPO_4, 20 g glucose, 120 mg thiamine, 2.0 g asparagine, 12 g agar.

Note added in proof:

In vitro completion of the sexual life cycle of *U. maydis*

Ruiz-Herrera *et al.* (87) recently demonstrated that the whole life cycle of *U. maydis* can be completed *in vitro* by growing mixtures of mating-compatible strains on porous cellulose-acetate membranes (Millipore type HA, 25 mm in diameter, 0.45 μm pore size) bridged by a liquid droplet to a growing callus of *Zea mays*. Mycelial masses formed which slowly acquired a dark colour by the formation of teliospore-like cells appearing to be engaged in recombination and germination processes. Following recovery of fungal mass from the membrane and plating on charcoal minimal medium, fuzzy and smooth colonies appeared, representing diploid and haploid colonies, respectively. Haploid colonies could be subcultured on suitable media to test for recombination of genetic markers.

87. Ruiz-Herrera, J., Léon-Ramírez, C., Cabrera-Ponce, J. L., Martínez-Espinoza, A. D., and Herrera-Estrella, L. (1999). *Mol. Gen. Genet.*, **262**, 468.

Chapter 6
Genome analysis in filamentous fungi

Mark L. Farman
Department of Plant Pathology, University of Kentucky, Lexington, KY 40546–0091, USA.

1 Introduction

Fungal genome analysis is an extremely broad topic in the sense that the isolation of most fungal genes cloned to date was made possible by using techniques to analyse and manipulate the fungal genome. In this chapter I have focused on methods for genetic and physical mapping of chromosomes and large chromosomal fragments, as might be applied in the analysis of chromosome structure and for map-based cloning of genes. The first sections outline approaches and methods for mapping chromosomes and analysing intact chromosomal DNAs, while the later sections cover the construction and screening of bacterial artificial chromosome (BAC) libraries as a means of isolating and propagating large chromosome fragments for more detailed analyses. Many of the protocols provided are useful not only in chromosome walking studies but can also be applied or adapted for other uses, including isolation of genes from libraries.

2 Genetic mapping

For the purpose of fungal genome analysis, construction of genetic maps is a useful endeavour in and of itself. It provides information about genome organization, particularly as it relates to genetic recombination in meiosis or mitosis and enables the locations of important chromosomal landmarks to be determined, such as the centromeres, telomeres, and nucleolar organizer region. In the process of constructing a map, the researcher may also gain insight into genome/gene evolution through the identification of translocations, inversions, gene duplications, and regions of genetic instability. Finally, genetic maps have practical applications in many areas of molecular biology, such as for positional cloning of genes (1, 2) and as scaffolds for reconstructing genomes with BAC (3) or cosmid clones (4) as a prerequisite to whole genome sequencing.

Genetic maps can be constructed in a number of ways, using different types of genetic markers. Deciding on the best mapping strategy requires prior consideration of the likely uses of a map and an awareness of problems that may be encountered. This section considers the relative merits of different markers for typical uses related to genome analysis, such as for gaining entry into chromosome walks, or for comparative genome mapping.

2.1 Whole genome mapping versus targeted mapping

One decision that should be made before starting a mapping project is whether to construct a genetic map of the whole genome or to focus efforts on identification of markers surrounding a gene of interest. Generally speaking, in those fungi that are eminently amenable to genetic analysis, and where map-based cloning is likely to be used routinely for gene isolation, complete genetic maps are extremely valuable resources and should be given serious consideration.

A major advantage of having a genetic map of the whole genome is that once each chromosome is well represented by markers, genes controlling newly identified traits can be mapped very rapidly (5). This is because the mapping process no longer requires segregation analysis to be performed for every locus. Instead, a subset of markers spaced evenly throughout the genome is used first to determine the approximate location of the gene of interest. Additional markers from the target region are then used for fine mapping.

An alternative to whole genome mapping is to create a regional map around a locus containing a gene of interest. The usual purpose of this is to find linked molecular markers that can be used to initiate efforts to isolate the gene by chromosome walking. With this kind of targeted approach, one is not usually interested in unlinked loci and high-throughput screens are therefore desirable for rapid identification of useful markers. The most efficient way to screen large numbers of markers for linkage to specific loci is to perform bulked segregant analysis of mixed DNA samples, pooled according to the phenotype of interest (6). Markers that are linked to either allele of the corresponding gene should appear in only one of the pools. Candidate markers identified in the pooled DNAs are then mapped by analysing segregation in a larger population of individual progeny to determine their actual genetic distance from the target locus. If closer markers are desired, pools should be reconfigured to include recombinant progeny, as analysis of such pools will ensure that new markers are closer than those identified previously.

It is important to consider that if, at a later date, bulked segregant analysis is used to identify markers linked to another gene, every marker will need to be re-examined. Clearly, this is very inefficient and is a good argument for whole genome mapping.

2.2 Genetic markers and genome analysis

Any genetic difference between two compatible fungal strains, resulting from variation at a single locus can be used as a marker for mapping studies. The choice of which markers will be employed should be based on a careful consideration of how a map might eventually be used because certain kinds are more appropriate for particular applications. For application of a map to genome analysis, it is desirable that the markers provide a straightforward means of identifying the corresponding chromosomal regions in genomic libraries. DNA-based markers are obviously favourable in this regard. There are a number of different types of DNA markers that are suitable for fungal mapping studies including restriction fragment length polymorphisms (RFLPs) (7), randomly amplified polymorphic DNAs (RAPDs) (8, 9), and amplified polymorphic DNAs (AFLPs) (10).

2.2.1 Constructing genetic maps based on RFLP markers

When constructing a complete genetic map, RFLPs are one of the better choices of markers. Although the initial labour investment in an RFLP mapping study is greater than when using RAPDs or AFLPs, practical application of an RFLP map is much more straightforward than one based on these other markers. In the first instance, RFLP markers are identified using cloned DNA fragments as probes. Therefore, an RFLP marker that is linked to a chromosomal region of interest can be used immediately to identify the corresponding sequence(s) in a genomic library. Secondly, because an RFLP marker is co-dominant, it can usually be used to follow segregation of a particular locus in crosses between any pair of fungal strains. This enables a single map to be used as a framework for analysis of several different genomes. This property is extremely useful for comparative mapping studies (11) and also enables markers to be used for isolation of sequences from fungal strains other than those used for mapping.

In principle, almost any DNA fragment can be used to identify RFLPs. However, longer fragments generally are labelled more efficiently and have a higher probability of identifying polymorphisms. Random genomic DNA probes in the size range of 5–10 kb can be generated by cloning size-selected genomic restriction fragments into a standard plasmid vector.

Protocol 1

Cloning size-selected genomic DNA fragments for use as RFLP probes

Reagents

- PCI: phenol/chloroform/isoamyl alcohol (25:24:1)
- 1 × loading dye: 0.04% bromophenol blue, 5% glycerol, dissolved in TE buffer
- CI: chloroform/isoamyl alcohol (24:1)
- 10 × Freezer medium: 44% glycerol, 0.36 M K_2HPO_4, 0.132 M KH_2PO_4, 17 mM sodium citrate, 4 mM $MgSO_4$, 68 mM $(NH_4)_2SO_4$

Method

1 Digest 2 μg of plasmid DNA with a restriction enzyme present in the polylinker. Check that the digestion is complete by performing a test transformation of *E. coli*.

Protocol 1 continued

2 Treat the digested plasmid with calf intestinal phosphatase (CIP) according to the supplier's instructions and then extract the CIP with PCI, followed by extraction with CI. Finally, ethanol precipitate the vector DNA and redissolve in TE at a concentration of 0.2 μg/μl.

3 Digest 5–10 μg of genomic DNA with the same restriction enzyme as in step 1. Check that the digest is complete by running an aliquot (~ 100 ng) on an agarose gel. Ethanol precipitate the genomic DNA and redissolve in 20 μl of 1 × loading dye in TE. Load into a single lane of a 0.7% agarose gel in 0.5 × TBE. Perform electrophoresis at 1 V/cm for 24 h, using a 1 kb DNA ladder (Gibco BRL) as a size marker.

4 Stain the gel for 30 min in 5 μg/ml ethidium bromide solution in dH_2O. View the gel on a transilluminator using low intensity, long wavelength (312 nm) UV light. Excise the portion of the genomic DNA digest corresponding to the 5 kb and 10 kb size range.

5 Extract the DNA from the gel using a Qiagen Gel extraction kit (or equivalent).

6 Ligate 0.5 μg of the size-selected DNA to 1 μg of plasmid DNA in a reaction volume of ≤ 10 μl. Use 1 μl of ligation mix to transform *E. coli* cells by electroporation.

7 The following day, inoculate white colonies into microtitre plate wells containing LB + 1 × freezer medium + antibiotic. Grow overnight at 37 °C with gentle shaking. The following day, use a replicator tool to array the colonies on nylon membranes (*Protocol 9*) and then store the microtitre plates at −80 °C.

Cosmid clones with inserts of up to 40 kb also make excellent RFLP probes because their large size dramatically improves the probability that RFLPs will be detected. All plasmid and cosmid probes should be pre-screened to eliminate clones that contain repetitive DNA. These pre-screens can be easily performed by using labelled genomic DNA to probe the colony blot arrays prepared in *Protocol 1*. Clones containing repetitive DNA hybridize very strongly to a total genomic DNA probe, while single copy clones hybridize weakly, if at all (12).

2.2.2 cDNAs as RFLP probes

Some fungal genomes may be too heavily populated by repetitive DNAs to make analysis of RFLPs based on random probes unfeasible. In such cases, RFLP probes based on random cDNA clones could be used because protein-coding genes in fungi are usually present at single or low copy number. An advantage to mapping cDNAs in these 'problem' fungi is that the abundance of repetitive elements within their genomes will tend to increase the incidence of RFLPs.

2.2.3 Identification and screening of RFLPs

Each potential marker fragment should first be used to survey for polymorphisms between the parental strains used for mapping. It is an inefficient use of resources to digest the parental DNAs with many restriction enzymes to try and find polymorphisms with each and every marker. Instead, the parental DNAs should be digested with three or four enzymes to identify polymorphic probes. Non-polymorphic probes can be discarded or saved for other purposes. The restriction enzymes can be chosen arbitrarily, however, due to the large number of restriction digestions required in an RFLP mapping project, and it is wise to select ones that are insensitive to the impurities that often occur in fungal genomic DNA preparations. After the digests have been fractionated by electrophoresis and transferred to membranes, they should be hybridized with the intact plasmid/cosmid probes, using a hybridization buffer containing SDS as a blocking agent (*Protocol 11*). Such buffers enable membranes to be stripped and re-probed many times without an appreciable increase in background. Efficient labelling of intact plasmid or cosmid clones can be achieved using a commercially available nick translation kit (Gibco BRL) or they can be linearized for random-primed radioactive or non-radioactive labelling. It should be noted that use of radioactive probes generally allows blots to be probed more times before background signals become problematic. Once RFLPs have been identified in the survey blots, each polymorphic marker is used to monitor segregation of the corresponding locus in progeny DNAs digested with the 'polymorphic' enzyme. Preparation of several sets of survey and progeny blots enables markers to be screened with great efficiency. As long as the DNAs of the parental strains are reasonably polymorphic, it should be possible to screen 18 markers in a single week and map another 12.

2.2.4 Assembling genetic maps

Construction of genetic maps of fungi is performed conveniently using the program *MAPMAKER* V3.0 (http://waldo.wi.mit.edu/ftp/distribution/software/mapmaker3/), which is suitable for mapping in both haploid and diploid organisms. Its use is explained explicitly in the accompanying manual.

Like most mapping programs, *MAPMAKER* makes a basic assumption that the data it is provided with is suitable for mapping purposes. Most importantly, the markers should identify alleles that segregate in a 1:1 ratio. If this requirement is not met, false linkages are likely to occur. Therefore, this should be tested for each marker. *MAPMAKER* also lets the user define mapping parameters such as the maximum recombination frequency (θ) and the L.O.D. value (log of the odds) for map construction. The default values in *MAPMAKER* are not suitable for most fungal mapping studies. Parameters should be selected based on the number of markers and the size of mapping population. As a guideline, in an analysis of segregation among 50 progeny, markers separated by more than 30% recombination should be considered unlinked at the $P < 0.05$ level of significance. Recommended parameters for a new fungal mapping project are a L.O.D. value of 4.0 and a maximum recombination value (θ) of 0.2.

2.3 Other types of markers

In some cases it may be difficult to use RFLPs as genetic markers. One such instance is if the parental strains used to generate the progeny population are not very polymorphic. This situation poses a problem for mapping with any type of marker; and, if possible, other strains should be chosen. Failing this, detection of suitable polymorphisms may require use of methods such as RAPD or AFLP analysis, which can enable several markers to be detected in a single experiment. However, the reader should be aware that there are significant disadvantages to using dominant markers such as RAPDs and AFLPs for creating a map, especially if it will be used routinely for genome analyses.

2.3.1 RAPDs and AFLPs

RAPDs and AFLPs are best used in targeted mapping strategies which employ bulked segregant methods to identify markers linked to a specific gene of interest. They are particularly suited to these applications because they are easy to screen and each experiment usually yields several amplified fragments. These properties enable many loci to be screened rapidly and efficiently, until a marker is found that is linked to a gene of interest. If the sole purpose of the study is to clone the gene for further study of its function, then the disadvantages of RAPD and AFLP markers are usually compensated by the speed with which they can be identified.

Complete genetic maps of filamentous fungi have been created using RAPDs and AFLPs (8, 9). However, it should be noted that because these markers are dominant, many of the RAPDs and/or AFLPs appearing on one genetic map may not be useful in crosses between fungal isolates other than the parents of the mapping population. In some cases, it may be possible to overcome this problem by cloning RAPD and AFLP fragments and converting them to RFLP probes or sequence characterized amplified regions (SCARs). However, cloning of RAPD and AFLP bands can often prove problematic, leading to isolation of many different DNA fragments. The actual marker fragment may be hard to identify, especially if the fragments contain repetitive DNA and cannot be used for RFLP analysis (13). Finally, if dominant markers are used as markers in crosses between different strains, it may be necessary to perform cross-hybridization experiments to confirm that fragments of the same size are indeed allelic. This is especially important for RAPDs. As a result of these problems, routine application of RAPD and AFLP maps for most types of genome studies is usually laborious and inefficient.

3 Electrophoretic karyotyping

Fungal chromosomes are generally too small to be clearly visible under the microscope. Indirect visualization of chromosomes in most fungi is possible using contour-clamped homogeneous electrophoretic field techniques (CHEF) to resolve intact chromosomal DNA in agarose gels (14). CHEF involves applying an alternating electric field at a transverse angle to the main electrophoretic axis, forcing the DNA molecules to be periodically reoriented during electrophoresis. This results in a better relationship between mobility and molecular size.

Resolution of chromosomes in agarose gels results in an electrophoretic karyotype which can then be used for several purposes including estimation of overall genome size (15), assignment of RFLP markers/cloned genes to particular chromosomes (7), characterization of translocations (16), and analysis of chromosome polymorphisms (17).

3.1 Preparation of intact chromosomal DNA

Intact chromosomes are fragile molecules. To avoid damage from shearing, whole chromosomes are prepared by lysing cells in solidified agarose and then removing associated proteins by enzymatic digestion. This procedure releases the chromosomes into small aqueous pockets in the agarose left by the lysed cells, where they are protected from shearing forces.

To obtain clean chromosome separations with low background smearing, it is usually necessary to remove filamentous fungal cell walls prior to embedding cells in agarose. This is achieved by forming protoplasts (Chapter 3) and filtering them away from the undigested mycelium. However, protoplast formation is not necessary for some filamentous fungi, such as *Ustilago hordei*, which has a yeast-like phase in its life cycle (18). This is worth investigating, as elimination of this step can save a lot of time and effort.

Protocol 2

Preparation of intact chromosomal DNA in agarose microbeads

Note: clear resolution of chromosomal bands with minimal background smearing depends on having an optimal protoplast density. When first preparing intact chromosomal DNAs, it is therefore advisable to perform experiments to determine this density empirically.

Reagents

- EX buffer: 0.5 M EDTA, 0.01% Triton X-100

Method

1. Prepare protoplasts using a protocol appropriate for the fungus of interest (see Chapter 3 for examples). Osmotic stabilizers containing Mg^{2+} salts should, however, be avoided. After filtering the protoplasts away from the cell debris and washing to remove the protoplasting enzymes, resuspend the protoplasts at a density of approx. 2×10^9/ml in 20% (w/v) sorbitol or sucrose (the composition and concentration of the osmoticum should be adjusted to suit the fungus under study).
2. Take up to 4 ml of protoplast suspension and pre-warm to 37 °C. Add an equal volume of 1.2% InCert agarose (FMC Bioproducts) in 20% sucrose, maintained at 45 °C. Mix well by gentle pipetting and then pour the protoplast suspension into an equal volume of pre-warmed (50 °C) mineral oil in a 25 ml conical flask. Vortex at full speed using a benchtop unit, holding the flask horizontally to allow the liquid to impact forcefully on the glass walls. After approx. 30 sec, transfer the flask to salted ice water to solidify the agarose as rapidly as possible.

Protocol 2 continued

3 Add an equal volume of EX buffer and transfer to a 15 ml centrifuge tube. Pellet the microbeads at 4000 g for 5 min. Discard the supernatant, being careful not to dislodge the beads. Add 4 ml of fresh EX buffer and use a pipette with a cut-off tip to resuspend the beads. Centrifuge as before. Repeat this wash step two times, then decant the EX buffer, and store the microbeads at 4°C as a slurry in tightly-sealed tubes.

Protocol 3

Loading agarose microbeads onto gels

Reagents

- TEX buffer: TE buffer with 0.01% Triton X-100

Method

1 Equilibrate 100 μl of beads in 10 vol. of TEX buffer. Pellet beads by brief centrifugation and carefully remove supernatant. Lay the gel comb horizontally on the bench and load the microbeads onto the teeth. Remove excess TEX from the bead suspension by wicking onto a paper tissue. Do not try to remove too much liquid as this will draw DNA out of the beads.

2 Glue beads into place with molten 0.6% InCert agarose in 0.5 × TBE. Allow to harden at 4°C. Insert the comb into the gel tray and carefully pour molten (≤ 50°C) 0.8% SeaKem GOLD agarose (FMC Bioproducts) around it to form the gel. Allow the gel to set and then place at 4°C for 20 min to harden.

3 Carefully remove comb and fill the empty well with molten 0.8% agarose in 0.5 × TBE. Immerse gel in running buffer and allow to equilibrate to the running temperature (14°C).

3.2 CHEF electrophoresis

Allow the gel to sit in 0.5 × TBE buffer in the CHEF unit and let it equilibrate to the running temperature of 14°C. Perform the CHEF electrophoresis in a running buffer of 0.5 × TBE and maintain the temperature of the running buffer at 14°C throughout the experiment. Examples of CHEF conditions suitable for resolving different sized DNA fragments are given in *Table 1*.

After electrophoresis, the chromosomes should be visualized by staining the CHEF gel in a 0.5 μg/ml ethidium bromide solution, followed by destaining for 1 h in distilled water. Assignment of genes and other sequences to particular chromosomes can be achieved by Southern blotting the resolved chromosomal DNAs to nylon membranes using standard protocols (19). It is recommended that the depurination wash is performed for 30 min, to ensure that the large chromosomal DNA molecules are fragmented sufficiently for efficient transfer.

Table 1 Examples of conditions for CHEF gel electrophoresis

Condition	Size range	Voltage gradient	Switch time(s)	Run time
1	5–50 kb	6 V/cm	1–6 sec	12 h
2	50–1000 kb	6 V/cm	50–90 sec	22 h
3	200–2000 kb	6 V/cm	60–120 sec	24 h
4	1 Mb–4 Mb	3 V/cm	250–900 sec	48 h
5	3 Mb–10 Mb	1 V/cm	5400 sec followed by 3600 sec	120 h 48 h

4 Physical mapping

It is often necessary to determine the physical distance between two DNA sequences on the same chromosome or between one sequence and a chromosomal landmark such as a telomere. Such measurements are used to gain prior knowledge of the length of a chromosome walk, to measure gaps between contigs of overlapping clones, and to determine sizes of chromosomal deletions.

4.1 Procedures for physical mapping in filamentous fungi

Rare cutting restriction enzyme analysis is a straightforward and often used physical mapping procedure and in some cases requires little or no prior knowledge of the loci/genome of interest. A restriction enzyme that cuts very infrequently in the genome of interest is used to digest chromosomal DNA into large fragments. These are then resolved by CHEF electrophoresis, transferred to hybridization membranes, and probed with two, or more, sequences to determine if they lie on the same restriction fragments. Unfortunately, physical mapping with rare cutting enzymes has limited resolution as it is possible only to determine if two sequences reside on the same restriction fragment (or at least on fragments of the same size). Unless these sequences contain restriction sites for the enzyme used, such analyses provide no information on the actual distance between them.

Frequently, it may be necessary to measure physical distances that are too great to be determined by restriction analysis, or a greater precision may be required. A strategy that has been applied successfully for long-range mapping in filamentous fungi is the use of targeted transformation to engineer unique cleavage sites into specific chromosomal loci (20, 21). Unfortunately, this usually requires a significant investment of labour. Furthermore, physical mapping usually entails measurement of distances between two loci, requiring engineering of two cleavage sites. For this reason, physical mapping experiments are easier to perform using RecA-assisted Achilles' cleavage techniques.

4.2 RecA-assisted Achilles' cleavage (RecA-AC)

RecA-AC (22) is a highly effective physical mapping method, applicable to any fungus for which it is possible to make intact chromosomal DNA. It requires minimal knowledge of the target loci and, unlike rare cutting restriction analysis, it enables measurement of the actual distance between them. RecA protein from *E. coli* catalyses strand exchange between single-stranded DNA and a complementary double-stranded DNA, in an ATP-driven reaction. Addition of RecA protein to chromosomal DNA, in the presence of an oligonucleotide and a non-hydrolysable analogue of ATP results in formation of a stable triplex at a region homologous to the oligonucleotide. RecA-AC takes advantage of this property by using large, synthetic oligonucleotides (62-mers) to direct RecA protein to specific restriction sites, affording them steric protection from methylases. Unprotected sites are then inactivated by methylation, whereas sites protected by the RecA-oligo complex, remain susceptible to subsequent enzyme digestion.

Protocol 4

Protocol for equilibration of microbeads in enzyme reaction buffers

Reagents

- TEX (see *Protocol 3*)
- 1 × reaction buffer

Method

1. Place 100 μl of microbeads in a microcentrifuge tube. Add 10 vol. of TEX. Mix by flicking the tube and allow to equilibrate for 1 min. Centrifuge for 10 sec at full speed in a microcentrifuge and then remove the supernatant by careful pipetting.
2. Add 10 vol. of 1 × reaction buffer, mix by flicking tube and allow to equilibrate for 1 min. Pellet beads by centrifugation as above and remove the supernatant. Repeat the reaction buffer equilibration step two more times.
3. After removing as much supernatant as possible, add enzymes and other reagents as required by the specific protocol.

Protocol 5

RecA-assisted Achilles' cleavage of intact chromosomal DNA (22)

Note: fresh DTT is critical for the success of RecA-AC.

Reagents

- 1 M DTT: made from a previously unopened vial, stored frozen in aliquots; do not re-freeze
- $RecA_4$ buffer: 25 mM Tris acetate pH 7.5, 4 mM magnesium acetate, 10 mM fresh DTT
- 10 × $RecA_{10}$ buffer: 250 mM Tris acetate pH 7.5, 100 mM magnesium acetate, 100 mM fresh DTT
- 1 mg/ml BSA in TE
- 0.4 mM ATPγS: store at −80 °C in 10 μl aliquots; do not re-freeze after thawing
- 120 mM magnesium acetate stock solution
- 1.6 mM S-adenosyl methionine: just before use dilute stock solution supplied by manufacturer in 5 mM H_2SO_4
- ES: 0.1 M EDTA, 1% SDS

Protocol 5 continued

Method

1 Following *Protocol 4*, equilibrate 100 μl of microbeads in TEX and then equilibrate the microbeads three times with $RecA_4$ buffer. Remove the buffer above the beads, add 10 μl of BSA, and then store on ice.

2 Form an oligo-RecA nucleoprotein complex by mixing 1 μl of oligonucleotide (0.165 μg/μl) with 6.25 μg of RecA protein (NBL). Add 1 μl of 10 × $RecA_{10}$ buffer and water to 9 μl. Place in a 37 °C water-bath for 1 min. Then add 1 μl of a 0.4 mM stock of ATPγS. Incubate at 37 °C for 10 min (when using RecA-AC to cleave two sites simultaneously, form the nucleoprotein complex by mixing 1 μl of each oligonucleotide and doubling the volume of the other reagents).

3 Warm the bead solution to 37 °C in water-bath and then add 10 μl of nucleoprotein solution. Adjust the magnesium acetate concentration to restore 4 mM and place on ice overnight.

4 In the morning, incubate beads at 37 °C for 30 min. Then add fresh DTT to 10 mM and magnesium acetate stock solution to bring the Mg^{2+} concentration up to 12.5 mM. Finally, add 1 μl (40 U) of *Eco*RI methylase (NBL) and start the reaction by adding 5 μl of 1.6 mM S-adenosyl methionine. Incubate at 37 °C for 1 h.

5 Inactivate the methylase and dissociate the nucleoprotein complex by adding 1 vol. of ES and heat to 52 °C for 30 min.

6 Equilibrate microbeads in TEX, followed by equilibration in *Eco*RI restriction buffer (NBL buffer 3). Add 10 μl of BSA and finally 1 μl (20 U) of *Eco*RI. Digest for 1 h at 37 °C.

7 Add 1 vol. of ES and heat to 52 °C for 15 min to inactivate the enzyme. Equilibrate beads in TEX and load on a CHEF gel, along with appropriate controls. After electrophoresis, stain the gel as described in *Protocol 1*.

When performing RecA-AC for the first time, or at a new site, it is advisable to digest the cleaved chromosomal DNA with a second restriction enzyme such as *Not*I. Southern hybridization analysis can then be used to demonstrate that the target *Not*I fragment is efficiently cleaved by the RecA-AC treatment. Once the investigator is confident that the reactions are working, cleaved chromosomal DNAs should be resolved by CHEF electrophoresis initially using CHEF conditions 2 (*Table 1*). However, additional gels may need to be run using other conditions because in most cases there will be little or no prior knowledge of the expected sizes of cleavage products. In all these experiments, it is important to note that large cleavage products may not be visible in the stained gel but are usually clearly visible upon hybridization (*Figure 1*).

5 Construction and screening of genomic DNA libraries

Genomic libraries are essential resources for genome analysis and gene isolation. There are a number of vectors available for such purposes, including λ, P1,

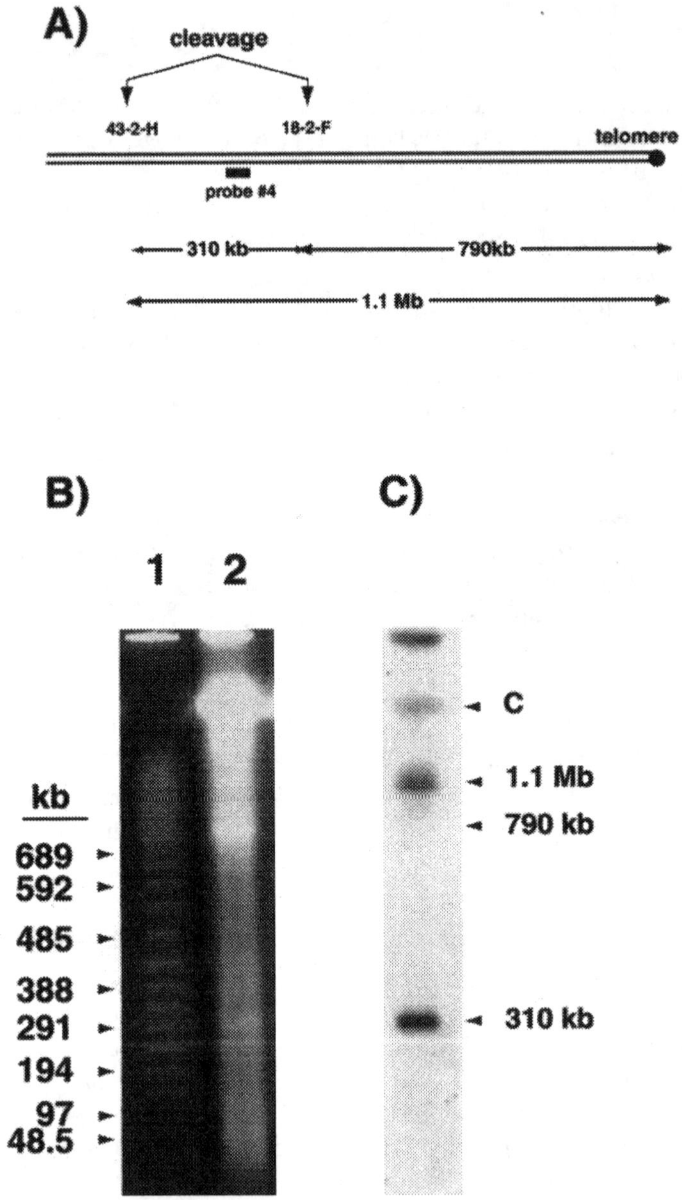

Figure 1 Physical mapping of a *Magnaporthe grisea* chromosome by RecA-AC. (A) Organization of the chromosomal region under study, as determined by genetic mapping. The chromosome was cleaved in two locations by designing oligonucleotides overlapping *Eco*RI sites in cosmid markers 43-2-H and 18-2-F. RecA-AC directed to these chromosomal locations resulted in the release of a 310 kb fragment representing the distance between these two markers. (B) This fragment is clearly visible in the photograph of RecA-AC treated DNA. (C) Autoradiograph showing the result of Southern hybridization after probing with a subclone of the gel purified 310 kb cleavage product (indicated as probe No. 4 in A). This probe also hybridized to a 1.1 Mb cleavage product, representing the telomeric fragment resulting from incomplete protection from methylation (and hence incomplete cleavage) of sites in 18-2-F.

cosmids, fosmids, bacterial artificial chromosomes (BACs), and yeast artificial chromosomes (YACs). For the purpose of fungal genome analysis, the most useful vector systems are based on λ, which will accept inserts up to 23 kb (23); cosmids (24) or fosmids (25), which can accommodate up to 45 kb; and BACs with maximum cloning capacities of around 300 kb (26).

A genomic library will usually be used to isolate many different sequences from a given fungal genome. It is, therefore, very important that a library represents the genome adequately. This is ensured by maintaining enough clones such that the probability of occurrence of any given sequence is greater than 0.99. This number may be estimated using the following equation:

$$N = \frac{\ln(1 - P)}{\ln(1 - F)}$$

where N is the number of clones required; P is the desired probability of identifying any given sequence; and F is the fraction of the genome contained in an individual clone.

Construction of λ and cosmid libraries was covered in a previous *Practical Approaches* volume (27). Therefore the following section focuses on construction of BAC libraries.

5.1 Bacterial artificial chromosome (BAC) libraries

Bacterial artificial chromosomes (BACs) are excellent vectors for construction of filamentous fungal genomic DNA libraries. They are maintained a very low copy number of one to two per cell, which reduces instability caused by repetitive DNA sequences (26). This is an important consideration as most filamentous fungal genomes contain repetitive DNA. The 300 kb cloning capacity of the BAC vectors is optimal for most fungal genome studies. For example, a typical filamentous fungal genome of 40 Mb, should be 99% represented by only 1840 clones, assuming an average clone size of 100 kb. Although YACs can accommodate much larger inserts, meaning fewer clones would be required, the comparatively small size of fungal genomes does not justify the additional challenges associated with manipulation of yeast and YAC DNAs.

5.1.1 Preparation of BAC vector DNA

The BAC vector, pBeloBAC11 (http://www.tree.caltech.edu/protocols/BAC_lib_construction.html) and its derivatives is recommended for these experiments, because it has a lacZ gene which enables blue/white colour screening to be performed for identification of recombinant clones. Due to the low copy number of BAC vectors, preparation of vector DNA typically requires culturing several litres of *E. coli*. When working with such large culture volumes, special attention is required to minimize the possibility of the BAC vector preparations becoming contaminated with bacterial DNA.

Protocol 6

Preparation of BAC vector DNA

Equipment and reagents

- Qiagen-tip 500 DNA purification cartridges (Qiagen)
- Solution I: 25 mM Tris-HCl pH 8.0, 50 mM glucose
- Solution II: 0.2 M NaOH, 1% SDS
- Solution III: 5 M potassium acetate pH 4.8

Method

1 Start an overnight culture with a single *E. coli* colony expressing the lacZ$^+$ phenotype. Use the culture to inoculate 3 litres of LB + chloramphenicol (12.5 μg/ml). Shake vigorously at 37 °C for 20 h.

2 Harvest the cells by centrifugation in twelve 250 ml plastic bottles. Use a single 75 ml volume of solution I to resuspend all of the pellets and then divide the suspension into 3 × 250 ml bottles. Add 60 mg of lysozyme to each bottle and mix by inversion.

3 Add 50 ml of solution II to each bottle, mix by inversion, and then place on ice for 10 min. Then add 40 ml of solution III to each bottle, mix by inversion, and allow to stand on ice for another 10 min.

4 Pellet cellular debris by centrifugation at 8000 g for 30 min at 4 °C. Collect the supernatant and filter it through cheesecloth to remove remaining debris. Add RNase A to a concentration of 0.1 mg/ml.

5 Purify the BAC DNA using four Qiagen-tip 500 columns, following the manufacturer's directions.

6 Redissolve the DNA pellets obtained from the Qiagen preps in 18.6 ml of TE, and then dissolve 20.5 g of CsCl in the DNA solution. Add 0.4 ml of ethidium bromide (10 mg/ml) and perform ultracentrifugation for two to three days at 45 000 r.p.m. in a Beckman 70.1Ti rotor (or equivalent).

7 Isolate the lower DNA band from the CsCl density gradient and extract the ethidium bromide with water-saturated *n*-butanol until there is no visible colour in the organic phase (to avoid loss of sample, extract by removing the *n*-butanol layer), then extract one more time. Finally, dialyse the aqueous layer against TE overnight.

8 Precipitate the BAC DNA with 0.1 vol. 3 M sodium acetate pH 5.2 and 2 vol. EtOH, wash with 70% ethanol, air dry, and finally, redissolve in 20 μl of TE. Measure DNA concentration and adjust to 0.5 μg/μl by addition of TE or, if necessary, by re-precipitation.

5.1.2 Preparation of BAC insert DNA

BAC insert DNA is prepared by partially digesting intact chromosomal DNA to obtain a large number of fragments in the 100–300 kb size range. Use of agarose plugs for partial digestion can be rather challenging because, while the enzyme diffuses into the plug, excessive digestion occurs in the periphery. Although this can be accomplished by allowing a preliminary 4°C diffusion step, preparation of partially digested DNA is more easily performed using chromosomal DNA prepared in agarose microbeads (*Protocol 2*). The large surface area to volume ratio in microbeads ensures that the diffusion of buffers and enzymes is almost instantaneous.

Protocol 7

Partial digestion of intact chromosomes in agarose microbeads

Reagents

- See *Protocol 3* and *4*

Method

1 Following the steps in *Protocol 4*, equilibrate 500 µl of agarose microbeads with 10 vol. of TEX and then equilibrate in 1 × restriction enzyme buffer.

2 Remove excess restriction buffer, leaving enough to just cover plug or microbeads. Then add 0.1 vol. of 1 mg/ml BSA and then dispense into one aliquot of 100 µl and eight aliquots of 50 µl (priming the pipette tip with TEX before each aliquot helps prevent the beads from sticking).

3 Add 1 U of enzyme to the first tube only, mix by stirring with the pipette tip and flicking the tube gently. Then transfer 50 µl of beads to the second tube and mix by stirring/flicking. Repeat the serial dilution process down to the ninth tube. Digest for 1 h at the optimal temperature for the enzyme employed.

4 Stop the reaction by adding 10 vol. ES to each tube and heat to 52°C for 15 min. Then equilibrate beads one time with 10 vol. TEX, pellet by centrifugation, and remove supernatant.

5 Load microbeads onto the well comb as described in *Protocol 3*. Electrophorese the partial digests using CHEF conditions 2 (*Table 1*), using a concatemer of λ DNA as a size marker.

6 After electrophoresis stain and photograph the gel to determine which enzyme concentration yielded the most fragments in the desired size range.

7 Scale up the digestion for a preparative partial digest.

After performing the preparative digest, it is necessary to select fragments in the desired size range. Size-selection is performed by resolving fragments on a CHEF gel, followed by excision of fragments in the correct size range. However, any given region of a CHEF gel is usually contaminated with smaller fragments that were 'caught up' in the agarose plug and released throughout the electrophoresis. Although these fragments are vastly outnumbered by the larger molecules, the extreme size bias inherent in transformation will favour cloning of the smallest fragments. Therefore, a second electrophoresis is usually necessary to eliminate these small, contaminating fragments.

Protocol 8

Size selection, recovery, and cloning of BAC insert DNA

Reagents

- See *Protocol 1*
- SeaPlaque agarose (FMC Bioproducts)
- TAE buffer
- TBE buffer
- β-agarase (NBL)

Method

1 Use CHEF conditions 2 to fractionate the partially digested preparative DNA sample in 1% SeaPlaque agarose made in 0.5 × TAE buffer. In the next lane but one, run a λ DNA concatemer as a size marker.

2 Do not expose the genomic DNA to UV light. Instead, separate the lane containing the λ size marker and stain it with ethidium bromide (*Protocol 1*). Place the gel slice on a transilluminator next to a ruler. Expose to UV light and take a photograph. Determine the position of fragments in the 200–300 kb size range and excise the corresponding region from the lane(s) containing digested chromosomal DNA.

3 Place the gel slice(s) into a well of a 0.8% SeaKem GOLD agarose gel in 0.5 × TBE and electrophorese for 15 min using conditions 1 to eliminate contaminating small fragments. Recover the gel slice(s) from the well.

4 Transfer the gel slice(s) to a microcentrifuge tube, place in a 65 °C water-bath for 5 min to melt the slice(s). Add 0.1 vol. of 10 × β-agarase buffer and 1 U of β-agarase per 200 μl of molten agarose. Incubate at 42 °C for 1 h.

5 In a total reaction volume of < 100 μl, ligate between 100–500 ng of CIP-treated BAC vector to ≤ 50 μl of the size-selected DNA.

6 Use 1–2 μl of the ligation mix to transform *E. coli* cells by electroporation. Strains that are recommended for propagating BAC clones are DH10B (Gibco BRL) and XL10-Gold-kan^R cells (Stratagene). After electro-transformation, plate cells on LB containing 12.5 μg/ml chloramphenicol, 50 μg/ml X-gal, and 0.2 mM IPTG. Recombinant BAC clones will grow as white colonies on these plates.

7 **Before** picking colonies to microtitre plates for storage and screening, isolate DNA from a number of clones using a standard alkaline lysis protocol (19) and check the size distribution of BAC inserts.

5.2 Picking colonies and library replication

Once clones have been picked to liquid culture, grown, and then stored at −80 °C, repeated thawing of the library should be avoided as this will compromise clone viability. Therefore, it is desirable that library replication, clone pooling, and

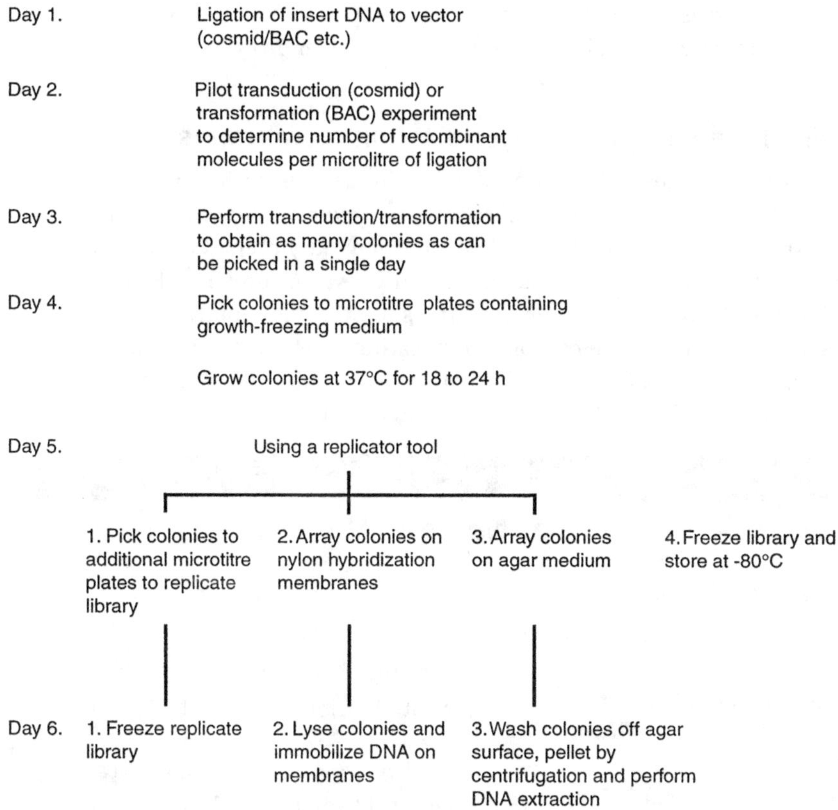

Figure 2 Organizational scheme for creating, replicating, arraying, pooling, and storing DNA libraries.

colony arraying is performed immediately after the library has been picked from the primary transformation/transduction plates and grown in liquid medium. A scheme for these efforts is suggested in *Figure 2*.

Colonies containing recombinant clones are most economically grown and stored in 96- or 384-well microtitre plates (Nalge Nunc International Corp—NNI). If the primary transformants are picked by hand, they should be inoculated to just one set of plates. Attempts to replicate libraries manually invites the introduction of errors. Instead, replication should be performed using tools designed specifically for this purpose. Inclusion of freezer medium (*Protocol 1*) in the LB broth enables the bacteria to be grown and frozen in the same culture medium without additional amendments.

Almost invariably, colonies on the primary plates will display a wide range in colony size. There may be a tendency to avoid picking very small colonies, which may lead to under- or non-representation of certain sequences in the library. It is recommended that at least 500 small colonies are picked to microtitre plates dedicated specifically to slow-growers. The similar, albeit slow, growth rates of these colonies means that they will be roughly equally represented in the DNA pools and arrays, increasing the likelihood of isolating clones which would be missed if small colonies were avoided or cultured with faster growers.

5.3 Arraying library clones on hybridization membranes

In most applications, clones will be identified in libraries using hybridization-based screens. This means that colony blot arrays will eventually be required to identify individual clones. These blots are most conveniently prepared after the primary colonies have been picked and cultured (*Figure 2*). Commercially available 96- and 384-pin replicator tools enable arrays to be made rapidly and with great accuracy. Also available are complete replicating systems which facilitate manual preparation of high-density arrays containing up to 9 × 96-well plates or 4 × 384-well plates on a single 11.5 cm × 7.5 cm membrane.

Protocol 9

Arraying clones on hybridization membranes

Equipment and reagents

- 96-pin and/or 384-pin replicator tool (NNI, 250520 and 250393)
- 11.5 cm × 7.5 cm hybridization membranes (Pall Biodyne B—NNI, or equivalent)
- (Optional) Omnitray copier and Omnitrays (all from NNI)
- Whatman 3MM paper
- Denaturation solution: 0.5 M NaOH
- Neutralization buffer: 0.5 M Tris-HCl pH 7.5, 1.5 M NaCl
- 20 × SSC: 175.4 g NaCl and 88.2 g sodium citrate dihydrate per 1 litre of dH_2O

Method

1. Prepare LB + antibiotic agar in large, sterile Pyrex dishes, being very careful to avoid air bubbles. Alternatively use agar poured in Omnitrays (NNI).
2. Sterilize nylon membranes by autoclaving. Number the membranes and place them on the surface of the solidified agar. Do not allow bubbles to form under the membranes.
3. Use a sterilized 96- or 384-pin replication tool to transfer cultures from each microtitre plate onto membranes. It is advisable to make at least two replicates at a time.
4. Between uses, decontaminate the replicator pins in 10% bleach; then rinse in H_2O, flame-sterilize with 95% EtOH; and, finally, cool the pins in sterile H_2O before placing in the next microtitre plate.
5. Invert the culture dishes and leave colonies to grow at 37 °C, overnight (20 h).
6. Use tweezers to lift the membranes from the agar surface. Allow them to air dry, with colonies up, on paper towels.
7. Place the membranes, with colonies facing up, for 10 min on Whatman 3MM paper

Protocol 9 continued

soaked in denaturation solution. Then transfer the membranes to 3MM paper soaked in neutralization buffer for 10 min.

8 Remove the membranes from the second sheet of paper and shake gently for 10 min in 2 × SSC. Transfer membranes to 2 × SSC, 0.5% SDS and use a clean tissue to gently massage the membrane surface to remove cell debris. A discoloured spot will remain under each colony.

9 Fix the DNA to the membranes using a UV crosslinker and then allow them to air dry.

10 Hybridize membranes following *Protocol 11*. It is important to avoid the use of hybridization buffers which include salmon sperm DNA and Denhardt's solution as blocking agents, as these tend to cause increased background with repeated probing.

5.4 Pooling library clones

Identification of target sequences in genomic libraries can be performed in a number of ways including genetic complementation, hybridization, and PCR-based methods. Most of these applications are performed more efficiently if the library is screened by analysing pools of clones. The simplest pooling strategy is to combine colonies from an individual 96- or 384-well microtitre plate and to isolate DNA from this mixture. In this case, second, and sometimes third round screens are necessary to identify the target clone. However, use of an alternative pooling strategy enables a target clone to be identified in the primary screen (28). When constructing DNA pools, care must be taken to minimize over- or under-representation of particular clones. This can be achieved by culturing each colony on agar medium, where visual inspection of colonies enables one to be sure that each clone is adequately represented in the pool.

Protocol 10

Pooling clones for DNA preparation

Equipment

- Omnitrays (NNI) or sterilized empty pipette tip boxes
- 96- or 384-pin replicator tool

Method

1 Prepare LB + antibiotic agar in sterile Omnitrays or pipette tip boxes. Use a sterilized 96- or 384-pin replication tool to spot colonies onto the agar surface. Rinse and re-sterilize the replicator pins between each use as described in *Protocol 9*.

2 Invert plates/boxes and allow the colonies to grow at 37 °C overnight (20 h).

Protocol 10 continued

3 When colonies have reached 2–4 mm in size, pipette 5 ml of LB onto the surface of the agar and dislodge the bacteria by gentle massaging with a sterile spreader. Pipette the cell suspension into a centrifuge tube and add 5 ml fresh LB back to the plate to recover the remaining cells. Pellet cells by centrifugation and discard the supernatant.

4 Either, prepare DNA from the pooled colonies immediately using a standard alkaline lysis protocol (19), or freeze the cell pellets and extract DNA later.

5.5 Library storage

Long-term storage of the library should be at −80 °C. Clones should be thawed only when the library is replicated or when new membranes/pools are required. Thawing should not be allowed when clones are recovered (see Section 6.2).

6 Screening libraries

Approaches for screening libraries depend on whether the libraries are ordered or not. In unordered libraries, amplified or unamplified clones are maintained in a common pool. This makes identification and recovery of desired clones more difficult. In ordered libraries, however, the clones are stored individually in microtitre plates, or some other manner, where each clone is in a known location and can be uniquely identified. For most experimental applications, identification of clones is most easily and efficiently performed using ordered libraries as these lend themselves well to screening strategies that analyse pools of clones. Although pooling is also possible in unordered libraries, in the majority of cases, it is not possible to make sub-pools for zeroing in on a positive clone.

6.1 Screening libraries by hybridization

In most fungal genome studies, such as chromosome walks, genome reconstruction experiments, and analyses of genome instability and evolution, genomic DNA libraries are most likely to be screened by hybridization using RNA or DNA probes.

The quickest means of identifying positive clones by hybridization is to screen the whole library arrayed on membranes by colony blotting. However, colony blots are labour-intensive to make and each membrane has a limited life span. For this reason, it is useful to perform a primary screen using DNA pools. Each pool sample is digested with a restriction enzyme and fractionated in an agarose gel. Also included is a lane of total genomic DNA digested with the same enzyme. This gel is blotted and then hybridized with the probe. Pools containing the target sequence should yield hybridizing fragments of the same size as observed in the genomic DNA lane. Positive clones are then identified in secondary screens of the colony blots representing the positive pools. In this manner, colony blots are only subjected to hybridization if they contain a positive clone. The following protocol may be used in library screens based on Southern and colony blotting.

Protocol 11

Southern hybridization analysis of DNA pools and clone arrays

Reagents

- Hybridization buffer: 50% formamide, 0.125 M $NaHPO_4$ pH 7.2, 7% SDS, 10 mM EDTA
- 0.5 M $NaHPO_4$ stock solution: make up a 0.5 M solution of Na_2HPO4 and adjust pH with phosphoric acid
- Denatured RNA or DNA probe
- SSPE: add to 800 ml H_2O, 175.3 g NaCl, 27.6 g NaH_2PO_4, and 7.4 g EDTA; adjust to pH 7.4 with NaOH, and add H_2O to 1 litre

Method

1. Pre-hybridize membranes for at least 15 min at 42 °C in an excess of hybridization buffer. Then pour off the hybridization buffer and replace with 5 ml of fresh buffer. Allow the temperature of the fresh buffer to equilibrate to 42 °C.
2. Add denatured probe to the hybridization solution, being careful not to apply it directly to the membrane surface. Mix the probe by swirling and then allow the solution to cover the membrane(s). Hybridize at 42 °C for $\geq$ 16 h.
3. Wash the membrane(s) two times at 42 °C for 10 min in 2 × SSPE, once at 65 °C for 15 min in 2 × SSPE, and finally at 65 °C for 15 min in 0.1 × SSPE, 0.1% SDS.
4. (a) Blot the membranes dry with paper towels, seal in plastic wrap, and expose to photographic film or a PhosphorImager screen.
   (b) Alternatively, process the membranes according to the appropriate non-radioactive detection protocol and expose to photographic film.
5. Develop the film/scan the screen and either:
   (a) Determine which pools contain positive clones. Perform secondary screens of colony blots of the positive pools.
   (b) Align the developed film/screen with the membrane(s) to determine which colonies contain the desired clone.
6. Strip the membranes *immediately* by heating at 100 °C for 1 h in 0.1 × SSPE, 0.1% SDS. Rinse in fresh 0.1 × SSPE, 0.1% SDS, then allow membranes to air dry, and store between paper towels.

6.2 Recovery of clones from frozen library stocks

Frozen *E. coli* stock cultures should ALWAYS be streaked prior to starting liquid cultures. Failure to do so is poor microbiological practice and may result in poor yields of BAC or cosmid DNA. When recovering clones, do not thaw the frozen stocks and do not use a hot loop. Instead, scrape a sliver of ice from the surface with a sterile toothpick and transfer to agar plus antibiotic. Then use a sterilized loop to streak for single colonies.

7 Chromosome walking

Chromosome walking is a method used to isolate a gene or other chromosome structure using a cloned DNA fragment, or marker, that is near to the target sequence on the same chromosome. The cloned DNA fragment is used to identify overlapping clones in a genomic DNA library and this process is reiterated in a systematic manner using the ends of newly identified clones to identify the next overlaps. The net result is a gradual progression, or walk, along the chromosome. Eventually, a successful walk will lead to isolation of a clone that contains the required sequence(s).

7.1 Synthesis of probes from insert ends

Chromosome walks proceed quickest when overlaps are identified using probes derived from the endmost piece of the insert in the current clone. A number of BAC and cosmid vectors have been designed specifically to help generate such probes. These vectors incorporate bacteriophage RNA polymerase promoters flanking the cloning site (29–31), allowing labelled RNA transcripts to be generated from the DNA sequences at the very ends of the inserts. These labelled RNAs can then be used as probes to screen a library for identification of overlapping clones.

Protocol 12

Synthesis of RNA probes from BAC or cosmid ends

Note: all these experiments should be performed in an RNase-free environment (19).

Reagents

- BAC/cosmid clone
- RNasin (Promega)
- 5 mM rATP, rGTP, rTTP
- 100 μM rCTP
- [α-^{32}P]rCTP (3000 Ci/mmol)
- Polymerase (T7, T3, or SP6)

Method

1. Create linear template DNA by restriction digestion of BAC/cosmid clone.
2. Mix together the following at room temperature (**not on ice**):
   - 2 μl of 10 × polymerase buffer
   - 0.5 μl RNasin
   - 2 μl of 0.1 M DTT
   - 2 μl of 5 mM rATP, rGTP, rTTP

Protocol 12 continued

- 2.5 μl of 100 μM rCTP
- 2 μl [α^{32}P]rCTP (3000 Ci/mmol)
- 1 μg of linearized DNA
- H_2O to 19 μl

3 Add 1 μl of the appropriate polymerase (T7, T3, or SP6) and incubate at 37 °C for 1 h.

4 Add 1 μl RQ1 DNase (Promega) and incubate at 37 °C for 15 min.

5 (Optional) Purify unincorporated nucleotides away by ethanol precipitation or spin column chromatography through P-30 gel (Bio-Rad).

6 Denature probe by boiling for 5 min and then use for hybridization. Use the hybridization conditions outlined in *Protocol 11*.

Most vectors used for cloning large inserts also incorporate restriction sites for rare cutting enzymes surrounding the cloning site. This enables the insert DNAs to be excised from the vector backbone, thereby facilitating their subsequent analysis. Probes for the ends of insert DNAs may also be generated efficiently by first creating an 'endclone'-containing derivative. The steps are outlined in *Protocol 13*.

Protocol 13

Creating 'enclone' derivatives from large insert clones

Reagents

- Cosmid/BAC DNA
- Ligase buffer
- T4 DNA ligase (NBL)

Method

1 In a total reaction volume of 10 μl, digest 100 ng of cosmid/BAC DNA with an enzyme that does not cut the vector.

2 (a) For heat resistant enzymes (http://www.neb.com/neb/tech/tech_resource/restriction/properties/heat_inactivation.html): remove enzyme by PCI extraction and ethanol precipitate the DNA. Redissolve in 10 μl of TE.

(b) For heat sensitive enzymes: transfer the digest to 65 °C for 10 min to kill the enzyme.

3 Add H_2O to bring the volume of the DNA solution to 45 μl, then add 5 μl of 10 × ligase buffer and 40 NBL units of T4 DNA ligase. Incubate overnight at 16 °C.

4 Use the whole ligation mix to transform *E. coli* cells using a standard protocol (or 1–2 μl for electroporation).

When analysing 'endclone' derivatives, DNA should be prepared from at least three colonies. Digestion with the enzyme used for digestion/religation and a rare cutting enzyme which cuts on each side of the cloning site should release the vector, along with two additional fragments derived from the ends of the insert DNA (29). It should be noted that some clones may have sites for the rare cutter near to the insert ends, in which case, there may be more than three fragments. In any event, all three clones should be identical. If not, the most likely cause is that internal fragment(s) were re-cloned into a derivative. This should be easily detectable by comparing the restriction pattern with the original clone. Usually, at least two of the clones will have the same restriction pattern and these will usually be the correct derivatives.

The distal fragment in the endclone derivative is readily identified by Southern hybridization analysis using the previous clone in the walk as a probe (29). The required fragment will not hybridize and can then be isolated by elution from a gel and labelled using a standard protocol.

7.2 Determining the correct direction of a walk

Initial entry into a chromosome walk is gained by isolating a BAC or cosmid containing a marker linked to the target gene. At this point it is useful to know in which direction to walk. If the target gene has been mapped in a large progeny population, it may be separated from the initial marker by several well-distributed crossover points. In this case it is possible that the insert of the BAC/cosmid may span one or more crossover points. This information can be used to establish the orientation of the BAC/cosmid and reveal which end of the insert is closest to the target gene.

If crossovers between the insert ends are not detected, a decision may be made to begin walking in both directions, until RFLPs detected by the new clones identify a crossover. Depending on how many recombinant progeny are available and how evenly distributed the crossovers are, this may result in walking several steps.

If markers are available which flank the target gene on both sides, RecA-AC can be used to establish cosmid orientation using a physical approach. Simultaneous cleavage of the genome at restriction sites present in BACs/cosmids containing the flanking markers will liberate a fragment containing the target gene. Only one end of the BAC/cosmid insert will reside on this fragment and this is the end that should be used for walking. These may be identified readily by probing Southern blots of CHEF-resolved RecA-AC products with probes from each end of the insert.

7.3 Monitoring progress

Continual monitoring of the progress of a walk is absolutely essential. This provides information on progress and indicates when the walk arrives at the target locus. Most importantly, however, it ensures than the walk remains 'on track'. Fungal genomes are often riddled with middle repetitive or duplicated sequences which, if undetected, can cause chromosome walks to 'jump' onto other chromosomes. Similar diversions can occur if some library clones are chimeric.

Progress should be monitored by mapping new RFLPs detected using a BAC or cosmid identified in each 'step'. It is not usually necessary to include every progeny isolate in the mapping population for this purpose. However, it is important to include all the recombinant progeny to indicate when crossovers have been traversed. A number of non-recombinant progeny should also be examined. Otherwise, it is possible that a single step will appear to traverse a number of crossover points when, in fact, the new cosmid is actually from an unlinked locus.

7.4 Overcoming obstacles to walking

The most likely obstacles to be encountered in chromosome walks are regions of repetitive DNA. Many repetitive DNAs are not too much of a problem unless they occur at the ends of inserts, where they will result in hybridization of end-specific probes to many clones in a library. If the repeat is an isolated element, it is usually possible to use an internal probe from the insert to find a BAC/cosmid clone that will jump over the offending sequence. However, clusters of repetitive DNA can prove a little more problematic. Even in these cases, it is usually possible to identify single copy fragments, interspersed with repeats. These fragments enable overlapping clones to be identified which enable gradual progression out of highly repetitive region.

These approaches may require fairly extensive characterization of repetitive clones to identify suitable probes for walking. Clearly, navigation of long tracts of highly repetitive DNA would convert a walk into a crawl. Some obstacles are potentially capable of stopping a walk completely however. This would be the case for genomic regions that are not present, or are uncloneable in genomic libraries (1). Such obstacles can be circumvented by using RecA-AC to excise the target region of the chromosome. The cleavage products can then be subcloned to produce probes enabling re-entry into the walk at a point beyond the problem region (1).

A less common problem is presented by a duplicated chromosome segment. In this case it may be necessary to walk through both copies of the duplication until one emerges on the other side of each duplicate. Walking can then be refocused on the correct locus, identified by mapping RFLPs.

7.5 Identification of clones containing a target gene

It is important to be able to walk to a gene in an efficient manner. Equally important is knowing when to stop. If new clones have been mapped diligently throughout the walk, one should be aware of when it arrives at the target locus because RFLPs identified by the new clones will co-segregate with the gene. However, it would be premature to stop walking as soon as the walk arrives at the target locus because the locus may span many kilobases. Instead, the walk should be continued until a crossover point is traversed on the other side of the locus. Ultimately, proof that the gene has been cloned is demonstrated when a BAC/cosmid clone can be shown to confer the desired phenotype in transformation experiments. In the event that no clone confers the expected phenotype, it is important to have walked through the entire locus so that one can be reasonably certain that the desired sequences were covered by the walk. In this case, failure of transformation could be because the gene was truncated when cloned in the BAC/cosmid, or that it is recessive in action. Having confidence that a walk spans the entire locus should provide the impetus to test these possibilities.

8 Resources for fungal genome analysis

Clearly it was not possible in the space provided to cover all protocols used for fungal genome analysis. The following web sites provide additional information and protocols that are relevant to such studies:

- http://www.cals.ncsu.edu/fungal_genomics/index.html
- http://www.gene.genetics.uga.edu/
- http://www.genome.ou.edu/fungal.html
- http://www.fgsc.net/

References

1. Farman, M. L. and Leong, S. A. (1998). *Genetics*, **150**, 1049.
2. Wan, Y., Liu, H., Li, C., and Schmidhauser, T. J. (1997). *Fungal Genet. Biol.*, **21**, 329.
3. Zhu, H., Blackmon, B. P., Sasinowski, M., and Dean, R. A. (1999). *Genome Res.*, **9**, 739.
4. Prade, R. A., Griffith, J., Kochut, K., Arnold, J., and Timberlake, W. E. (1997). *Proc. Natl. Acad. Sci. USA*, **94**, 14564.
5. Zhu, H., Whitehead, D. S., Lee, Y.-H., and Dean, R. A. (1996). *Mol. Plant Microbe Interact.*, **9**, 767.
6. Michelmore, R. W., Paran, I., and Kesseli, R. V. (1991). *Proc. Natl. Acad. Sci. USA*, **88**, 9828.
7. Skinner, D. Z., Budde, A. D., Farman, M. L., Smith, J. R., Leung, H., and Leong, S. A. (1993). *Theor. Appl. Genet.*, **87**, 545.
8. Whisson, S. C., Drenth, A., Maclean, D. J., and Irwin, J. A. (1995). *Mol. Plant Microbe Interact.*, **8**, 988.
9. Van der Lee, T., De Witte, I., Drenth, A., Alfonso, C., and Govers, F. (1997). *Fungal Genet. Biol.*, **21**, 278.
10. Vos, P., Hogers, R., Bleeker, M., Reijans, M., van de Lee, T., Hornes, M., *et al.* (1995). *Nucleic Acids Res.*, **23**, 4407.
11. Nitta, N., Farman, M. L., and Leong, S. A. (1997). *Theor. Appl. Genet.*, **95**, 20.
12. Landry, S. and Michelmore, R. W. (1985). *Plant Mol. Biol. Rep.*, **3**, 174.
13. Arnau, J., Housego, A. P., and Oliver, R. P. (1994). *Curr. Genet.*, **25**, 438.
14. Chu, G., Vollrath, D., and Davis, R. W. (1986). *Science*, **234**, 1582.
15. Orbach, M. J., Vollrath, D., Davis, R. W., and Yanofsky, C. (1988). *Mol. Cell. Biol.*, **8**, 1469.
16. Talbot, N. J., Salch, Y., Ma, M., and Hamer, J. E. (1993). *Appl. Env. Micriobiol.*, **59**, 585.
17. Zolan, M. E. (1995). *Microbiol. Rev.*, **59**, 686.
18. McCluskey, K., Russell, B. W., and Mills, D. (1990). *Curr. Genet.*, **18**, 385.
19. Sambrook, J., Fritsch, E. F., and Maniatis, T. (ed.) (1989). *Molecular cloning: a laboratory manual.* Cold Spring Harbor Laboratory Press, Cold Spring Harbor, NY.

20. Ahn, J. H. and Walton, J. D. (1996). *Plant Cell*, **8**, 887.
21. Farman, M. L. and Leong, S. A. (1995). *Genetics*, **140**, 479.
22. Koob, M., Burkiewicz, A., Kur, J., and Szybalski, W. (1992). *Nucleic Acids Res.*, **20**, 5831.
23. Murray, N. (1983). In *Lambda II* (ed. R. W. Hendrix, J. W. Roberts, F. Stahl, and R. A. Weisberg), pp. 395–432. Cold Spring Harbor Laboratory Press, Cold Spring Harbor, NY.
24. Collins, J. and Hohn, B. (1978). *Proc. Natl. Acad. Sci. USA*, **75**, 4242.
25. Kim, U. J., Shizuya, H., de Jong, P. J., Birren, B., and Simon, M. I. (1992). *Nucleic Acids Res.*, **20**, 1083.
26. Shizuya, H., Birren, B., Kim, U.-J., Mancino, V., Slepak, T., Tachiiri, Y., *et al.* (1992). *Proc. Natl. Acad. Sci. USA*, **89**, 8794.
27. Frischauf, A.-M. (1991). In *Essential molecular biology: a practical approach* (ed. T. A. Brown), Vol. 2, pp. 15–38. Oxford University Press, Oxford, UK.
28. Metzenberg, R. L. and Kang, S. (1987). *Fungal Genet. Newsl.*, **34**, 44.
29. An, Z., Farman, M. L., Budde, A. D., Taura, S., and Leong, S. A. (1996). *Gene*, **176**, 93–6.
30. Orbach, M. J. (1994). *Gene*, **150**, 159.
31. Taylor, J. L. and Borgmann, I. E. (1996). *Fungal Genet. Newsl.*, **43**, 52.

Chapter 7
Cell biology of the duplication cycle in fungi

Michelle Momany
Department of Botany, 2502 Plant Sciences, University of Georgia, Athens, GA 30602, USA.

1 Introduction

Filamentous fungi undergo an ordered series of events during growth and division. These events include mitosis, periods of isotropic and polar growth, cytoskeletal rearrangements, and partitioning of cytoplasmic contents via septa (1). Because the cells within a hypha do not physically separate from one another these events are known as a 'duplication cycle' rather than a cell cycle.

Microscopy and classical genetics is a powerful combination for exploring the duplication cycle of filamentous fungi. Using microscopy to observe mutant phenotypes, a surprising variety of important questions can be answered.

One of the most compelling examples of the power of microscopic examination of mutant phenotypes comes from the *Saccharomyces cerevisiae* cell division cycle (*cdc*) literature (2). The *cdc* mutants are temperature-sensitive (ts) and arrest with uniform terminal phenotypes. Hartwell recognized that the morphological and biochemical arrest phenotypes could serve as reference points, or landmarks, in the cell cycle. Determining the order of these landmarks and their dependency relationships formed the basis for understanding the cell cycle of budding yeast.

To identify landmarks, one must be able to stop the cell at a specific point in the cycle and examine the phenotype. While the most common way to arrest the cycle is to incubate Ts mutants at restrictive temperature, chemical inhibitors have also been used. An arrest phenotype furnishes a 'snapshot' of an intermediate in a multistep process. Landmarks are ordered relative to each other to generate a pathway, or multiple related pathways. Ordering landmarks is often accomplished through epistasis analysis, the construction of double mutants and observation of their phenotypes at restrictive temperature (see Chapters 4 and 5).

Many aspects of nuclear division of the filamentous fungus *Aspergillus nidulans* have been clarified through the careful examination of arrest phenotypes (3). By combining nuclear staining and mutant studies, Bergen and Morris (4) showed

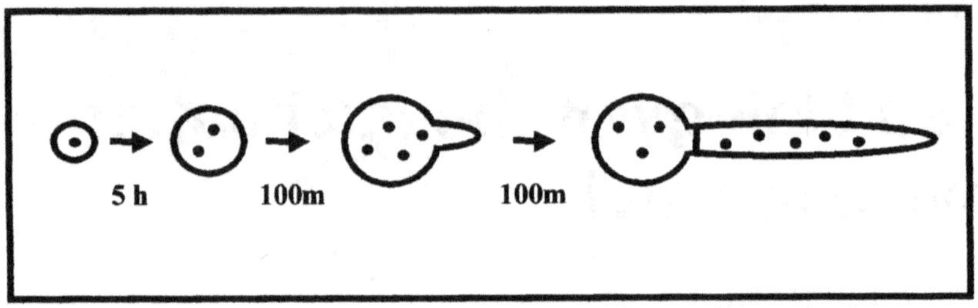

Figure 1 Landmarks in the early duplication cycle of *A. nidulans*. After an initial incubation period of about 5 h, nuclear divisions occur every 100 min. Between the first and second nuclear division the germ tube emerges. After the third nuclear division the first septum is laid down.

that the nuclear division cycle in *A. nidulans* takes about 100 minutes. Recently, other researchers have begun to correlate cytoplasmic landmarks with nuclear divisions as shown in *Figure 1*.

2 Staining nuclei and cell walls

When inoculated to rich medium, the conidia of *A. nidulans* break dormancy and maintain a degree of synchrony for the first several mitotic divisions. Examining wild-type germlings at hourly intervals using a simple protocol for staining nuclei and septa (*Protocol* 1), allowed Harris *et al.* (5) to show that the first septum in *A. nidulans* is laid down after the third nuclear division. Similarly, this protocol has been used to show that the switch from isotropic to polar growth occurs between the first and second nuclear divisions (6, 7). *Figure 2* shows a micrograph of germlings stained using *Protocol* 1. This protocol may also be used in conjunction with temperature-shift experiments to determine when the temperature-sensitive step occurs relative to other landmarks. It may also be combined with inhibitor treatments (see *Table* 1) to give insight into dependency relationships.

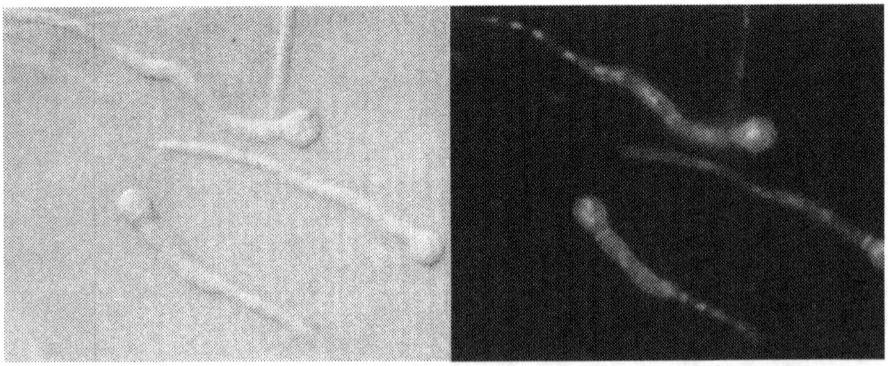

Figure 2 Visualization of nuclei and chitin. *A. nidulans* germlings were prepared according to *Protocol 1*. The left panel shows a Nomarski differential interference contrast image. The right panel shows the corresponding fluorescent image stained with Hoechst and Calcofluor.

Table 1 Common inhibitors used in filamentous fungi

Inhibitor	Mode of action	Effect	Reference
Benomyl	Destabilizes microtubules	Blocks nuclei in mitosis	11, 12, 13
Hydroxyurea	Prevents synthesis of dNTPs	Blocks nuclei in S phase	4
Cytochalasin A	Destabilizes actin	Abnormal tip growth; blocks septation	5, 14

Protocol 1

Staining nuclei, cell walls, and septa[a]

Equipment and reagents

- Tweezers with flat ends for handling coverslips (Sigma)
- Small Coplin staining jars, coverslip size (Baxter)
- Parafilm
- Fixer: 3.7% formaldehyde, 0.2% Triton X-100, 50 mM phosphate buffer pH 7
- Hoechst solution: 1 mg/ml Hoechst 33258 (Sigma) in H_2O; store at 4 °C protected from light
- Calcofluor solution: 1 mg/ml Blankophor BBH standard SV-2460 (Bayer) in H_2O; add 1–2 drops 1 M KOH to get into solution
- Hoechst/Calcofluor stain: 2 μl Calcofluor solution, 1 μl Hoechst solution in 5 ml H_2O; make fresh and cover with foil
- Mounting solution: 50% glycerol, 10% phosphate buffer pH 7, 0.1% *n*-propyl-gallate

Method

1 With flame-sterilized forceps place a clean coverslip[b] in the bottom of a small, sterile Petri dish.

2 Place 10 ml of appropriate medium in a sterile disposable 15 ml centrifuge tube. Add approx. 10^6 conidia. Invert tube several times to mix.

3 Pour medium with conidia over the coverslip. To prevent growth of hyphae on the underside of the coverslip, press down on the coverslip with the sterile end of the tube (the screw cap end). This will remove any air bubbles trapped under the coverslip and seal it to the bottom of the Petri dish.

4 Incubate at temperature and time appropriate for specific experiment.

5 Fill a small Coplin jar with fixer. Place coverslip in jar and incubate coverslip for 30 min.[c]

6 Rinse briefly in a small beaker of water.

7 On a piece of Parafilm place a 450 μl drop of Calcofluor/Hoechst staining solution.[d] Place coverslip on drop, with hyphae facing down. Cover with a box lid or piece of aluminium foil to protect from light. Incubate at room temperature for 5 min.

8 Rinse briefly in a small beaker of water.

9 Place 7 μl of mounting solution on a slide. Carefully place coverslip on mounting solution. Try not to introduce bubbles.

Protocol 1 continued

10 Cover with a box or foil and let air dry for 1 h.[e] Seal the edges of the coverslip with nail polish.

11 View with UV light using a fluorescence microscope.

12 Store at 4 °C protected from light. Stain should be visible for several months.

[a] From ref. 5.

[b] Some investigators pre-clean coverslips with 95% EtOH, 0.1 M HCl, and then autoclave. We have not found this treatment to be necessary.

[c] It is important to know which side of the slide your hyphae are on. We always place coverslips so that the hyphae face forward.

[d] We place our Parafilm on a piece of Plexiglass (Perspex) and then rub it with a straight edge to seat it. This makes a smooth working surface. DAPI may be used instead of Hoechst.

[e] If coverslips are left longer before sealing, place at 4 °C so that they will not dry out.

3 Indirect immunofluorescence

Another very useful microscopy technique for studying the duplication cycle is indirect immunofluorescence (*Protocol 2*). In indirect immunofluorescence, an unconjugated primary antibody binds to the epitope of interest. A secondary antibody coupled to a fluorescent molecule such as fluorescein isothiocyanate (FITC) binds to the primary antibody. Because of the fluorescent tag, the secondary antibody may be visualized using a fluorescence microscope (see also Chapter 12). To use immunofluorescence to view molecules within the cell, the cell wall must first be partially digested to allow antibodies in. In *A. nidulans* this is usually accomplished by incubating the germling with cell wall-degrading enzymes such as Novozyme234 or Glucanex. Indirect immunofluorescence has also been used without enzymatic digestion to localize cell wall epitopes (*Figure 3*) (6). Because no digestion is required, antibodies to external epitopes may be applied to living cells. This allows the antibody to be used as label in a pulse-chase experiment. Here, antibody is applied and then the cell is allowed to grow further. Growth occurring prior to the antibody treatment will be labelled while that occurring after antibody treatment will be unlabelled.

3.1 Commercially available antibodies

Several laboratories have used commercially available monoclonal antibodies against actin and tubulin (N350 from Amersham; C4 from ICN Biochemicals; DM1A from Sigma) (5, 7, 8).

Many anti-mouse and anti-rabbit antibodies conjugated to fluorescent dyes are commercially available for use as secondary antibodies. Sometimes these secondary antibodies recognize *A. nidulans* proteins and give high background levels. Therefore it is important to include a control with secondary antibody

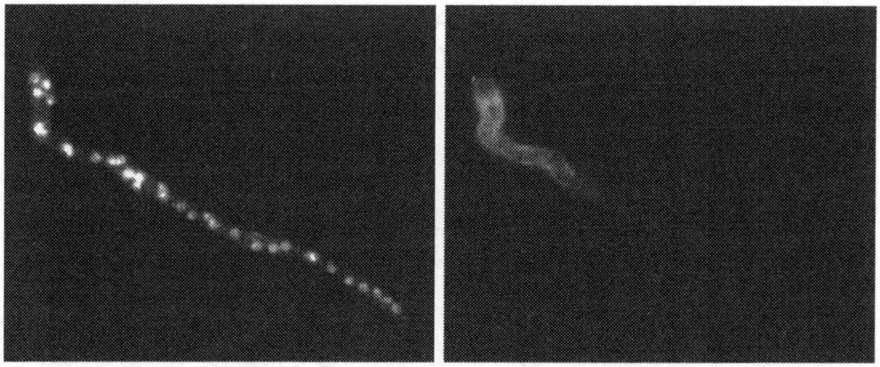

Figure 3 Immunolocalization of a cell wall epitope. *A. fumigatus* germlings were prepared according to *Protocol 2* without the cell wall digestion. The primary antibody was a monoclonal against a cell wall epitope. The secondary antibody was goat anti-mouse coupled to FITC. The left panel shows immunofluorescence. The right panel shows the corresponding image stained with Hoechst and Calcofluor.

only (no primary) whenever a new secondary is being used. We have had very good results with affinity purified secondary antibodies from Jackson ImmunoResearch (West Grove, PA).

3.2 Raising antibodies

In cases where an antibody against the molecule of interest is not available, an investigator can have antibodies made in rabbits or mice. Most researchers hire either a special University facility or a private company for antibody production and the researcher need only provide purified protein to the facility. Generally, the easiest way to do this is to make a fusion between the protein of interest and a purification tag and then to isolate the fusion protein making use of the tag. Popular fusion partners include glutathione S-transferase and His-tag (9). Several companies sell kits that include both the vector and purification matrix.

Protocol 2

Immunofluorescence staining[a]

Equipment and reagents

- All equipment and reagents from *Protocol 1*, except for fixer
- IF fixer: 3.7% formaldehyde, 5% DMSO, 25 mM EGTA pH 7, 5 mM $MgSO_4$
- Digestion solution: 50 mg/ml Novozyme 234[b] in PEM buffer. Mix with equal volume egg white just before use. Store Novozyme and egg white aliquots at −20 °C.
- PEM: 50 mM Pipes pH 6.7, 25 mM EGTA pH 7, 5 mM $MgSO_4$
- Extraction solution: 100 mM Pipes pH 6.7, 25 mM EGTA pH 7, 10% DMSO, 0.01% Nonidet NP-40
- PBSA: phosphate-buffered saline + 0.1% bovine serum albumin

Protocol 2 continued

Method

1 Follow *Protocol 1*, steps 1–5 for growing and fixing germlings, except use IF fixer.

2 Wash coverslips for 5 min in PEM.[c] Repeat wash three times.

3 On a piece of Parafilm, place a 200 μl drop of digestion solution. Place coverslip on drop, with hyphae facing down. Incubate at room temperature for 15–90 min.[d]

4 Wash coverslips for 5 min in PEM. Repeat wash three times.

5 Incubate coverslips in extraction solution for 2 min.[e]

6 Wash coverslips for 5 min in PEM. Repeat wash three times.

7 On a piece of Parafilm, place a 100 μl drop of primary antibody diluted in PBS. Place coverslip on drop, with hyphae facing down. Incubate at room temperature for 1–2 h.

8 Wash coverslips for 5 min in PBSA. Repeat wash three times.

9 On a piece of Parafilm, place a 100 μl drop of secondary antibody diluted in PBSA. Place coverslip on drop, with hyphae facing down. Cover with a box or foil to protect from light. Incubate at room temperature for 1–2 h.

10 Wash coverslips for 5 min in PBSA. Repeat wash three times.

11 Follow *Protocol 1*, steps 7–12 to stain chitin and nuclei.

[a] From ref. 10.

[b] Novozyme 234 has been the enzyme of choice for several years, however the manufacturer has recently stopped making it. Other researchers report good results with Glucanex, although we have not yet tried this enzyme.

[c] Coplin jars are used for all washes and incubations unless otherwise stated.

[d] Novozyme 234 differs from batch to batch. We generally set up several identical samples and incubate for 15, 30, 45, 60, and 90 min to find optimal conditions.

[e] For anti-actin staining include an incubation in −20 °C methanol for 10 min after extraction and PEM wash.

References

1. Harris, S. D. (1997). *Fungal Genet. Biol.*, **22**, 1.
2. Pringle, J. R. and Hartwell, L. H. (1981). In *The molecular biology of the yeast Saccharomyces: life cycle and inheritance* (ed. J. N. Strathern, E. W. Jones, and J. R. Broach). Cold Spring Harbor Press, Cold Spring Harbor.
3. Aist, J. R. and Morris, N. R. (1999). *Fungal Genet. Biol.*, **27**, 1.
4. Bergen, L. G. and Morris, N. R. (1983). *J. Bacteriol.*, **156**, 155.
5. Harris, S. D., Morrell, J. L., and Hamer, J. E. (1994). *Genetics*, **136**, 517.
6. Momany, M., Westfall, P. J., and Abramowsky, G. (1999). *Genetics*, **151**, 557.
7. Harris, S. D., Hofmann, A. F., Tedford, H. W., and Lee, M. P. (1999). *Genetics*, **151**, 1015.
8. Oakley, B. R., Oakley, C. E., Yoon, Y., and Jung, M. K. (1990). *Cell*, **61**, 1289.
9. Guan, K. and Dixon, J. E. (1991). *Anal. Biochem.*, **192**, 262.

10. Oakley, B. R. and Osmani, S. A. (1993). In *The cell cycle: a practical approach* (ed. P. Fantes and R. Brooks), pp. 127–42. Oxford University Press, Oxford.
11. Orr, E. and Rosenberger, R. F. (1976). *J. Bacteriol.*, **126**, 895.
12. Sheir-Neiss, G., Lai, M. H., and Morris, N. R. (1978). *Cell*, **15**, 639.
13. Oakley, B. R. and Morris, N. (1980). *Cell*, **19**, 255.
14. Allen, E. D., Aiuto, R., and Sussman, A. S. (1980). *Protoplasma*, **102**, 63.

Chapter 8
Investigations of Oomycete cell biology

Adrienne R. Hardham
Plant Cell Biology Group, Research School of Biological Sciences, Building 46, The Australian National University, GPO Box 475, Canberra, ACT, 2601, Australia.

1 Introduction

1.1 Oomycete phylogeny

The Oomycetes are a distinct group of organisms which, like many true fungi (Basidiomycetes, Ascomycetes, and Zygomycetes), form filamentous hyphae during vegetative (somatic) growth. They are also heterotrophs with absorptive nutrition. Many characteristics of Oomycete organisms, however, indicate that they are not true fungi but are most closely related phylogenetically to the Chromophyte algae and other flagellate protists which possess tripartite flagellar hairs (1–3).

One of the main distinguishing features of the Oomycetes is the production of motile, biflagellate zoospores during asexual sporulation. For many species, the zoospores serve as the major infective agent for disease dissemination and establishment. Other distinguishing Oomycete characters include the possession of tubular rather than plate-like cristae in the mitochondria, the occurrence of cellulose rather than chitin as the predominant cell wall microfibrils, and the occurrence of stacks of Golgi cisternae similar to those in plants and algae rather than the simple Golgi cisternae seen in true fungi. Oomycetes are diploid throughout most of their life cycle, with meiosis occurring in the developing gametangia; they also have a number of differences in biochemical pathways to those in true fungi (see ref. 3). Comparisons of nucleic acid sequences support the phylogenetic relationship of the Oomycetes with heterokont algae and other protists (4, 5).

The phylogenetic position of the Oomycetes has important implications for cell biological investigations of Oomycetes. It means, for example, that microtubules in Oomycete cells may be resistant to drugs used to inhibit microtubule function in true fungi (for example, the benzimidazoles), but may be susceptible to drugs which inhibit microtubules in plants and algae: colchicine (6), nocodazole (7), and oryzalin (8). Differences in wall composition (for example the absence of chitin) may also be an important factor to be taken into consideration in studies of cell wall synthesis or degradation, or of cell growth.

1.2 General biology and life cycle of Oomycetes

Many of the aquatic species of Oomycetes are saprobes which feed on dead plant and animal material. A few cause diseases in algae (some species of *Pythium*), insects (species of *Lagenidium*), and fish (some species of *Saprolegnia*). Most of the terrestrial species of Oomycetes are facultative parasites (for example species of *Phytophthora*) or obligate biotrophs (species of *Bremia* or *Plasmopara*) which cause diseases in vascular plants. The growth conditions required to culture species of Oomycetes vary widely and reference to texts such as Erwin and Ribeiro (9), Fuller and Jaworski (10), or Dick (3), and to original publications should be made for instructions on culturing protocols and media.

Oomycetes may reproduce both sexually, through the production of oogonia and antheridia, or asexually through the production of chlamydospores, sporangia, zoospores, and cysts. During sexual reproduction, a thick-walled, resistant oospore is formed which is capable of surviving for long periods of time under adverse conditions. Germination of the oospore gives rise to a diploid mycelial thallus. During asexual reproduction, multinucleate chlamydospores or sporangia develop. Chlamydospores germinate to give rise to two or three germ tubes which typically form terminal sporangia.

1.2.1 Induction of asexual sporulation

Oomycete species grow throughout most of their life cycle as filamentous hyphae (*Figure 1*). When in fresh media containing a complete supply of nutrients, the hyphae are usually in a vegetative (somatic) state but when nutrients are depleted (or perhaps when staling factors accumulate), the mycelium is induced to sporulate, usually asexually. Sporangia form on sporangiophores or at hyphal apices by expansion of the hyphal tip and the inflow of cytoplasm from the subtending hypha (*Figure 1*). When mature (*Figure 2*), the cytoplasm in the sporangium or chlamydospore is sealed off by a basal septum. Sporangia may remain dormant for

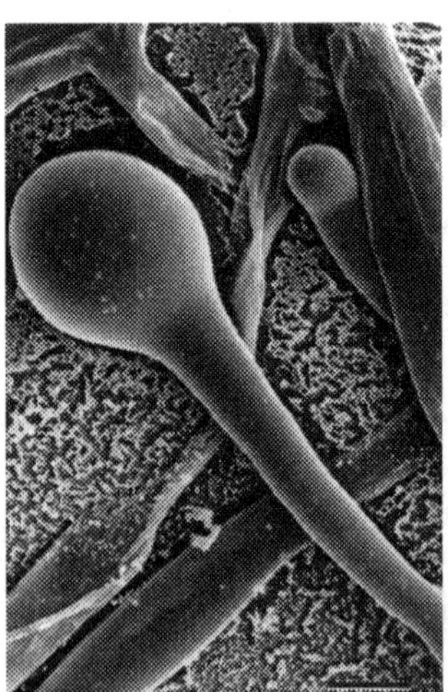

Figure 1 Scanning electron microscope (SEM) image of a frozen, hydrated hypha and young sporangium of *Phytophthora cinnamomi*. Reproduced with permission from *Protoplasma*, **181**, 213 (1994). Bar = 2 μm.

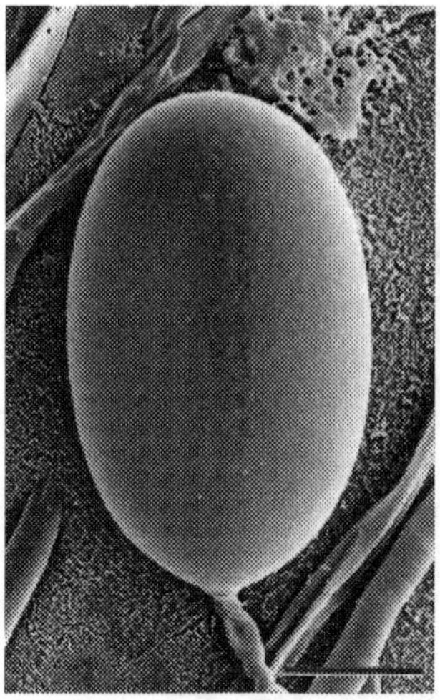

Figure 2 SEM image of a frozen, hydrated mature sporangium of *P. cinnamomi*. Reproduced with permission from *Protoplasma*, **181**, 213 (1994). Bar = 10 μm.

many days or even weeks, but a number of factors, such as a reduction in temperature, will induce their germination either directly via production of a germ tube or indirectly via cytoplasmic cleavage and the formation and release of uninucleate spores.

With regard to studies of vegetative hyphae, it is important to note that changes in gene expression and cell constituents characteristic of the reproductive state can occur relatively rapidly as the composition of the growth medium changes. For example, vesicles which are synthesized during asexual sporulation in *Phytophthora* species appear not only six to seven hours after transfer of the culture into a nutrient-free, mineral salts medium, but will also appear after two to three days in nutrient broth or in older regions of a colony on nutrient agar plates (11) (A. R. Hardham, unpublished observations). In order to be confident the hyphae are in a vegetative stage, the cultures should be transferred to fresh nutrient medium 12–17 hours before sampling.

1.2.2 Production of zoospores, cysts, and germinated cysts

Once sporangial cleavage is complete, one or multiple pores in the wall of the sporangium rupture and the spores are released into the medium. In some species, the spores are non-motile cysts, but usually they are zoospores (*Figure 3*) which will remain motile for relatively long periods of time (hours or even days) if released into water or a suitable buffer and held under optimal conditions. Some species of Oomycetes produce two types of zoospores: a primary zoospore which has its two flagella inserted into the apex of the cell and which is a weak swimmer, and a secondary zoospore which has its flagella inserted midway along a lateral groove in the kidney-shaped cell (*Figure 3*). Most species of Oomycetes produce zoospores of the second type only.

Zoospores can be induced to encyst by a number of means, such as agitation, centrifugation, addition of nutrients, or plant extracts (*Figure 3*). The propensity of zoospores to encyst varies between different species, and is also influenced by

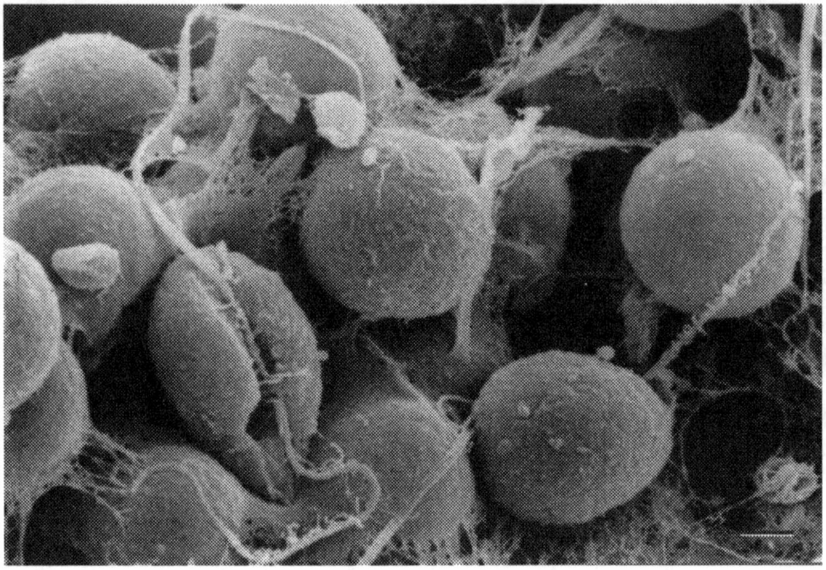

Figure 3 SEM image of chemically fixed and dried zoospore and cysts of *P. cinnamomi*. Micrograph courtesy of Entomology CSIRO. Reproduced with permission from *Protoplasma*, **181**, 213 (1994). Bar = 2 μm.

zoospore density and by the volume of the zoospore suspension. For example, small volumes (10–15 ml) of *P. cinnamomi* zoospores can be induced to encyst by vortexing for 20 seconds or by centrifugation at 1500 *g*. However, centrifugation or vigorous shaking of large volumes (50–200 ml) produces only a relatively low percentage of cysts. A better method for the production of large volumes (200–500 ml) of *Phytophthora* cysts is to place the zoospore suspension in a beaker at −20 °C and monitor the temperature until it reaches 0–1 °C. This induces encystment of over 95% of the zoospores.

Germination of Oomycete cysts may occur in distilled water but is usually inefficient and growth of the germlings is limited. Recently-encysted cells may also lyse. Better rates of germination and germling growth are achieved by suspending the cysts in nutrient medium.

2 Observations of living cells

2.1 Assessment of hyphal health

In investigations of the cell biology of Oomycetes, as in those of filamentous fungi and other organisms, it is often helpful to commence studies at the light microscope level before proceeding to electron microscopy. Similarly, it is important, whenever possible, to observe the organism in its living state before studying the structure of fixed and dead material. An understanding of the structure and behaviour of organisms at the light microscope level will give a better appreciation of its biology, a better perspective for ultrastructural observations, and a better chance of recognizing artefacts that might arise as a result of fixation and processing for electron microscopy.

The health of hyphae can, for example, be assessed by observing hyphal growth and the stratification of organelles within the cytoplasm near the hyphal tip. The spitzenkörper, a cluster of small vesicles near the hyphal tip in most true fungi, is not observed in Oomycete hyphae, but they do show zonation of organelles in the cytoplasm along the hypha (12). The extreme tips of Oomycete hyphae appear homogeneous, and usually lack large organelles. Sub-apically, there are clusters of mitochondria (2–20 μm from the tip) and then nuclei (approx. 15 μm from the tip). Cytoplasmic streaming and the rapid bidirectional movement of small refractile vesicles should be visible. Further back ($>$45 μm from the tip), large vacuoles span almost the entire width of the hypha and are crossed by transvacuolar strands. Oomycete hyphae are very fragile and growth and normal morphology are easily perturbed. Cellular damage may be indicated by cessation of tip growth, reduction of cytoplasmic movement, and by loss of organelle stratification (13).

2.2 Video microscopy of living cells

The analysis of living and growing Oomycetes and filamentous fungi has been greatly assisted by the availability of high quality video equipment and digital enhancement of video images. Video microscopy is particularly valuable in observations of cell movement and analysis of experimental perturbations of intracellular dynamics. For example, Oomycete zoospores contain a water expulsion vacuole which is responsible for zoospore osmoregulation and which forms after the sporangial cytoplasm is cleaved. Details of the behaviour of the water expulsion vacuole during the onset of its operation, steady-state function, and decline have been obtained from analysis of video recordings of sporangia, zoospores, and encysting cells, respectively (14). Analysis of pharmacological treatments showed that KNO_3, an inhibitor of vacuolar-type H^+-ATPase, increased the cycle time of the water expulsion vacuole from six seconds to ten seconds (14). Video microscopy can also be used to study the characteristics of Oomycete zoospore motility (for example swimming speed, the amplitude and wavelength of flagellar beat) (15).

2.3 Introduction of fluorescent probes into living cells

A range of highly specific fluorescent probes is now available for use in studies of living cells. Tubulin and phalloidin conjugated to fluorescent markers such as FITC or rhodamine can localize microtubules and filamentous actin, respectively; many other proteins may be labelled by fluorescently-conjugated antibodies; and a number of dual ratio dyes can measure cytoplasmic pH and Ca^{2+} concentration. Some of these probes are able to cross the cell wall and plasma membrane and can be loaded into cells simply by placing the cells in a solution of the probe. Other probes will not cross the plasma membrane/cell wall boundary and must be loaded into the cells by microinjection. A description of membrane-permeable and membrane-impermeable dyes and methods for loading them into plant cells is given in ref. 16. The use of these probes in studies of the cell biology of Oomycete cells is still relatively limited.

2.3.1 Membrane-permeable dyes

Nuclear and mitochondrial DNA can be labelled in living Oomycete cells by staining with DAPI (4,6-diamidino-2-phenylindole), a dye which can freely enter cells and which fluoresces blue when it reacts with DNA. A stock solution of 1 mg/ml DAPI in distilled water should be prepared and can be stored at 4°C in the dark for several months. For labelling, incubate cells in a solution of 0.2 μg/ml DAPI (diluted in PBS) for 25 min and then place the cells in dye-free medium and examine with ultraviolet excitation.

Other membrane-permeable dyes which can be used to study an intracellular component of Oomycete hyphae are 6-carboxyfluorescein diacetate and Oregon Green 488 carboxylic acid diacetate (see *Protocol* 1). These molecules are taken up by hyphae and accumulate in a vacuolar reticulum. When excited with blue light and examined in an epifluorescence microscope, the organization and dynamics of the reticulum can be observed (*Figure* 4) (13). It is important to note that the vacuolar reticulum is easily disrupted and great care must be exercised not to damage the hyphae and alter the reticulum.

2.3.2 Membrane-impermeable probes: microinjection

Charged or large molecules will not pass across the plasma membrane/cell wall and must be introduced into the cell cytoplasm by microinjection. The basic equipment required for microinjection of plant or Oomycete cells is described in refs. 16 and 17. In ionophoretic microinjection, small charged molecules are moved out of the micropipette on application of an electric current. The advantage of this method is that the process of injection does not lead to an increase in

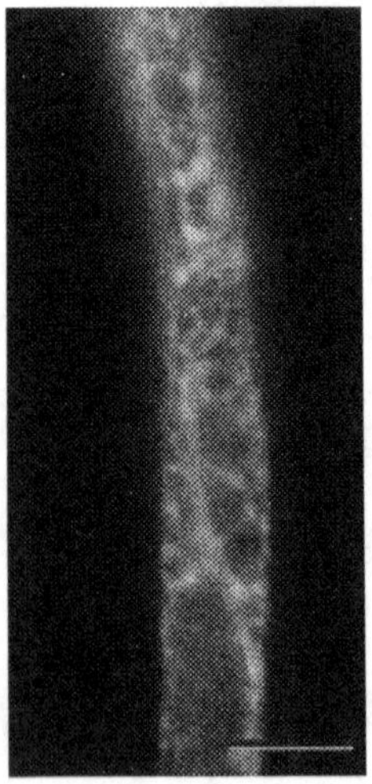

Figure 4 Fluorescence micrograph showing the sub-apical region of a living hypha of *Saprolegnia diclina* that has been loaded with Oregon Green 488 carboxylic acid diacetate. The dye has accumulated within the vacuolar reticulum in the cells, anterior to the region containing large vacuoles. Micrograph courtesy W. G. Allaway and A. E. Ashford. Bar = 10 μm.

pressure or volume in the cell. In pressure microinjection, increased hydraulic pressure is used to force molecules out of the micropipette into the infected cell. This method has been successfully used to microinject *Phytophthora* sporangia with rhodamine–phalloidin to localize F-actin (18), with Fura2–dextran to monitor changes in Ca^{2+} (19), and with BCECF–dextran to monitor changes in pH (20) during sporangial cleavage. Dextran-conjugated probes are used because the chance of the dye moving into vacuoles or other membrane-bound cell components is reduced.

Protocol 1

Loading living Oomycete hyphae with 6-carboxyfluorescein diacetate or Oregon Green 488 carboxylic acid diacetate[a]

Equipment and reagents

- 10 mg/ml stock of 6-carboxyfluorescein diacetate (Molecular Probes, Inc.) in acetone or 10 mg/ml Oregon Green 488 carboxylic acid diacetate (Molecular Probes, Inc.) in dimethyl sulfoxide
- Coverslips
- Microscope slides prepared with strips of Teflon to act as spacers
- Growth medium

Method

1. Place a small sample of mycelium into nutrient medium and allow to grow until the colony is 0.5–1.0 cm in diameter.[b]
2. Keeping the mycelium submerged and as undisturbed as possible, replace the growth medium with the same medium containing 10 μg/ml 6-carboxyfluorescein or Oregon Green dyes which has been freshly added from a 10 mg/ml stock. Incubate for 10 min without disturbance.
3. With minimal disturbance, gently remove dye solution by three partial rinses over 10 min in fresh growth medium, again keeping the mycelium submerged at all times.
4. Carefully slide a coverslip under the mycelium, lift it away from the medium, and invert onto Teflon spacers on a microscope slide.[c] Add more growth medium to fill the space between coverslip and slide and allow the preparation to recover until hyphal growth has resumed.
5. On a fluorescence microscope, use differential interference contrast to locate hyphal tips which are growing and in which organelle zonation characteristic of healthy cells is evident.
6. To observe labelling of the vacuolar reticulum, irradiate with blue light (epifluorescence filter combination: BP450–490, FT510, LP520). Photograph immediately.[d]

[a] After method given in ref. 13.

[b] The smaller the colony, the easier it is to transfer.

[c] The spacers prevent the hyphae from being squashed.

[d] On prolonged exposure, the vacuole system may leak fluorochrome and/or break up, and a range of artefactual fluorochrome distributions will result.

3 Investigations of cell ultrastructure

3.1 The flagellar apparatus of Oomycete zoospores

The close phylogenetic relationship of the Oomycetes to the heterokont algae is highlighted by comparisons of the structure of the two flagella of the zoospores with those of the algae. Both classes of organisms have two, morphologically different flagella. One flagellum is shorter than the other and is covered by two rows of tubular hairs called mastigonemes. The longer flagellum is sparsely covered by fibrous hairs. In Oomycete zoospores, the shorter, tinsel flagellum projects forwards and provides propulsion; the longer, posterior flagellum projects backwards and appears to act like a rudder to steer the cells as they swim.

The tubular and fibrous hairs on the surface of zoospore flagella are usually either lost or are not visible on cells that have been processed by routine resin embedding and sectioning for transmission electron microscopy, although addition of ruthenium red to the fixative helps in this regard (21). Details of the ornamentation of the flagellar surface can be observed on cells air dried onto a formvar-coated grid. The flagella can be negatively stained with uranyl acetate on the grid (*Protocol 2*) or can be rotary shadowed with carbon–platinum in a vacuum coating unit (15, 22). When examining in the microscope, care should be taken to avoid focusing the electron beam on the body of the zoospore as this will cause the sample to heat up and disintegrate.

Protocol 2

On-grid whole mount negative staining of zoospore flagellar hairs

Equipment and reagents

- Double-strength fixative: e.g. 2% glutaraldehyde in 100 mM Pipes buffer pH 6.8–7.2
- Formvar-coated grids
- 1% uranyl acetate in water

Method

1. Place a 2 μl aliquot of zoospore suspension onto a formvar-coated grid held in fine forceps.
2. Add 2 μl aliquot of double-strength fixative.
3. Allow the cells to settle onto the formvar surface for 10–20 min.
4. Rinse the grids gently in a stream of distilled water.
5. Add a drop of 1% uranyl acetate. Wick away the solution with filter paper and allow to air dry.

3.2 CryoSEM: scanning electron microscopy of frozen, hydrated cells

One of the characteristic features of the scanning electron microscope is its ability to reveal details of the surface of cells. In the past, most scanning electron microscopy was done on material that had been chemically fixed and dehydrated by freeze-drying or critical point drying. Both fixation and drying can introduce artefacts in the structure and appearance of the sample. To avoid these potential problems, whenever possible, view unfixed cells in their hydrated state. In studies of Oomycetes, this technique has been used for the preparation and observation of sporulating mycelia (*Figures 1* and *2*) and early stages of plant infection (21). The procedure is simple and follows the same protocols as those described for other organisms (23, 24). In brief, infected material or small samples of mycelium growing on cellophane, or dialysis or Millipore membranes are mounted on a cryo-stub and frozen by plunging into liquid nitrogen. The specimen is inserted into the chamber of the microscope, and surface water allowed to sublime away before the sample is sputter-coated with gold. The sample is examined in the microscope between $-100\,°C$ and $-175\,°C$.

3.3 Chemical fixation and processing for transmission electron microscopy

The basic protocols for fixing and processing Oomycete cells for transmission electron microscopy are generally similar to those employed for plant or fungal material (Chapter 10). This is especially so for walled cells such as hyphae, sporangia, and cysts, however, the wall-less zoospores have specific requirements for fixation and processing. Zoospores and cysts are easily lost during processing. A convenient way to handle them is to embed them in low melting point agarose after primary fixation (*Protocol 3*). In step 2 of this protocol, it is better not to resuspend the cells completely, but to allow small clumps of cells to become dispersed in the agarose. This means that small clusters of cells can be later selected for sectioning.

Protocol 3

Embedding zoospores and cysts in agarose

Reagents

- Fixative (see *Protocols* 4, 5, or 6)
- 2% low melting point agarose (Sigma Type VII) in water or buffer

Method

1 Fix cells as described in *Protocols* 4, 5, or 6, and then wash by three cycles of gentle centrifugation and resuspension in buffer.

2 Pellet the cells in an Eppendorf tube, remove as much liquid as possible, then

Protocol 3 continued

partially resuspend in a small volume of 2% low melting point agarose held at 37–40 °C.

3 Solidify the agarose by placing the tube in ice for 10–20 min.

4 Carefully cut the Eppendorf tube away from the block of agarose and place the agarose block into buffer. The block should be cut into small pieces at this stage or before resin embedding.

Because Oomycete zoospores lack a cell wall, they are unable to osmoregulate by building up turgor pressure, and instead have a contractile, water expulsion vacuole which pulses cyclically to expel water from the cell. The wall-less zoospores are very sensitive to the osmotic pressure of their surrounding solution, including that of any chemical fixative being used to kill and preserve the cells.

Studies of the effect of fixatives based on 1% or 2% glutaraldehyde in buffers with different osmolarities indicate that the optimal osmolarity of the fixative (as determined by the freezing point depression method) is about 500 mOsm (25). Fixatives with an osmolarity close to this value include 1% glutaraldehyde in 100 mM Pipes buffer, and 2% glutaraldehyde in 50 mM Pipes, 100 mM cacodylate, or 100 mM phosphate buffer. At low fixative osmolarities (e.g. 1–2% glutaraldehyde in 25 mM phosphate buffer: 200–300 mOsm), the zoospores are dilated and the plasma membrane and peripheral vesicle membranes are broken. At high fixative osmolarities (e.g. 2% glutaraldehyde in 100–200 mM Pipes buffer: 750–1000 mOsm), the zoospores appear shrunken and the peripheral vesicles bulge extensively from the cell surface. It should be noted that even with the most careful attempts to attain an optimal fixation, the zoospores often still show signs of perturbations such as protrusion of the large peripheral vesicles and dilation of the rim of the peripheral cisternae.

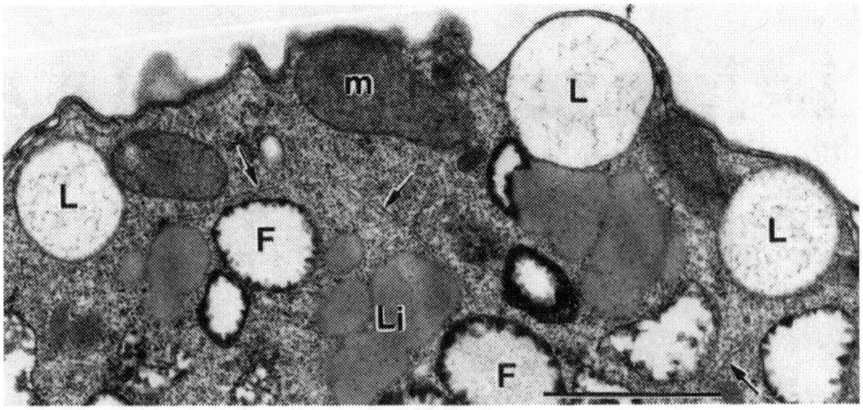

Figure 5 Transmission electron micrograph (TEM) of a zoospore of *P. cinnamomi* fixed in 2% glutaraldehyde in 50 mM Pipes buffer and post-fixed in 1% OsO_4. L, large peripheral vesicles; F, fingerprint vesicles; Li, lipid; m, mitochondria; arrows, endoplasmic reticulum. Bar = 1 μm.

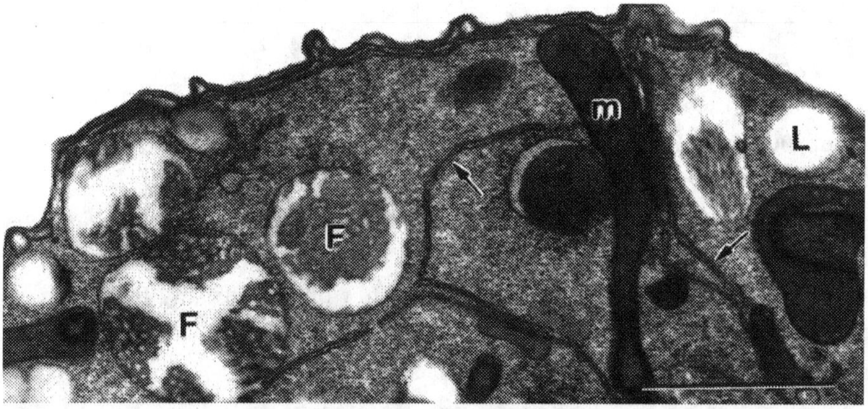

Figure 6 TEM of a zoospore of *P. cinnamomi* fixed in 2% glutaraldehyde in 100 mM cacodylate buffer and post-fixed in OsFeCN. Membranes and the mitochondria are highly electron dense. F, fingerprint vesicles; m, mitochondria; arrows, endoplasmic reticulum. Bar = 1 μm.

Oomycete zoospores may be fixed following routine procedures (primary fixation in glutaraldehyde and secondary fixation in 1–2% OsO_4) (*Figure 5*); alternatively, secondary fixation may be in OsFeCN (26) (*Protocol 4*). In walled cells, OsFeCN fixation results in electron dense deposits in the lumen of the ER (26); in zoospores, this procedure gives increased membrane contrast (*Figure 6*). Samples should be dehydrated and embedded in Spurr's or LR White resin following standard protocols (e.g. see ref. 23).

Protocol 4

Fixation of zoospores in OsFeCN[a]

Reagents

- 25% glutaraldehyde
- 500 mM sodium cacodylate buffer pH 6.8
- 100 mM $CaCl_2$
- 2% OsO_4 in water
- 8% KFeCN
- 100 mM sodium cacodylate buffer pH 6.8

Method

1 Prepare 10 ml of double-strength primary fixative (4% glutaraldehyde in 200 mM cacodylate buffer pH 6.8 containing 10 mM $CaCl_2$) by mixing:
- 1.6 ml of 25% glutaraldehyde
- 4 ml of 500 mM cacodylate buffer
- 1 ml of 100 mM $CaCl_2$
- 3.4 ml distilled water

2 Prepare 5 ml OsFeCN fixative (1% OsO_4, 0.8% KFeCN, 100 mM cacodylate buffer, and 5 mM $CaCl_2$) by mixing:
- 2.5 ml of 2% OsO_4

Protocol 4 continued

- 500 μl of 8% KFeCN
- 1 ml of 500 mM cacodylate buffer
- 250 μl of 100 mM $CaCl_2$
- 750 μl distilled water

3 Add an equal volume of glutaraldehyde fixative (made in step 1) to cells in suspension. Incubate for 1 h at room temperature.

4 Wash cells three times in 100 mM cacodylate buffer by centrifuging for 2 min at 1500 g to pellet the cells and resuspending in 3 ml of buffer for 5 min.

5 Embed the cells in agarose as described in *Protocol* 3.

6 Place agarose-embedded cells in osmium fixative (made in step 2) for 1 h at room temperature.

7 Wash in three changes of 100 mM cacodylate buffer, 5 min each.

[a] After method in ref. 26.

3.4 Cryofixation and freeze-substitution for transmission electron microscopy

The fact that chemical fixation can cause artefacts in cell preservation, particularly of membranes, has been recognized for a considerable time (27, 28). Rapid freezing can be used to avoid chemical fixation, however, difficulties in attaining consistently good ultrastructural preservation, free of ice-crystal damage, are largely responsible for the relatively infrequent use of cryofixation and freeze-substitution. Because rapid freezing rates and low levels of ice-crystal formation occur only in the outer 10–50 μm of the sample, large cells or multicellular tissue are not preserved at all well. Apical cells of fungal hyphae, however, are sufficiently small to freeze well and cryofixation and freeze-substitution have been applied successfully to a number of species of filamentous fungi (see ref. 29). Hyphae of *Saprolegnia* (12, 30, 31), sporangia of *Phytophthora* (32), and zoospores of *Phytophthora* (14, 33, 34) have also been successfully preserved by cryofixation and freeze-substitution.

3.4.1 Slam-freezing and plunge freezing hyphae and sporangia

For cryofixation, hyphae should be grown on cellophane or dialysis membrane on nutrient agar and pieces containing the growing edge of the colony cut out, allowed to recover, and then frozen. Adhesion to the dialysis membrane can be enhanced by pre-treatment with locust bean gum (35). In a freeze-substitution study of *Saprolegnia* hyphae, the best results were obtained by slam-freezing the hyphae against a liquid helium cooled copper block (36). Freeze-substitution in 2.5% OsO_4 in acetone followed standard protocols.

Sporangia of *P. cinnamomi* at different stages of cleavage have been successfully preserved by plunge freezing tufts of mycelium on formvar-covered loops into liquid propane (32, 37). This freeze-substitution study demonstrated that aldehyde fixation of cleaving sporangia produced artefactual vesiculation of the cleavage membranes.

3.4.2 Plunge, propane jet, and spray freezing zoospores

Oomycete zoospores present other problems for freeze-substitution because they are difficult to prepare for freezing without inducing encystment and because they are easily lost during processing after freezing. *Phytophthora* zoospores have, however, been successfully frozen and freeze-substituted by placing an aliquot of zoospores onto a piece of filter paper and plunging into liquid propane (34), by spraying zoospores from an atomizer into liquid propane (33), and by dipping a formvar-coated mesh grid into a zoospore suspension and rapidly freezing at −180 °C in a propane jet freezer (14). Whichever method is used, it is imperative to be cognisant of the likelihood of inducing zoospore encystment before freezing.

3.4.3 High pressure freezing

Good quality cryopreservation in large cells and multicellular tissues can be attained by freezing under high pressure (38–40). Hyperbaric freezers freeze samples at high hydrostatic pressure (2100 atm) and have been used to study *Phytophthora* sporangia, cysts and germlings infecting roots. In a study of *P. cinnamomi* and *P. palmivora* sporangia, it was found that high pressure freezing caused one type of vesicle in the cells, the so-called large peripheral vesicles, to rupture (41). Good preservation of cysts and germ tubes infecting roots of *Eucalyptus seiberi* has been obtained by high pressure freezing and freeze-substitution (*Figure 7*) (A. R. Hardham and K. Mendgen, unpublished observations).

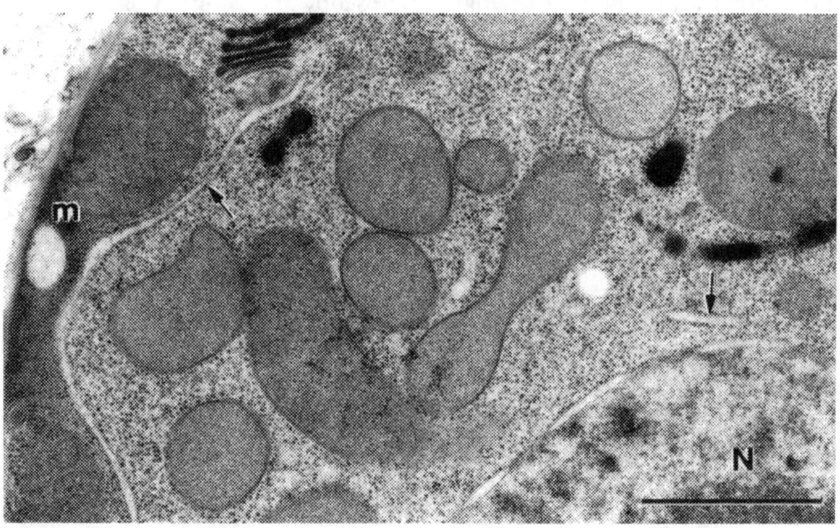

Figure 7 TEM of cyst of *P. cinnamomi* frozen under high pressure and freeze-substituted in acetone before embedding in Araldite/Epon resin. Vesicles are more spherical than they are after chemical fixation. At least some are likely to be fingerprint vesicles. m, mitochondria; arrows, endoplasmic reticulum; n, nucleus. Bar = 1 μm.

4 Immunolabelling of Oomycete cells

4.1 Fixation

The discussion of fixation in the previous section focused on the goal of obtaining as optimal preservation of cell structural as possible. In immunolabelling studies, this goal must be balanced against a number of other criteria, most importantly retention of antigenicity, accessibility of antigens to antibodies and, in the case of immunofluorescence microscopy, avoidance of high levels of background autofluorescence. The following sections address these issues as they pertain in particular to studies of Oomycete cells. More general aspects of principles and procedures of fixation regimes that are applicable to most organisms may be found elsewhere (see, for example, ref. 42).

4.1.1 Minimizing aldehyde-induced fluorescence

Many plant cells contain molecules (such as polyphenolics or chlorophyll) which are autofluorescent. This is not usually the case with Oomycete cells, although it may be an issue when studying infected plant material. Autofluorescence of Oomycete cells can, however, be induced during fixation. Glutaraldehyde is recognized as yielding superior structural preservation to that resulting from fixation in formaldehyde, however, when used alone, even at low concentrations glutaraldehyde may induce fluorescence of a sufficiently high level to obscure the signal from subsequent immunofluorescence labelling. In studies of *Phytophthora* cells, it has been found that this problem can be reduced by adding 4% formaldehyde to low concentrations of glutaraldehyde (e.g. 0.2%) in the fixative (25) (see *Protocol 5*). The presence of the glutaraldehyde gives structural preservation of a quality adequate for the light microscopy and addition of the formaldehyde keeps the autofluorescence to a sufficiently low level for subsequent immunolabelling.

Protocol 5

Fixation of zoospores for immunofluorescence labelling of surface antigens

Reagents

- 25% glutaraldehyde
- 16% formaldehyde
- 500 mM Pipes buffer pH 6.8
- 100 mM Pipes buffer pH 6.8

Method

1 Prepare 10 ml of 0.4% glutaraldehyde/8% formaldehyde in 100 mM Pipes buffer pH 6.8 by adding:
   - 160 μl of 25% glutaraldehyde
   - 5 ml of 16% formaldehyde

Protocol 5 continued

- 2 ml of 500 mM Pipes buffer pH 6.8
- 2.84 ml distilled water

2 Add equal volumes of fixative to motile zoospores in a 15 ml centrifuge tube, seal, and invert gently to mix. Fix for 30 min at room temperature.

3 Centrifuge cells at 1500 g for 2 min. Remove fixative and resuspend pellet in 3 ml of 100 mM Pipes buffer. Leave for 5 min and repeat wash twice.

4.1.2 Fixation of zoospores for immunofluorescence labelling

As discussed in Section 3.3, Oomycete zoospores are sensitive to the osmolarity of their surrounding medium and are easily damaged during fixation. This fact has important bearing on the fixative chosen for immunolabelling studies. If the zoospores are fixed in 4% formaldehyde alone (without addition of glutaraldehyde), the plasma membrane and internal membranes are poorly preserved and fragmented. This means that antibodies will have access to intracellular antigens in the zoospores (*Figures 8* and *9*), but that surface labelling is poor. On the other hand, fixation of zoospores in 0.2% glutaraldehyde/4% formaldehyde at optimal osmolarity gives good preservation of the plasma membrane and antibodies do not enter the cell. This means that only surface antigens can be labelled in glutaraldehyde/formaldehyde-fixed zoospores (*Figure 10*). These two different fixation regimes can be exploited to differentiate between surface and intracellular labelling of Oomycete zoospores. The best procedure is to prepare the

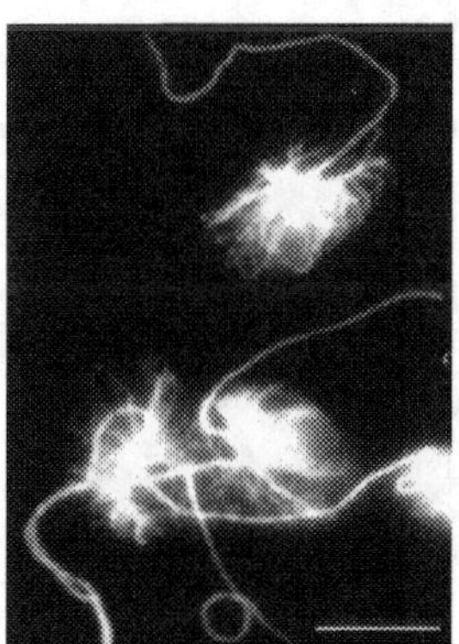

Figure 8 Immunofluorescence image of zoospores of *P. cinnamomi* fixed in 4% formaldehyde in 50 mM Pipes buffer and labelled with anti-tubulin. Cytoplasmic and flagellar microtubules are labelled. Bar = 10 μm.

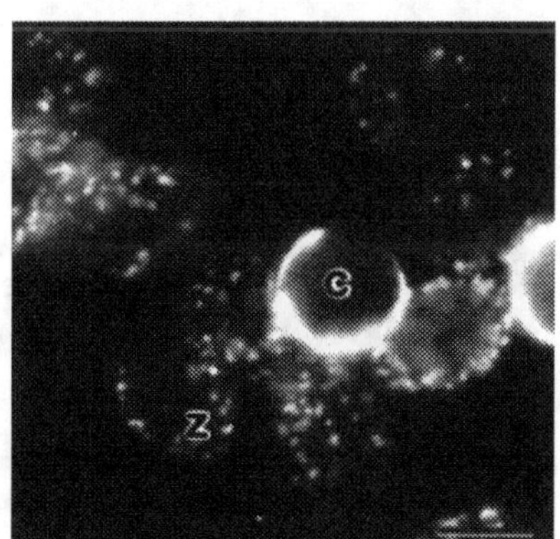

Figure 9 Immunofluorescence image of zoospores and cysts of *P. cinnamomi* fixed in 4% formaldehyde in 50 mM Pipes buffer and labelled with Cpa-2, an antibody which reacts with the contents of dorsal vesicles in the zoospores (z). This material is secreted on encystment and coats part of the cyst (c) surface. Bar = 10 μm.

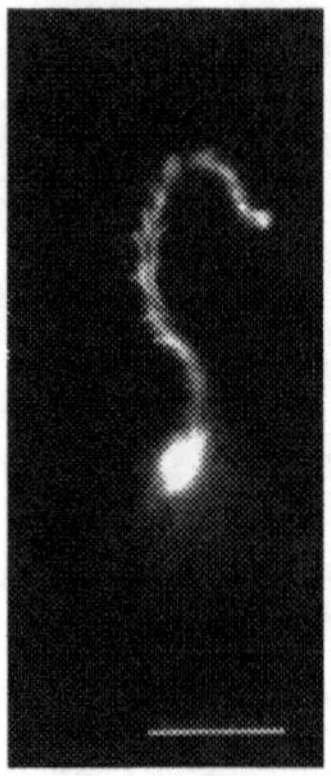

Figure 10 Immunofluorescence image of a *P. cinnamomi* zoospore fixed in 0.2% glutaraldehyde and 4% formaldehyde in 50 mM Pipes buffer and labelled with Zg-1, an antibody which reacts with tubular hairs on the anterior flagellum and at the flagellar base. Bar = 10 μm.

fixative at double strength and add it to an equal volume of the motile zoospores in water (*Protocols* 5 and 6).

Protocol 6

Fixation of zoospores for intracellular, immunofluorescence labelling

Reagents

- 16% formaldehyde
- 500 mM Pipes buffer pH 6.8
- 100 mM Pipes buffer pH 6.8

Method

1 Prepare 10 ml of 8% formaldehyde in 100 mM Pipes buffer pH 6.8 by adding:
- 5 ml of 16% formaldehyde
- 2 ml of 500 mM Pipes buffer
- 3 ml water

2 Add equal volumes of formaldehyde fixative to motile zoospores in a 15 ml centrifuge tube, seal, and invert gently to mix. Fix for 30 min at room temperature.

3 Centrifuge cells at 1500 g for 2 min. Remove fixative and resuspend pellet in 3 ml of 100 mM Pipes buffer. Leave for 5 min and repeat washing procedure twice.

4.2 Cryosectioning for immunofluorescence microscopy

A crucial consideration for all immunolabelling studies is to ensure that antibodies have access to their antigens. Antibodies are large proteins (M_r 160–900 kDa) and are unable to penetrate cell walls or intact membranes. Many early studies of intracellular antigens in plant cells used wall-degrading enzymes and mild detergent treatment to render the cell wall and plasma membrane permeable to the antibodies. This approach has been used to label microtubules and actin filaments in *Saprolegnia* hyphae (43) but use of a range of wall-degrading enzymes failed to make the wall of *Phytophthora* sporangia permeable to anti-tubulin antibodies, except in the region of the apical papillum (44).

An alternative approach which has been used successfully to label both cytoskeletal and vesicular components in hyphae, sporangia, and chlamydospores of *Phytophthora* is to cut and label cryosections of the sample (*Figures 11* and *12*) (*Protocol 7*). In this technique, the sample is frozen in cryoprotectant and sectioned while frozen at a thickness of 10–20 μm. Antibodies are then able to enter cells through their cut surface.

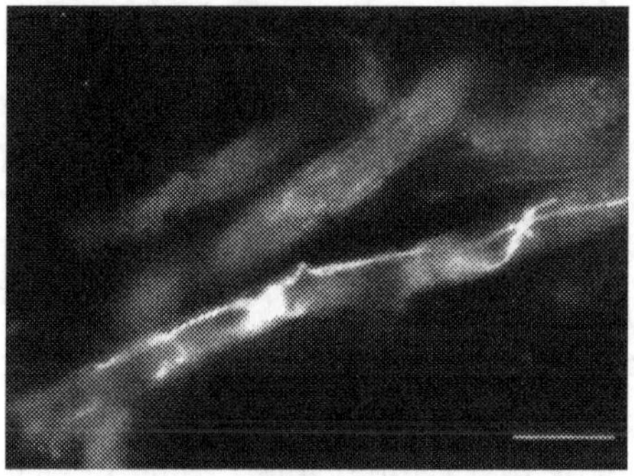

Figure 11 Immunofluorescence image of a cryosection of *P. cinnamomi* hyphae fixed in 4% formaldehyde in 50 mM Pipes buffer and labelled with anti-tubulin. Micrograph courtesy of M. Cope. Bar = 10 μm.

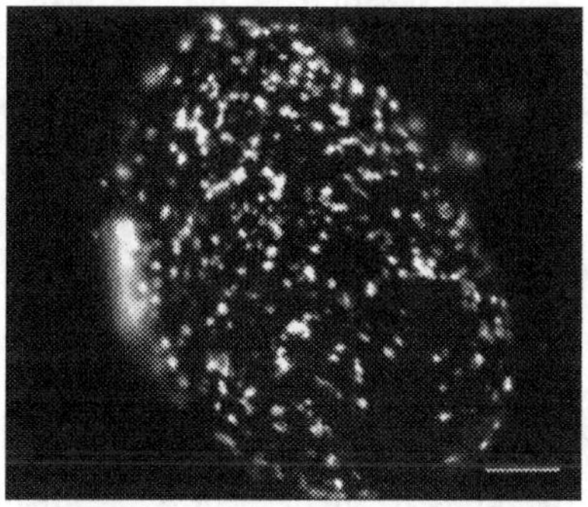

Figure 12 Immunofluorescence image of a cryosection of *P. cinnamomi* sporangium fixed in 4% formaldehyde in 50 mM Pipes buffer and labelled with Cpa-2 to localize the dorsal vesicles. Bar = 10 μm.

Protocol 7

Cryosectioning for *in situ* labelling of hyphae, sporangia, and chlamydospores

Equipment and reagents

- Cryo-moulds (Miles Inc.)
- Tissue-Tek OTC compound (Miles Inc.)
- 16% formaldehyde
- 500 mM Pipes buffer pH 6.8
- 100 mM Pipes buffer pH 6.8
- PBS: 20 mM sodium phosphate buffer, 150 mM NaCl pH 6.8–7.2
- Liquid N_2

Method

1 Prepare 10 ml of 4% formaldehyde in 50 mM Pipes buffer pH 6.8 by adding:
   - 2.5 ml of 16% formaldehyde
   - 1 ml of 500 mM Pipes buffer
   - 6.5 ml distilled water

2 Remove a tuft of mycelium from the culture and place in fixative solution in small glass vial. Fix for 1 h room temperature.

Protocol 7 continued

3 Remove fixative and rinse sample in three changes of 100 mM Pipes buffer, about 5 min each.

4 Place a drop of OTC Tissue-Tek in the base of a cryo-embedding mould and, after blotting very briefly, transfer sample to the drop of Tissue-Tek in the mould. Fill the mould with more Tissue-Tek. The moulds can be labelled on their flat surface using a felt-tip marker pen.

5 Freeze the sample in the mould by plunging into liquid N_2. The samples can then be stored in sealed plastic bags at −20 °C or −80 °C.

6 *Cryosectioning.* Freeze a drop of Tissue-Tek on a specimen holder. Remove the sample from the embedding mould, and mount on the frozen Tissue-Tek on the holder using more Tissue-Tek.

7 Place the specimen holder in the microtome and trim the block. Cut 10–16 μm sections with the specimen at about −15 °C and the chamber at about −20 °C.

8 Pick up sections by gently lowering a glass microscope slide onto the section as it sits on the knife.[a] The slides should be pre-coated with 1 mg/ml poly-L-lysine or gelatin subbing solution (see *Protocol 8*).

9 Allow the slides to sit in a moist atmosphere for 10–15 min before rinsing in PBS buffer to remove the Tissue-Tek. Do not allow the sections to dry out.[b]

10 Immunolabel and mount in Mowiol mounting medium containing 0.1% phenylene-diamine (see *Protocol 14*) under a coverslip.

[a] Standard microscope slides or multiwell slides may be used. In the former case, it is helpful for later labelling steps to encircle the sections on the other side of the slide using a diamond pencil or permanent marker pen. Multiwell slides allow labelling with a number of different antibodies.

[b] This may result in increased levels of background fluorescence and loss of antigenicity.

Protocol 8

Gelatin subbing of glass microscope slides[a]

Reagents

- Gelatin
- Chromium potassium sulfate
- Standard or multiwell glass microscope slides

Method

1 Dissolve 5 g of gelatin in 1 litre of warm distilled water. Stir slowly over low heat until dissolved. Turn off heat.

2 Add 0.5 g of chromium potassium sulfate and stir until dissolved.

3 Filter as solution cools.[b]

Protocol 8 continued

4 When cool, dip clean microscope slides into the adhesive solution and allow to dry thoroughly in a vertical position.[c]

[a] After method in ref. 45.

[b] The subbing solution can be stored at 4 °C for several weeks if 0.02% sodium azide is added.

[c] If using multiwell slides, apply adhesive solution to wells individually.

4.3 Methacrylate embedding

An alternative approach to cryosectioning for immunofluorescence labelling of intracellular components is to embed the sample in a methacrylate or acrylic resin and to label sections of the resin-embedded cells (*Figure 13*). The advantage of this approach is that the preservation of cell structure is much better than that in cryosections in which the sample is damaged by ice-crystal formation during freezing and thawing. However, one of the limitations of immunolabelling sections of cells embedded in Lowicryl K4M (a methacrylate resin) or LR White (an acrylic resin) either for fluorescence or electron microscopy, is that the antibodies only have access to antigens at the cut surface of the section. If the antigen is highly abundant this may not be a problem but for less abundant proteins the chances of the protein being exposed at the surface of the section will be low and consequently the intensity of immunolabelling will also be low.

4.3.1 Lowicryl K4M

Lowicryl K4M and LR White (or LR Gold) are the resins most often used for post-embedding labelling. Fixation should be as mild as possible while achieving adequate structural preservation. In most cases, 1–2% glutaraldehyde in 50–100 mM Pipes buffer at pH 6.8–7.2 is appropriate. Osmication is avoided because it often destroys antigenicity and may also inhibit resin polymerization by ultraviolet light.

Lowicryl K4M, like other methacrylate resins, has a low viscosity and therefore

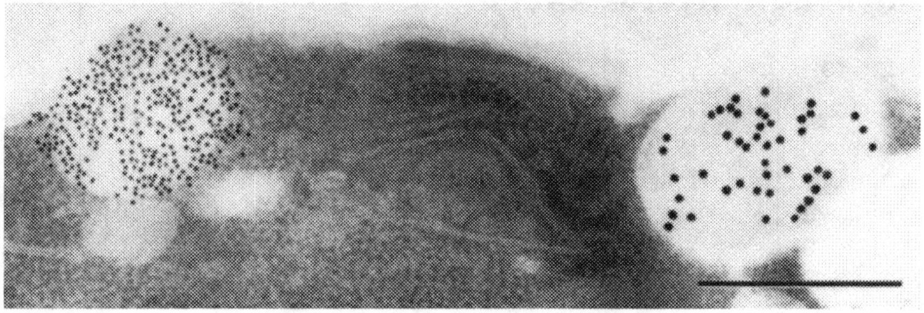

Figure 13 Section of a *P. cinnamomi* zoospore embedded in Lowicryl K4M and labelled with Cpa-2 conjugated to 10 nm gold particles and with Lpv-1 conjugated to 18 nm gold particles. Cpa-2-Au_{10} reacts with the contents of dorsal vesicles and Lpv-1-Au_{18} reacts with the contents of large peripheral vesicles. Micrograph courtesy of F. Gubler. Bar = 0.5 μm.

infiltrates into most samples quite well. The resin is supplied as three components, which must be mixed prior to use (*Protocol 9*). Procedures for embedding in Lowicryl K4M are given in *Protocol 10*. Resin polymerization is achieved by exposure to ultraviolet light, can occur at low temperatures, but is inhibited by O_2. O_2 may be excluded by embedding the samples in gelatin or Beam capsules. Alternatively, samples may be embedded in open dishes by constructing a sealed polymerization chamber which is equipped with inlet and outlet ports for flushing with gaseous N_2. The polymerization box should contain a bank of ultraviolet lights and, ideally, a refrigeration unit capable of cooling to approximately −30 °C. Polymerization of the resin at low temperatures is believed to help retain antigenicity. Note that when sectioning Lowicryl K4M resin, lower the water level in the boat as much as possible while still maintaining contact of the water with the knife edge. This reduces the chances of the water wetting the block face.

Protocol 9

Preparation of Lowicryl K4M resin

Reagents

- Lowicryl resin kit[a] (Polysciences, Inc.)

Method

1. In a fume hood weigh out 10.8 g of solution A (crosslinker), 69.2 g of solution B (monomer), and mix by rocking bottle gently. Avoid formation of air bubbles.
2. Add 0.4 g of the initiator (benzoin methyl ether).
3. Mix resin by bubbling N_2 through the solution (still in a fume hood). Use a flow rate of approx. 0.4 litres/min.

[a] Methacrylate resins may cause eczema therefore wear gloves and use only in fume hood.

Protocol 10

Embedding in Lowicryl K4M resin

Equipment and reagents

- Aluminium foil trays or planchettes
- Polymerization box, equipped with refrigeration unit, ultraviolet lights,[b] and ports for flushing with N_2
- 1% glutaraldehyde in 100 mM Pipes buffer pH 6.8–7.2
- 10%, 25%, 50%, 75%, 95%, and 100% ethanol
- Lowicryl K4M resin[a] (see *Protocol 9*)

Method

1. Fix samples in 1% glutaraldehyde in 100 mM Pipes buffer pH 6.8–7.2 for 2 h at room temperature in small glass vials.[c]
2. Rinse tissue three times for about 5 min each in 100 mM Pipes buffer at same pH as

Protocol 10 continued

fixative. If samples are zoospores or cysts, embed cells in agarose as described in *Protocol 3*.

3 Dehydrate samples by passing through a graded series of ethanol solutions, with 30 min steps in each solution, and two changes of 100% ethanol. The 10% and 25% steps should be done at 4 °C and the 50–100% steps should be at −20 °C.[d]

4 Prepare 2:1, 1:1, and 1:2 of ethanol/resin mixtures and infiltrate resin into samples by incubating samples in each mixture for approx. 2–3 h.[e] Finally transfer tissue to 100% resin. Leave for two days, changing the resin once during that time. Oxygen inhibits resin polymerization so avoid bubbling air through the resin whenever possible. When changing the solution, put the vials on ice and do the solution change in the fume hood.

5 2 h prior to embedding, turn on the refrigeration in the polymerization box and flush with N_2 at a flow rate of 0.6–1.2 litres/min.

6 Transfer the samples into fresh resin (2–3 mm deep) in aluminium foil trays or planchettes and place on the cooled base of the polymerization box. Close the lid of the box and turn on the ultraviolet lights. After 24 h check that the resin has polymerized. If hard, turn the refrigeration unit off and leave the blocks to cure under the ultraviolet lights for another 48 h. If resin is not hard, leave refrigeration until resin has set.

[a] Methacrylate resins may cause eczema therefore wear gloves and use only in fume hood.

[b] Long wavelength ultraviolet lights (e.g. TW6: Philips).

[c] Because the main reason for using Lowicryl K4M is to do immunolabelling, the fixation is kept as mild as possible to retain antigenicity, while still achieving acceptable levels of structural preservation.

[d] It is best to get the tissue into at least 75% ethanol (ideally the second change of 100% ethanol) before leaving overnight.

[e] If resin does not infiltrate thoroughly into samples, extend infiltration times.

4.3.2 Butyl-methyl methacrylate

As mentioned above, one of the main limitations of immunolabelling sections of samples embedded in Lowicryl K4M or LR White is that only antigens exposed on the surface of the section are accessible to the antibodies. A resolution to this problem, at least for light microscopy immunolabelling, came about with the introduction of butyl-methyl methacrylate resin for studies of plant cells (46). The key feature of this resin mixture is that the polymerized resin can be dissolved after sectioning by rinsing with acetone, a relatively mild treatment which usually does not destroy antigenicity. Removal of the resin means that antigens throughout the section of the sample may be labelled by the antibodies. The method has proved to be highly successful for labelling cytoskeletal elements in plant and fungal cells (e.g. refs 47–49). The resin is simple to prepare (*Protocol 11*) and can be polymerized at cool temperatures (4 °C) under long wavelength ultraviolet lights. Dehydration and embedding (*Protocol 12*) are conducted at low temperatures to help retain antigenicity.

Protocol 11

Preparation of butyl-methyl methacrylate resin[a]

Equipment and reagents

- Small (100 ml) disposable plastic bottle
- Dithiothreitol
- Butyl methacrylate (Merck or Aldrich)
- Methyl methacrylate (Merck or Aldrich)
- Bezoin ethyl ether (Aldrich)

Method

To prepare 50 ml of butyl-methyl methacrylate resin containing 80% butyl methacrylate, 20% methyl methacylate, and 10 mM dithiothreitol.[b]

1. Place 77 mg dithiothreitol into a small plastic bottle.
2. Add 40 ml butyl methacrylate and 10 ml methyl methacrylate to the dithiothreitol in the bottle.
3. Let the mixture stand undisturbed for 5–30 min while the dithiothreitol dissolves.[c]
4. Degas the resin mixture for 10–20 min by bubbling nitrogen through the mixture.[d] Use a flow rate of about 0.4 litres/min.
5. Add 250 mg bezoin ethyl ether (the catalyst). At this point the resin becomes light-sensitive and should be stored in the dark (e.g. wrap bottle in foil). Store at −20 °C when not in use, but at 4 °C during the few days of the embedding run.
6. Degas the resin mixture again just before using to embed the samples and polymerize the resin.

[a] After method in refs 46 and 47.

[b] Dithiothreitol is a reducing agent and is added to inhibit oxidation of the tissue during embedding. It causes the resin to become brittle. In some cases it can be omitted without reducing the quality of the immunolabelling results.

[c] Do not shake as the dithiothreitol will be likely to form an insoluble mass.

[d] This also serves to dissolve the dithiothreitol and thoroughly mix the contents.

Protocol 12

Embedding in butyl-methyl methacrylate resin[a]

Equipment and reagents

- Polymerization box equipped with long wavelength ultraviolet lights (TW6: Philips), and optionally with provision for nitrogen gas flow through the chamber, and temperature control
- 10%, 25%, 50%, 75%, 95%, and 100% ethanol
- Butyl-methyl methacrylate resin mixture (see *Protocol 11*)

Protocol 12 continued

Method

1. Fix sample in 4% formaldehyde or 0.2% glutaraldehyde/4% formaldehyde in 50 mM Pipes buffer at pH 6.8–7.2 (see *Protocols 5* or *6*).
2. Dehydrate samples in a 10–100% ethanol series for 30 min each step, with two changes of 100% ethanol. Ideally the samples should be placed on a rotator and the rotator placed at 4 °C during the 10% and 25% ethanol steps and at either 4 °C or −20 °C for the later steps in the ethanol series.
3. Infiltrate samples in 2:1, 1:1, and 1:2 ethanol/resin mixtures at 4 °C or −20 °C for about 2 h each change.
4. Transfer samples through three changes (2–3 h each) of 100% methacrylate resin (4 °C or −20 °C) and leave in the final change for at least 24 h.
5. Place samples in aluminium foil trays and cover with fresh methacrylate resin mixture that has been degassed prior to use (see *Protocol 11*). Alternatively place samples in gelatin or Beam capsules, fill with resin, and cap.
6. Place samples in the chamber of the polymerization box, close box, and turn on ultraviolet lights. If using foil trays, the box should be flushed with N_2 gas (a flow rate of 0.6–1.2 litres/min). If using sealed capsules, flushing with N_2 gas is not necessary. Cool the chamber to 0–4 °C if possible. The resin should be polymerized and hard in 4–8 h.

[a] After method in refs 46 and 47.

In a recently reported variation on this procedure, it has been found that embedding fungal or Oomycete cells in butyl-methyl methacrylate resin after cryofixation and freeze-substitution and removal of the butyl-methyl methacrylate resin, results in the permeabilization of walled hyphae (50) and sporangia (14) without the need to section the material.

4.4 Immunolabelling for light microscopy

Methods for immunolabelling Oomycete cells are similar to those for plants or fungi (see, for example, refs 51 or 52). The method described in *Protocol 13* is that employed in extensive studies of Oomycete cells (e.g. ref. 53). Aspects of the procedure which are especially relevant to Oomycetes are discussed below.

Use of multiwell slides is extremely helpful when labelling with a number of different antibodies, including, for example, when screening hybridoma supernatants during production of monoclonal antibodies. The multiwell slides can be purchased from a number of sources, including ICN, and are available with a number of different well sizes and configurations. The wells are defined by a thin PTFE coating on the slide and serve to hold the aliquot of antibody within the confines of the well during antibody incubation. Care should be taken, however, to avoid cross-contamination between wells during the rinsing step after primary antibody incubation. Rather than placing the slide straight into a container of washing solution, it is preferable to wick away the antibody solutions from each well and replace with a drop of rinsing buffer, which is then also wicked away before the slide is placed into a large volume of wash buffer.

Aliquots of zoospores and cysts can be loaded into the wells and allowed to air dry onto the slide. Although many of the cells are lost during subsequent processing, if high densities of cells are applied, there should be more than sufficient numbers of cells left at the end of the labelling process. If this is not the case, the slides can be pre-treated with an adhesive such as poly-L-lysine, gelatin subbing solution (*Protocol 8*), or a commercial adhesive product such as Vectabond (Vector Lab Inc.). Use of these pre-treatments may lead to increased levels of background fluorescence; blocking before antibody application may reduce this.

Protocol 13

Indirect immunofluorescence labelling

Equipment and reagents

- Multiwell microscope slides (ICN)
- Coplin jars
- PBS: 20 mM sodium phosphate buffer, 150 mM NaCl pH 6.8–7.2
- PBS-BSA: PBS containing 1% BSA
- Primary antibodies (diluted if necessary in PBS-BSA)
- Secondary antibodies conjugated to fluorescent marker such as FITC or Alexa dye (Molecular Probes, Inc.)

Method

1 If labelling zoospores or cysts, fix and wash cells as described in *Protocols 5* or *6*. Suspend the final pellet of cells in a small volume of water at a density of approx. 10^5 cells/ml and place a 10–15 μl aliquot of cells into each well of a multiwell microscope slide. Allow the samples to air dry either at room temperature or at 37 °C.[a]

2 Rehydrate zoospores or cysts, or wash away Tissue-Tek from cryosections of hyphae, by placing slides in a Coplin jar containing PBS for 5 min. After 5 min, shake or blow excess liquid off slide.

3 Place 10–15 μl of primary antibody in each well and incubate slides at room temperature or 37 °C for 45–60 min.[b] Use either PBS-BSA or a non-reacting primary antibody for a negative control.

4 Blot away the antibody solutions from each well on the slides using small pieces of filter or blotting paper cut into triangular or rectangular wicks. Replace with a drop of PBS. Blot the PBS away.

5 Place the slides into PBS in a Coplin jar and leave for 1–2 min before transferring to a fresh change of PBS for a further 5 min. Shake or blow excess liquid off slide.

Protocol 13 continued

6 Place 10–15 μl of secondary antibody in each well and incubate at room temperature or 37 °C for 45–60 min.

7 Rinse the slides twice in PBS (1–2 min then 5 min) in a Coplin jar. Rinse briefly in distilled water.

8 Place small drops of mounting medium (see *Protocol 14*) in the wells and along the slide and place a coverslip over all the wells containing sample.[c]

[a] This should take about 1 h at 37 °C. If the samples are dried completely, many cells will remain adherent to the slide through the processing, however, if the samples are not dried sufficiently, the cells will wash off the slides. The slides can be coated with an adhesive (see text). Do not over-dry as it may increase background fluorescence.

[b] The optimal time will vary depending on the antibody and sample being labelled.

[c] An aliquot of 80 μl of mounting medium is sufficient to cover all wells.

Antibody dilutions may be made in PBS-BSA or PBS-BSA containing 0.1% gelatin (fish scale gelatin: Sigma). For immunofluorescence microscopy, secondary antibodies may be conjugated to FITC, rhodamine, or a number of other fluorochromes such as the Alexa dyes (from Molecular Probes Inc.) which are especially suitable for confocal microscopy. An alternative approach for light microscopy immunolabelling is to use a secondary antibody conjugated to 1 nm or 5 nm colloidal gold particles and follow this with silver enhancement of the gold labelling (see ref. 51). This is particularly useful if autofluorescence of the sample is a problem. Silver deposits on the gold particles to produce a black deposit clearly visible in the light microscope.

Application of the coverslip in a suitable mounting medium after labelling serves three purposes.

(a) The medium solidifies after several hours and holds the coverslip in place, preventing movement when using an oil immersion lens.

(b) The mounting medium buffers the pH of the preparation. This is important because fluorescence of the marker may be pH sensitive; optimal fluorescence of FITC, for example, occurs at pH 8.5.

(c) Anti-fade agents in the mounting medium will inhibit bleaching of the fluorescent signal.

Protocol 14 describes preparation of an inexpensive mounting medium suitable for use. It is based on a mixture of polyvinyl alcohol and glycerol and includes *p*-phenylene diamine dihydrochloride (Sigma) as an anti-fade agent. Alternative mounting media are also available commercially (e.g. Citifluor, Citifluor Pty Ltd.).

Protocol 14

Mounting medium for fluorescence microscopy[a]

Equipment and reagents

- 50 ml plastic centrifuge tube
- Mowiol 4-88 (polyvinyl alcohol; Hoechst-Roussell, NJ)
- Glycerol
- 0.2 M Tris–HCl buffer: 2.42 g Tris/100 ml water adjusted to pH 8.5 with HCl
- 1% *p*-phenylenediamine dihydrochloride in water

Method

1 Place 6 g glycerol in a 50 ml plastic centrifuge tube.
2 Add 2.4 g Mowiol 4-88 and stir thoroughly to mix the Mowiol with the glycerol.[b]
3 Add 6 ml distilled water and leave the solution for 2 h at room temperature.
4 Add 12 ml of 0.2 M Tris buffer.
5 Incubate solution in a water-bath at 50 °C for 10 min with occasional stirring to dissolve the Mowiol.[c]
6 Clarify the mixture by centrifuging at 5000 g for 15 min.
7 Divide Mowiol mixture into 1 ml aliquots and store frozen at −20 °C (for up to 12 months).
8 Immediately before use, thaw and add 10 μl of 1% *p*-phenylenediamine to 90 μl of Mowiol mounting medium.

[a] After method in ref. 54.

[b] At this stage the Mowiol swells but doesn't dissolve.

[c] At this stage most, but not all, the Mowiol will dissolve.

4.5 Post-embedding immunogold labelling

Cells that have been embedded in Lowicryl K4M or LR White resin can be labelled, after sectioning, for electron microscopy by using gold-conjugated antibodies. The routine procedure (*Protocol 15*), especially for labelling sections with a single antibody, is to use a gold-conjugated secondary antibody (such as goat anti-mouse immunoglobulin) which can be purchased from a commercial source. However, in order to label a section with two different antibodies (*Figure 13*), it is often necessary to purify the primary antibodies and conjugate them directly to gold particles of different sizes. This would be required, for example, if both antibodies were raised in the same animal and were of the same immunoglobulin type. It is not difficult to make colloidal gold particles of selected sizes (see ref. 55) or to conjugate antibodies or other proteins to the colloidal gold (51). Gold particles 8–10 nm and 15–20 nm in diameter are suitable sizes for this double labelling procedure.

When sectioning cells for immunogold labelling, pick up the sections on gold or nickel grids. Do not use copper grids because the copper reacts with the phosphate in the buffers, resulting in a black precipitate being deposited over the sections.

Protocol 15

Indirect immunogold labelling

Equipment and reagents

- Ultra-thin sections (70–100 nm in thickness) of samples embedded in Lowicryl K4M or LR White on formvar-coated gold or nickel grids
- PBS: 20 mM sodium phosphate buffer, 150 mM NaCl pH 6.8–7.2
- PBST: PBS containing 0.2% Tween 20
- PBS-BSA-gelatin: PBS containing 1% BSA and 0.1% gelatin
- Primary antibodies
- Secondary antibodies conjugated to colloidal gold (5–20 nm diameter)

Method

1. Place drops of PBS-BSA-gelatin in a row on a clean piece of Parafilm, a drop for each grid to be labelled.
2. Float grids, section-side down, on the drops of PBS-BSA-gelatin for 10 min. During incubations, cover the Parafilm with a box or plastic lid to reduce evaporation and stop dust settling onto the grids.
3. Place a row of 10 μl aliquots of primary antibody on the Parafilm beneath the drops of blocking solution (the PBS-BSA-gelatin).
4. Transfer each grid to a drop of primary antibody and incubate for 1–3 h at room temperature.[a]
5. Place a row of 10 μl aliquots of secondary antibody diluted in PBS-BSA-gelatin on the Parafilm beneath the grids.
6. Wash each grid with PBST by holding the grid in forceps[b] and directing a gentle stream of washing solution from a squeeze bottle onto the forceps just above the grid. Remove excess liquid from grid and from between the arms of the forceps with filter paper.
7. Transfer each grid after it is washed to the drop of secondary antibody. Incubate for 1 h at room temperature.
8. Wash with PBST as in step 6.
9. Wash with distilled water[c] as in step 6.
10. For samples of Lowicryl-embedded cells, stain the sections with uranyl acetate for 5 min and lead citrate for 2 min. For sections of LR White resin, stain in uranyl acetate for 20 min and lead citrate for 10 min.

[a] Alternatively, incubate at 4°C overnight, but take care to ensure that the antibodies do not dry out.

[b] A sliding band (e.g. made by cutting the end off a Beam capsule) to keep them closed can be helpful.

[c] If this is not done, the salt will crystallize out as the grids dry.

Acknowledgements

I wish to thank the many colleagues, in particular Dr Frank Gubler, who have worked with me in the development of methods for cell biological investigations of Oomycete cells. I also thank Prof. K. Mendgen for the material prepared using high pressure freezing, Ms J. Maleszka for technical assistance with the cryoSEM, and Dr F. Gubler, Dr M. Cope, Prof. W. G. Allaway, and Prof. A. E. Ashford for permission to use their unpublished micrographs.

References

1. Cavalier-Smith, T. (1989). In *The chromophyte algae: problems and perspectives* (ed. J. C. Green, B. S. C. Leadbeater, and W. L. Diver), p. 381. Clarendon Press, Oxford.
2. Patterson, D. J. (1989). In *The chromophyte algae: problems and perspectives* (ed. J. C. Green, B. S. C. Leadbeater, and W. L. Diver), p. 357. Clarendon Press, Oxford.
3. Dick, M. W. (1990). In *Handbook of protoctista* (ed. L. Margulis, J. O. Corliss, M. Melkonian, and D. J. Chapman), p. 661. Jones and Bartlett Publishers, Boston.
4. Gunderson, J. H., Elwood, H., Ingold, A., Kindle, K., and Sogin, M. L. (1987). *Proc. Natl. Acad. Sci. USA*, **84**, 5823.
5. Bruns, T. D., White, T. J., and Taylor, J. W. (1991). *Annu. Rev. Ecol. Syst.*, **22**, 525.
6. Heath, I. B. (1975). *Protoplasma*, **85**, 147.
7. Heath, I. B. (1982). In *Microtubules in microorganisms* (ed. P. Cappuccinelli and N.-R. Morris), p. 275. Marcel Dekker, Inc., New York.
8. Hyde, G. J. and Hardham, A. R. (1993). *Eur. J. Cell Biol.*, **62**, 75.
9. Erwin, D. C. and Ribeiro, O. K. (1996). *Phytophthora diseases worldwide.* APS Press, St. Paul, Minnesota.
10. Fuller, M. S. and Jaworski, A. (1987). *Zoosporic fungi in teaching and research*, p. 303. Southeastern Publishing Corporation, Athens, GA.
11. Dearnaley, J. D. W., Maleszka, J., and Hardham, A. R. (1996). *Mycol. Res.*, **100**, 39.
12. Heath, I. B. and Kaminskyj, S. G. W. (1989). *J. Cell Sci.*, **93**, 41.
13. Allaway, W. G., Ashford, A. E., Heath, I. B., and Hardham, A. R. (1998). *Fungal Genet. Biol.*, **22**, 209.
14. Mitchell, H. J. and Hardham, A. R. (1999). *Protoplasma*, **206**, 118.
15. Cahill, D. M., Cope, M., and Hardham, A. R. (1996). *Protoplasma*, **194**, 18.
16. Oparka, K. J. and Read, N. D. (1994). In *Plant cell biology: a practical approach* (ed. N. Harris and K. J. Oparka), p. 27. IRL Press, Oxford.
17. Jackson, S. L. (1995). *Can. J. Bot.*, **73** Suppl. 1, 435.
18. Jackson, S. L. and Hardham, A. R. (1998). *Fungal Genet. Biol.*, **24**, 24.
19. Jackson, S. L. and Hardham, A. R. (1996). *Eur. J. Cell Biol.*, **69**, 180.
20. Suzaki, E., Suzaki, T., Jackson, S. L., and Hardham, A. R. (1996). *Protoplasma*, **191**, 79.
21. Hardham, A. R., Cahill, D. M., Cope, M., Gabor, B. K., Gubler, F., and Hyde, G. J. (1994). *Protoplasma*, **181**, 213.
22. Hardham, A. R. (1987). *Protoplasma*, **137**, 109.
23. Hawes, C. (1994). In *Plant cell biology: a practical approach* (ed. N. Harris and K. J. Oparka), p. 69. IRL Press, Oxford.
24. Read, N. D. (1991). In *Electron microscopy of plant pathogens* (ed. K. Mendgen and D.-E. Lesemann), p. 17. Springer-Verlag, Berlin.
25. Hardham, A. R. (1985). *J. Histochem. Cytochem.*, **33**, 110.
26. Hepler, P. K. (1981). *Eur. J. Cell Biol.*, **26**, 102.
27. Mersey, B. and McCully, M. E. (1978). *J. Microsc.*, **114**, 49.

28. McCully, M. E. and Canny, M. J. (1985). *J. Microsc.*, **139**, 27.
29. Howard, R. J. and O'Donnell, K. L. (1987). *Exp. Mycol.*, **11**, 250.
30. Heath, I. B., Rethoret, K., Arsenault, A. L., and Ottensmeyer, F. P. (1985). *Protoplasma*, **128**, 81.
31. McKerracher, L. J. and Heath, I. B. (1985). *Protoplasma*, **125**, 162.
32. Hyde, G. J., Lancelle, S., Hepler, P. K., and Hardham, A. R. (1991). *J. Cell Sci.*, **100**, 735.
33. Shields, J. P. and Fuller, M. S. (1996). *Protoplasma*, **191**, 84.
34. Cho, C. W. and Fuller, M. S. (1989). *Can. J. Bot.*, **67**, 1493.
35. Heath, I. B. (1987). *Eur. J. Cell Biol.*, **44**, 10.
36. Heath, I. B. (1984). *J. Microsc.*, **135**, 75.
37. Lancelle, S. A., Callaham, D. A., and Helper, P. K. (1986). *Protoplasma*, **131**, 153.
38. Gilkey, J. C. and Staehelin, L. A. (1986). *J. Electron Microsc.*, **3**, 177.
39. Dahl, R. and Staehelin, L. A. (1989). *J. Electron Microsc.*, **13**, 165.
40. Mendgen, K., Welter, K., Scheffold, F., and Knauf-Beiter, G. (1991). In *Electron microscopy of plant pathogens* (ed. K. Mendgen and D.-E. Lesemann), p. 31. Springer–Verlag, Berlin.
41. Hyde, G. J., Lancelle, S., Hepler, P. K., and Hardham, A. R. (1991). *Protoplasma*, **165**, 1.
42. Griffiths, G. (1993). *Fine structure immunocytochemistry*, p. 458. Springer–Verlag, Berlin.
43. Kaminskyj, S. G. W. and Heath, I. B. (1994). *J. Histochem. Cytochem.*, **42**, 523.
44. Hyde, G. J. and Hardham, A. R. (1992). *Exp. Mycol.*, **16**, 207.
45. Jensen, W. A. (1962). *Botanical histochemistry*, p. 408. W. H. Freeman and Co, San Francisco.
46. Gubler, F. (1989). *Cell Biol. Int. Rep.*, **13**, 137.
47. Baskin, T. I., Busby, C. H., Fowke, L. C., Sammut, M., and Gubler, F. (1992). *Planta*, **187**, 405.
48. Baskin, T. I., Miller, D. D., Vos, J. W., Wilson, J. E., and Hepler, P. K. (1996). *J. Microsc.*, **182**, 149.
49. Czymmek, K. J., Bourett, T. M., and Howard, R. J. (1996). *J. Microsc.*, **181**, 153.
50. Bourett, T. M., Czymmek, K. J., and Howard, R. J. (1998). *Fungal Genet. Biol.*, **24**, 3.
51. Harris, N. (1994). In *Plant cell biology: a practical approach* (ed. N. Harris and K. J. Oparka), p. 157. IRL Press, Oxford.
52. Goodbody, K. C. and Lloyd, C. W. (1994). In *Plant cell biology: a practical approach* (ed. N. Harris and K. J. Oparka), p. 221. IRL Press, Oxford.
53. Hardham, A. R., Gubler, F., Duniec, J., and Elliott, J. (1991). *J. Microsc.*, **162**, 305.
54. Osborn, M. and Weber, K. (1982). In *Methods in cell biology* (ed. L. Wilson), Vol. 24, p. 97. Academic Press, New York.
55. Slot, J. W. and Geuze, H. J. (1985). *Eur. J. Cell Biol.*, **38**, 87.

Chapter 9
Signal transduction in filamentous fungi: regulation of protein kinases

Xiang S. Ye
Infectious Diseases Research, Lilly Research Laboratories, Lilly Corporate Center, Eli Lilly and Company, Indianapolis, IN 46582, USA.

Liping Wu
Millenium Pharmaceuticals, Inc., 75 Sidney Street, Cambridge, MA 02139, USA.

Stephen A. Osmani
Weis Center for Research, Pennsylvania State College of Medicine, Danville, PA 17822, USA.

1 Introduction

In eukaryotes, including fungi, protein kinases are key components of signal transduction pathways. Normally a cascade of reversible protein phosphorylation/dephosphorylation is the major mechanism of signal transduction, which provides both signalling specificity and diversity. Although second messengers, such as Ca^{2+}, IP3, cAMP, and G proteins have pivotal roles in many signalling events, their physiological functions are normally mediated through regulation of protein kinases.

Studies in model fungal systems, such as *Saccharomyces cerevisiae*, *Schizosaccharomyces pombe*, and *Aspergillus nidulans*, demonstrate that protein kinases play key regulatory roles in almost all cellular processes required for growth and development in response to both internal physiological and developmental cues, and external stimuli. These include highly conserved cellular processes such as cell cycle regulation, checkpoint control and regulation of transcription and translation, as well as fungal-specific functions that involve hyphal polarity and morphogenesis, cell wall metabolism, and sporulation. Complete genome sequencing of budding yeast reveals that about 2% of yeast genes encode protein kinases, representing more than 30 families, further highlighting the importance of protein kinases in cellular regulation and signal transduction (1). Although less than 20 protein kinases have so far been isolated from filamentous fungi (2), we can safely predict that filamentous fungi contain at least as many, probably more, protein kinases as the yeast.

In the last three years, genome and EST sequencing efforts have been initiated for several filamentous fungi. Many fungal protein kinase genes will soon be revealed by these sequencing efforts and it is just a matter of time before we will have the sequence of all protein kinases in several filamentous fungi. How to utilize the genomic resources to rapidly unravel signalling mechanisms in filamentous fungi will be a new challenge facing fungal signalling researches. Undoubtedly, fungal genome sequence databases and bioinformatics tools for their analysis will greatly accelerate the pace to identify key components of signalling pathways of interest. However, to fully establish a signalling pathway with a proper physiological context and to elucidate the underlying molecular mechanisms will still require a confluence of intelligent genetic, cell biological, and biochemical studies.

1.1 Scope of the chapter

Signal transduction research has been largely a study of the regulation of protein kinases. Therefore, in this chapter we will focus on major techniques and approaches generally applicable to studying protein kinases in filamentous fungi. We will discuss general properties of protein kinases, as all protein kinases share many conserved structural and biochemical characteristics. These properties serve as the basis for the protocols commonly used to study protein kinases. We will then describe the methods we have used to characterize the molecular and biochemical properties of the NIMA kinase and its regulation during the cell cycle. The NIMA kinase is probably the best-characterized protein kinase in filamentous fungi. We believe that relevant experiences with NIMA kinase will serve as a useful primer to study other novel protein kinases in filamentous fungi. The highly conserved kinases, such as PKA, PKC, $p34^{cdc2}$, and MAP kinases are not covered herein, as their regulation in signal transduction is well understood in yeasts and other eukaryotic systems. Protocols established for these highly conserved kinases are therefore directly applicable to filamentous fungi. In addition, many reagents for their detection and assay are available commercially.

2 Conserved properties of protein kinases

Protein kinases are enzymes that catalyse the transfer of the γ-phosphate from ATP to recipient serine, threonine, tyrosine, or histidine residues of a protein substrate. Based on the specificity of the amino acid receiving the phosphate, protein kinases are generally classified into four major groups: serine/threonine protein kinases, tyrosine protein kinases, dual specificity (serine/threonine and tyrosine) protein kinases, and histidine protein kinases (3, 4). Complete genome sequencing of the budding yeast, *S. cerevisiae*, confirming numerous earlier fruitless attempts to clone tyrosine kinases from fungi, shows that fungi lack true tyrosine kinases similar to those found in mammalian cells. Histidine kinases are structurally unrelated to serine/threonine or tyrosine protein kinases. Histidine kinases employ a phospho-relay system to transfer a phosphate from a phosphohistidine intermediate to an aspartate residue of a target substrate. Thus, histidine protein kinases are also called two-component systems. While the two-component signalling system is the prevailing signalling mechanism in bacteria in response to external stimuli, this family of protein kinases is not widely used for signalling in fungi although it is thought to play a role in osmo-regulation in yeast (5). Thus, signal transduction in fungi is mainly mediated through serine/threonine protein kinases (1).

The catalytic domain of all protein kinases can be divided into 12 subdomains with invariant and highly conserved residues (6). Moreover, serine/threonine and tyrosine protein kinases have different residues specifically conserved in their kinase domains. Thus, an unknown protein kinase can be quite reliably predicted to be either a serine/threonine or a tyrosine kinase based on its similarity to known protein kinases. Similarity in the catalytic domain of protein kinases has proven to be a good indicator for similarities in structure topology, regulation, and substrate specificity. Thus, a quick sequence homology search against known kinases in the database is a good way to start the characterization of new fungal protein kinases.

S. cerevisiae has been used as a model system to study mechanisms of signal transduction in eukaryotes because of its remarkable amenability to molecular genetic manipulations. Studies in yeast have provided many useful conceptual frameworks and paradigms for understanding signalling mechanisms in other eukaryotes, from fungi to humans. With a completely sequenced genome, budding yeast is likely to be the first eukaryotic organism for which we will have a complete understanding of all signalling mechanisms for a eukaryotic organism. A co-ordinated effort by an international consortium is well underway to analyse all 6000 plus ORFs found in the yeast genome systemically (7). In addition, DNA array and proteomics technologies have been developed since completion of the yeast genome. These new technologies will further facilitate the functional analysis of all yeast genes and help classify gene functions into specific metabolic and signalling pathways. The knowledge gained from analysis of yeast signalling mechanisms will be extremely useful to studies of signal transduction in filamentous fungi, as the majority of signalling events in filamentous fungi are likely to be very similar to yeast. However, given the multicellular nature and biochemical complexity of filamentous fungi, which is reflected in their larger genomes, it is likely that some signal transduction systems will be found in filamentous fungi that are lacking in *S. cerevisiae*.

Table 1 lists some useful Internet sites for comparative analysis of fungal protein kinases.

3 Protein extraction: importance of protecting state of phosphorylation

As protein kinases play pivotal roles in cellular regulation and signal transduction, their activities are often tightly regulated. The prevailing form of kinase

Table 1 Useful Internet sites for fungal protein kinase analysis

***Saccharomyces* Genome Database (SGD)**: http://genome-www.stanford.edu/Saccharomyces SGD is a comprehensive scientific database of the molecular biology and genetics of *Saccharomyces cerevisiae* with bioinformatics tools for yeast sequence analysis, search, and retrieval. Topics include comparative genomics, gene expression profiles, and yeast reference guide.
The Yeast Proteome Database (YPD): http://www.proteome.com/index.html YPD is an interactive knowledge database integrating accumulated data from the research literature on all yeast proteins with a complete reference list and links to PubMed, GenBank, SwissPro, etc.
Protein Kinase Resource: http://www.sdsc.edu/kinases The Protein Kinase Resource provides information on protein kinases with a compendium of kinase catalytic domains, tools for structural and computational analysis, and crystal structures of protein kinases or their regulatory subunits.
National Center for Biotechnology Information (NCBI): http://www.ncbi.nlm.nih.gov/ NCBI maintains the most comprehensive public DNA and protein databases as a national resource for molecular biology information with tools for sequence analysis and retrieval. Subdatabases include cluster of gene families, UniGene, Gene and Diseases, and protein 3D structures.
Prosite: http://www.expasy.ch/prosite/ Prosite is a database of protein families and domains. It contains biologically significant sites, patterns, and profiles. This database is particularly useful to reliably identify which protein family a new sequence may belong to.
Kyoto Encyclopedia of Genes and Genomes (KEGG): http://www.genome.ad.jp/kegg/ KEGG is a knowledge database of molecular and cellular biology organized on the basis of information pathways and protein–protein interactions. It provides links to other pathway-based databases.
Fungal Genetics Stock Center (FGSC): http://www.fgsc.net/ The FGSC at the University of Kansas and its mirror site in Europe is a repository for strains, clones, and libraries of filamentous fungi. For a nominal fee researchers can obtain fungal strains and clones from the centre. In addition, FGSC publish a newsletter online containing useful lab tips and protocols related to fungal research.
Fungal Genome Resource: http://fungus.genetics.uga.edu:5080/ The Fungal Genome Resource is a repository of information and services to assist the fungal genomics community and provides links to all the fungal genome sequencing centres.

regulation is phosphorylation, either by another protein kinase or through autophosphorylation. In a signal transduction pathway, protein kinases usually act through a phosphorylation cascade. Phosphorylation of a given protein kinase can be either a positive (activating) or negative (inhibitory) regulatory step. Moreover, the activation of many protein kinases requires phosphorylation. Thus, it is critical to protect phosphorylation during isolation when a protein kinase activity is followed by *in vitro* assays. Otherwise, incorrect interpretations can be easily made based on erroneous assay results caused by dephosphorylation of protein kinases during the isolation process. Buffers used to isolate protein kinases from fungal cell extracts should include potent protein phosphatase inhibitors. Often, it is necessary to include a combination of many different phosphatase inhibitors. For instance, we found that in order to fully stabilize

NIMA kinase activity during isolation it is necessary to add okadaic acid in the extraction buffer in addition to many conventional phosphatase inhibitors (8, 9). We find that this buffer satisfactorily protect both NIMA and NIMX^{cdc2} kinases against dephosphorylation and proteolysis during isolation.

Although most filamentous fungi can grow in minimal synthetic media, our experience with *Aspergillus* shows that fungal cells have much higher protease and phosphatase activities when grown in minimal synthetic medium as compared to those found in cells grown in rich complex medium. In addition, *Aspergillus* cells have higher activities of proteases and phosphatases as the cell culture ages. Thus, to follow protein kinase activities *in vitro*, we normally grow *Aspergillus* cells in yeast extract-supplemented medium (YG) and always harvest cells at early to mid log phase to minimize the effects of proteolysis and dephosphorylation.

Protocol 1

Protein extraction for kinase assays in *Aspergillus nidulans*

Reagents

- YG medium: 0.5% (w/v) yeast extract, 1% (w/v) dextrose, 20 mM magnesium sulfate, trace element mix (1 ml/litre), and any vitamin supplements (1000 × stock) as required by individual strains
- Trace element mix: $FeSO_4.7H_2O$ (1 g/litre), $ZnSO_4.4H_2O$ (8.8 g/litre), $CuSO_4.4H_2O$ (0.4 g/litre), $MnSO_4.4H_2O$ (0.15 g/litre), $NaB_4O_7.10H_2O$ (0.1 g/litre), $(NH_4)_6Mo_7O_{24}.4H_2O$ (0.05 g/litre).To make this stock (1000 ×), weigh and completely dissolve each chemical in the listed order. Then add 1 ml chloroform to each litre stock for long-term storage.
- Stop buffer: 160 mM sodium chloride, 1 mM sodium azide, 10 mM EDTA pH 8.0, 50 mM sodium fluoride
- Protein extraction buffer: 30 mM Tris–HCl pH 7.5, containing the following protease inhibitors: leupeptin (13 μg/ml), soybean trypsin/chymotrypsin inhibitor (13 μg/ml), aprotinin (13 μg/ml), *N*-tosylphenylalanine chloromethyl ketone (13 μg/ml), 3 mM Nα-*p*-tosyl-L-arginine methyl ester, 7 mM benzamidine, 0.3 mM phenylmethylsulfonyl fluoride (PMSF); the following phosphatase inhibitors: 100 mM β-glycerophosphate, 20 mM sodium fluoride, 1 mM sodium vanadate, 20 mM *p*-nitrophenyl phosphate, 10 nM microcystin, 1 μM/ml okadaic acid; and 25 mM EGTA pH 7.5, 7 mM EDTA pH 8.0, 0.2% (v/v) Nonidet P-40, 1 mM dithiothreitol (DTT). The extraction buffer should be prepared fresh and PMSF is added just at the time of homogenization. Okadaic acid is obtained from Upstate Biotechnology (New York, USA), microcystin from BRL Gibco, and all the other reagents from Sigma (St. Louis, USA).

Method

1. Grow *Aspergillus* cells (2×10^6 conidia/ml) in YG medium to early or mid log phase in a shaker incubator at 32 °C at 170 r.p.m.
2. Harvest hyphal cells by vacuum filtration through a layer of Miracloth and wash the cells immediately with ice-cold stop buffer.

Protocol 1 continued

3 Blot dry hyphal cells between paper towels and snap-freeze them in liquid nitrogen.

4 Grind the cells in extraction buffer with 2.5 ml/g of frozen cells for 5 min on ice with a mortar and pestle.

5 Centrifuge the homogenate at 14000 g for 10 min at 1 °C. Aliquot and then freeze the supernatant in liquid nitrogen for long-term storage.

6 Thaw the frozen protein extract with lukewarm water and centrifuge the extract again.

7 Quantify protein concentrations using Bio-Rad protein assay reagents before use.

4 Protein immunoprecipitation

Cellular levels of many protein kinases are highly regulated and are often low in abundance. In addition, protein kinases are somewhat promiscuous towards many artificial protein substrates routinely used for *in vitro* kinase assays. For instance, histone H1 has been widely used as an artificial substrate in many different kinase assays. Thus, to detect protein levels, or to assay kinase activities specifically in crude protein extracts, a quick affinity purification (normally by immunoprecipitation) using specific antibodies, is usually required. The protein level of NIMA kinase is highly cell cycle regulated and is in very low abundance. To follow the NIMA protein through the cell cycle, it is necessary to immunoprecipitate NIMA using NIMA-specific antibodies from >4 mg crude protein extract, and then detect NIMA by Western blotting of the immunoprecipitates (8, 9). Likewise, it is necessary to immuno-purify NIMA to specifically assay its kinase activity (8, 9).

With the advent of genome sequencing, the majority of fungal protein kinases will be identified through sequence homology searching. If a transformation system is established for a given fungus, epitope tagging is the method of choice to facilitate protein purification and cellular localization using commercial, epitope-specific antibodies. The tag is normally inserted by *in vitro* mutagenesis either at the N- or C-terminus. The tagged gene is then used to replace the endogenous gene or is expressed ectopically from an inducible promoter. If fungal cells expressing the tagged gene grow just as well as the wild-type under conditions requiring the function of the particular kinase under study, it indicates that the tag insertion does not interfere with the normal function of the gene. Alternatively, a large amount of fungal kinase protein can be produced in, and purified from, a heterologous expression system. Inducible expression in *E. coli* and baculovirus expression in insect cells are widely used protein expression systems. Many protein kinases produced in a heterologous system are active as kinases and retain proper intrinsic biochemical properties (such as $p34^{cdc2}$ and PKA expressed in insect cells by recombinant baculovirus and NIMA in *E. coli*). Heterologous expression systems are often a reliable source of pure kinase for biochemical characterization studies (see later sections). In addition, highly purified kinase obtained from heterologous systems can be excellent antigens to generate high titre specific antibodies.

4.1 Epitope tagging using the Quick Change mutagenesis kit

Recently a technique to rapidly introduce epitope tags to any region of a gene by PCR was developed by one of us using reagents of the Quick Change mutagenesis kit (Stratagene, La Jolla, CA) (10). This technique takes advantage of the fact that *Pfu* DNA polymerase lacks nucleotide replacement activity. As illustrated in *Figure 1*, the tagging oligonucleotide primers have two parts. The 5′ sequences of the primers encode the tag sequence on opposite strands (complementary). The 3′ sequences (25 to 45 base) of the primers are derived from the gene sequence, beginning at the insertion point on opposite strands so that the primers are juxtaposed to each other but do not overlap. The melting temperature of the gene-specific sequence of the primers should be about 10 °C above the extension temperature of 68 °C. The primers must be purified either by using an Oligonucleotide Purification Cartridge—OPC(™) from Perkin Elmer, or polyacrylamide gel electrophoresis (PAGE). The subsequent PCR amplification of the plasmid using the tagging primers, *Dpn*1 digestion of the parental plasmid, and bacterial transformation are essentially carried out according to the protocol

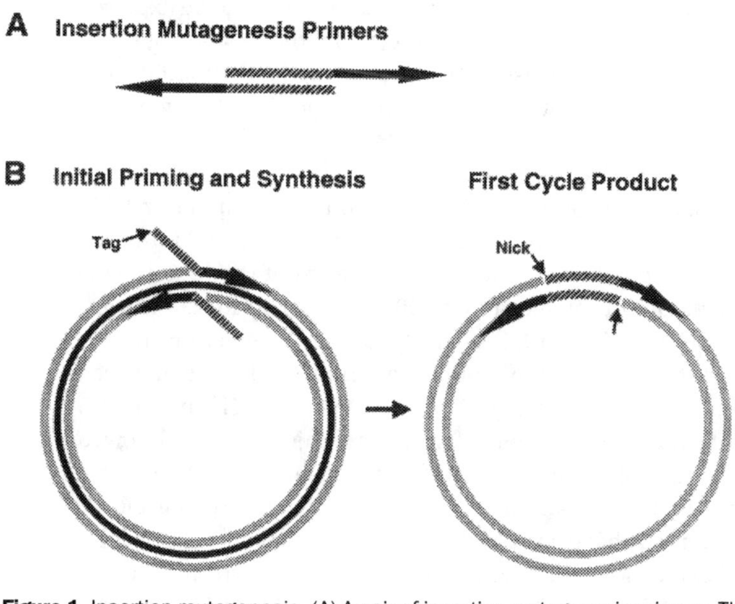

Figure 1 Insertion mutagenesis. (A) A pair of insertion mutagenesis primers. The hatched bars represent the same tag sequence on complementary strands; the solid black bars are non-overlapping, but immediately adjacent, gene-specific sequences also on complementary strands with the arrows pointing to the direction of DNA synthesis. (B) Insertion mutagenesis by polymerase chain reaction (PCR). The solid circle represents the parental plasmid carrying the gene to be tagged. Note the circular PCR products carry a nick point on each strand at the point of insertion. The nicks are repaired by bacteria after transformation.

provided by Stratagene for the Quick Change mutagenesis kit. Plasmids can be screened for tag insertion by colony PCR of the bacterial transformant using a primer derived from the tag sequence. In-frame insertion of the tag is confirmed by DNA sequencing. It is our experience that $>80\%$ of the plasmids have the correct tag insertion.

4.2 Heterologous expression and antibody production

For fungi lacking an efficient transformation system, epitope tagging of genes would be difficult. However, specific antibodies can be generated using proteins made in a heterologous system. Even with an efficient transformation system, as in *Aspergillus nidulans*, we find it advantageous to generate our own antibodies using bacterially-expressed *Aspergillus* proteins. Commercial antibodies are often quite expensive and have considerable batch variations. It is often necessary to calibrate the titre of new batches of commercial antibodies to produce consistent results.

There are several expression systems widely used to express proteins. We have used the pET inducible expression system (Novogen) to express our proteins due to the ease of affinity purifying proteins, using the His-tag, through nickel columns from both soluble and insoluble bacterial protein extracts (8, 11). We are able to purify highly active NIMA kinase from bacteria using this expression system. Often several expression systems and bacterial strains need to be explored to obtain optimal expression of some proteins. Furthermore, the function of many eukaryotic proteins requires post-translational processing and modification. Baculovirus insect cells are better suited to express biologically active forms of these proteins.

It is advisable to use several New Zealand white male rabbits for each antibody production. To generate antiserum suitable for immunoprecipitation (able to bind native form of the protein), as well as for Western blotting and immunostaining of fixed cells (able to recognize the denatured form of the protein), we use both native and denatured proteins (about 500 μg) at 1:1 ratio as antigens. The protein is denatured by boiling for 5 min in 1% SDS. One day before antigen injection, rabbits are pre-bled to collect pre-immune serum. Antigen is prepared in the RIBI adjuvant solution (RIBI ImmunoChem Research, Inc., Hamilton, MT) according to the manufacturer's instructions. Two booster injections are followed at 13 day intervals. The titre of antibodies is tested for each rabbit. We do find some variations in antibody titre among individual animals. The polyclonal antibodies made as described above are normally very specific and in high titre. We have successfully used them for immunoprecipitation, Western blotting, kinase assays, and indirect immunofluorescent microscopy often without the need to affinity purify the antibodies (8, 10).

The concentration of antibodies used in immunoprecipitation depends on the antibody titre and the concentration of the target protein present in the extract. This needs to be determined empirically. In cell cycle research it is necessary to follow cell cycle regulators throughout the cell cycle. To titre new antiserum we choose a cell extract containing the highest level of the target protein, for instance G_2 phase extract for NIMA kinase, and then use various amounts of antiserum for immunoprecipitation. The target protein present in the immunoprecipitates is quantified by Western blotting or enzyme assays. The lowest amount of antiserum able to bring down all of the target protein is then chosen for subsequent immunoprecipitations. At this concentration the antibody can quantitatively precipitate the target protein from different extracts and this minimizes contamination of non-specific proteins in the immunoprecipitates that could potentially be caused by excess amount of antiserum and protein A–Sepharose beads.

Protocol 2

Expression of NIMA in *E. coli* and His-tag affinity purification

Reagents

- BL21(DE3) pLysS cells containing pET21a-NIMA. pET21a-NIMA was created by cloning *nimA* ORF as a *BglII–NotI* fragment into pET21a plasmid in frame with the T7 leader sequence at its N-terminus and the His-tag at the C-terminus.
- HK binding buffer: 30 mM Tris–HCl pH 7.9, containing the following protease inhibitors: leupeptin (13 μg/ml), soybean trypsin/chymotrypsin inhibitor (13 μg/ml), aprotinin (13 μg/ml), *N*-tosylphenylalanine chloromethyl ketone (13 μg/ml), 3 mM *N*α-*p*-tosyl-L-arginine methyl ester, 7 mM benzamidine, 0.3 mM phenylmethylsulfonyl fluoride (PMSF); the following phosphatase inhibitors: 100 mM β-glycerophosphate, 20 mM sodium fluoride, 1 mM sodium vanadate, 20 mM *p*-nitrophenyl phosphate, and 0.2% (v/v) Nonidet P-40, 10 mM imidazole, 0.5 M NaCl.
- 1 M isopropyl β-D-thiogalactopyranoside (IPTG) in H_2O (sterilized by filter filtration and stored at −20 °C)
- Wash buffer: 30 mM Tris–HCl pH 7.9, 60 mM imidazole, 0.5 M NaCl
- Elution buffer: 20 mM Tris–HCl pH 7.9, 0.5 M imidazole, 0.5 M NaCl
- Column with 1 ml of packed nickel-charged His.bind resin, prepared as per Novagen's instruction
- NIMA storage buffer: 1 × PBS containing 15% glycerol and the following phosphatase inhibitors: 100 mM β-glycerophosphate, 20 mM sodium fluoride, 1 mM sodium vanadate, 20 mM *p*-nitrophenyl phosphate

A NIMA expression

1. Start a 50 ml culture of BL21(DE3) pLysS cells harbouring pET21a-NIMA in LB medium at 37 °C overnight from a single colony.
2. Collect bacterial cells by centrifugation at 5000 g for 5 min and wash once with fresh LB medium.
3. Inoculate the bacterial cells into 1 litre of fresh LB medium and incubate the culture (500 ml each in 2 litre flasks) at 37 °C until OD_{600} reaches about 0.5.

Protocol 2 continued

4 Transfer the culture to a 20 °C shaker incubator and incubate for another 10 min.

5 Induce NIMA expression by addition of 0.4 mM IPTG of the culture for 4 h at 20 °C.

6 Harvest cells by centrifugation at 5000 g for 5 min at 4 °C.

B NIMA purification

1 Resuspend bacterial cells in ice-cold HK binding buffer (20 ml for a 1 litre culture).

2 Sonicate on ice until the sample is no longer viscous.

3 Centrifuge the sample at 39 000 g for 15 min at 1 °C.

4 Load the supernatant from step 3 onto the charged His.bind resin column with a flow rate of 1 ml/min.

5 Wash with 15 ml of HK binding buffer.

6 Wash with 15 ml of wash buffer.

7 Elute NIMA with 15 ml of elution buffer and collect 1 ml fractions.

8 Exchange the elution buffer with storage buffer and concentrate NIMA using Centricon 30 (Millipore, USA).

Protocol 3

Protein immunoprecipitation (IP)

The following protocol was developed to immunoprecipitate NIMA either for protein detection by Western blotting or for kinase activity assays.

Reagents

- KAB buffer: 50 mM Tris–HCl pH 7.5, 10 mM $MgCl_2$, 1 mM dithiothreitol
- TENN buffer: 50 mM Tris–HCl pH 7.5, 5 mM ETDA pH 8.0, 0.5% Nonidet P-40, 150 mM NaCl, 0.02% sodium azide
- Protein A–Sepharose CL-4B beads: 1.5 g protein A–Sepharose beads (Pharmacia, NJ) are suspended in 10 ml of TENN buffer, bovine serum albumin (BSA) is then added to 0.1%. The beads are rocked on a rocking shaker overnight at 4 °C, and then washed 5 × by centrifugation with 8 ml of TENN buffer. The beads are finally resuspended in 10 ml of TENN buffer.

Method

1 Add 1 μl of NIMA specific antibody, E-14, to each milligram of protein extract in 1.5 ml microcentrifuge tubes in a total volume of 250 μl (make up the volume with fresh HK buffer) and mix.

2 Incubate on ice for 2 h with periodical mixing.

3 Add protein A–Sepharose beads to 1 mg/1 μl of the antibody and mix. Make sure to resuspend the beads well before pipetting.

4 Rock the samples for 30 min at 4 °C on a rocking shaker.

Protocol 3 continued

5. Wash the beads by centrifugation at 7000 r.p.m. in a refrigerated microcentrifuge, 3 × with HK buffer and 2 × with KAB buffer.
6. The immunoprecipitates are then either boiled in 2 × protein sample buffer for SDS-PAGE and Western blotting or their volume is properly adjusted for kinase assays.

Pre-clearance of protein extracts with pre-immune serum and protein A-Sepharose beads is normally recommended to reduce non-specific protein contamination. Based on our experiences with several *Aspergillus* proteins, however, this step may often be omitted. We find that all the β-casein kinase activity brought down by NIMA antibody is NIMA-specific (8). Omission of the pre-clearance step minimizes NIMA inactivation, as NIMA kinase activity is particularly sensitive to dephosphorylation (8, 9). In recent years, co-immunoprecipitation has been increasingly popular as a means to rapidly purify associated proteins or protein complexes, as new technologies in mass spectrometry have made it possible to accurately sequence minute quantities of proteins isolated from protein gels. These new technologies in mass spectrometry coupled with sequence databases allow rapid identification of genes from short peptide sequences (12). Thus, co-immunoprecipitation is particularly useful in signal transduction studies to identify additional signalling components as signalling pathway components often act as modules and physically associate with each other. For co-immunoprecipitation, it is advisable to pre-clear protein extracts to reduce potential false positives.

Expression of NIMA is toxic to bacterial cells. We tried several host bacterial strains and found that BL21(DE3) pLysS cells express NIMA well at 20 °C. Even with low temperature induction, NIMA still mostly accumulates in an insoluble form.

5 Gel-shift assay for detection of protein phosphorylation state

Activation and inactivation of protein kinases is often mediated by a series of phosphorylation and dephosphorylation events. It has been observed that phosphorylated proteins often exhibit a mobility shift during SDS-PAGE giving rise to a series of molecular weight species of protein bands caused by different levels of phosphorylation. Most phosphorylated proteins migrate more slowly during SDS-PAGE, seen as discrete higher molecular weight protein bands. However, a few phosphoproteins have been observed to migrate faster during SDS-PAGE. For instance, $p34^{cdc2}$ H1 kinase, when phosphorylated at T161, migrates faster than its non-phosphorylated form (11, 13). These mobility differences during SDS-PAGE have been exploited as a convenient means to rapidly determine phosphorylation states of protein kinases and to help dissect the role of phosphorylation events in kinase regulation and function.

As each phosphoprotein has subtle structural and chemical differences, they usually behave somewhat differently during SDS–PAGE. To achieve the best resolution of a protein kinase in different phosphorylation states, optimal electrophoresis conditions need to be empirically determined. Based on our experience the key factors affecting resolution are the gel concentration and the percentage of crosslinker (8, 11). Initially we detected the NIMA protein by Western blotting after electrophoresis with standard 10% SDS-acrylamide gels under normal electrophoresis conditions. NIMA in various phosphorylation states failed to separate from each other under these electrophoresis conditions. At the time we thus did not realize that NIMA has several distinct phosphorylation states during the cell cycle. By changing various parameters of the SDS–PAGE, we observed dramatic cell cycle-dependent phosphorylation events of the NIMA kinase. The best resolution was achieved with 7.5% (w/v) gels at a ratio of 1.2 g bisacrylamide to 30.8 g of acrylamide, 16 × 20 cm in size, cast with 1 mm spacers, and run with constant current of 25 mA.

Protocol 4

Gel-shift assay for NIMA phosphorylation states

Reagents

- TBST solution: 20 mM Tris pH 7.5, 500 mM NaCl, 0.05% (v/v) Tween 20, and 0.02% (w/v) sodium azide
- 5% BSA in TBST
- 3% non-fat milk in TBST
- PAP buffer: 25 mM Pipes pH 6.0, leupeptin (13 μg/ml), soybean trypsin/chymotrypsin inhibitor (13 μg/ml), aprotinin (13 μg/ml), *N*-tosylphenylalanine chloromethyl ketone (13 μg/ml), 3 mM *N*α-*p*-tosyl-L-arginine methyl ester, 7 mM benzamidine, 0.3 mM phenylmethylsulfonyl fluoride (PMSF)
- NIMA antiserum (E-14) diluted 1:200 in TBST containing 5% BSA
- Potato acid phosphatase (PAP): PAP suspension in $(NH_4)_2SO_4$ is obtained from Roche Molecular Biochemicals. PAP is precipitated from the suspension by centrifugation at 20 000 g for 1 min at 4 °C. PAP is resuspended in PAP buffer and dialysed against 1 litre of PAP buffer at 4 °C overnight. After addition of glycerol to 15%, PAP is aliquoted and stored at −70 °C.

Method

1. Immunoprecipitate NIMA from 4 mg protein extract prepared from cells at various stages of a synchronized cell cycle.
2. Complete SDS–PAGE of NIMA precipitates on 7.5%, 16 × 20 cm gels made with an acrylamide mix at a ratio of 1.2 g bisacrylamide to 30.2 g acrylamide, at a constant current of 25 mA. Note acrylamide is extremely toxic before being polymerized.
3. To further confirm NIMA phosphorylation, selected samples are dephosphorylated with 0.1 U of PAP for 60 min at 30 °C in the presence or absence of phosphatase inhibitors (100 mM β-glycerophosphate, 20 mM sodium fluoride, 1 mM sodium

Protocol 4 continued

vanadate, 20 mM *p*-nitrophenyl phosphate, 10 nM microcystin, 1 μg/ml okadaic acid, 25 mM EGTA pH 7.5, and 7 mM EDTA pH 8.0). As NIMA is present in an immunocomplex attached to protein A beads, we wash out the PAP with KAB by centrifugation before SDS–PAGE.

4 NIMA protein is electroblotted onto nitrocellulose membranes. Membranes are blocked in 5% (w/v) BSA in TBST solution for 2 h. Membranes are then incubated with NIMA antiserum (E-14) solution made at a 1:200 dilution in 5% (w/v) BSA in TBST for 2 h. Membranes are washed 4 ×, 10 min each, with TBST, incubated for 40 min with protein A–horse radish peroxidase conjugate at 1:25 000 dilution in 3% non-fat milk in TBST, and then washed again 4 × with TBST, 5 min each. NIMA protein is then visualized by enhanced chemiluminescence (ECL; Amersham) as per manufacturer's instructions.

In addition to phosphorylation many proteins, including protein kinases, also undergo other types of post-translational modifications, such as prenylation, ubiquitination, and acetylation. To confirm that protein mobility shifts during SDS–PAGE are caused by phosphorylation, phosphatase treatment of the protein is carried out. If a protein mobility shift is caused by phosphorylation, then dephosphorylation by phosphatase should completely eliminate the shift. However, certain phosphorylation sites of proteins in their native state are not accessible to phosphatases. Thus, to dephosphorylate these sites, the protein needs to be first denatured to expose such sites. For example, cyclin B binding protects the T161 phosphorylation site of $p34^{cdc2}$ kinase (11). To remove the phosphate from T161, $p34^{cdc2}$/cyclin B kinase complex is first boiled for 5 min in 1% SDS. The sample is then mixed with phosphatase buffer to dilute SDS to below 0.1%, followed by phosphatase addition. If phosphatase, even with the denatured protein, fails to reverse the mobility shift, then the protein is likely modified in some other way.

Many phosphatases are suitable for dephosphorylation assays. We have used potato acid phosphatase (PAP), because PAP is very potent and stable. PAP at 0.1–0.2 U can effectively dephosphorylate all the NIMA protein in the IP precipitates used for Western detection. As a negative control, a cocktail of phosphatase inhibitors is included.

6 Protein kinase assay and substrate specificity

Functional and biochemical characterization of a protein kinase requires an *in vitro* assay for kinase activity. Identification of endogenous substrates for protein kinases has been proven to be a very difficult task. *In vivo*, the specificity of protein kinases is largely determined by their cellular localization and temporal association with other proteins. However, *in vitro*, protein kinases can often phosphorylate many cellular proteins, most of which may have no physiological relevance. Thus, conventional biochemical fractionation methods, which have been successfully used to purify many physiological substrates of a wide range of enzymes, have limited success in identifying protein kinase substrates. As a result, endogenous substrates for most protein kinases are still unknown, even for the very well characterized kinases such as NIMA and $p34^{cdc2}$. However, this promiscuity towards substrates *in vitro* does often allow the use of artificial protein substrates to establish kinase activity assays. In fact, the majority of protein kinases are characterized and routinely assayed using artificial substrates.

A common approach to identify a suitable artificial substrate is to use the purified kinase to phosphorylate a variety of proteins known to be phosphorylated by other protein kinases. To identify a protein substrate suitable for NIMA kinase assay, NIMA kinase either purified from bacteria expressing NIMA or immunoprecipitated from *Aspergillus* cell extracts was used to phosphorylate a panel of proteins (14, 15). These proteins are dephosphorylated β-casein, phosvitin, MAP2, MBP, histone H1, H2A, and H2B, tau protein, α-tubulin, β-tubulin, dynein, kinesin, smooth muscle myosin light chain, BSA, MAP4, MAP1b, and IgG. These proteins are commercially available and can be obtained from several suppliers. Among the proteins tested β-casein was found to be the preferred substrate for NIMA kinase. NIMA also phosphorylates phosvitin, MAP2, and MBP but to a lesser extent. NIMA does not phosphorylate the other proteins.

Protein kinases can efficiently phosphorylate short peptide substrates. The use of peptide substrates has allowed the determination of structural requirements surrounding the phosphorylation sites for several protein kinases. In a screen of an assembled library of 50 synthetic peptides, each of which contains a phosphorylation site of a known protein kinase, PLM (54–72), DEEEGTFRSSIRRLSTRRR, was identified as an excellent substrate for NIMA kinase (16). Serine-10 is the only NIMA phosphorylation site on this peptide. A systematic amino acid replacement analysis around the phosphorylation site determined that the key structural determinant of NIMA kinase specificity is the phenylalanine residue –3 N-terminal to serine-10 (FXXT/S). This structural requirement for the phosphorylation site is found to be unique to the NIMA protein kinase family. PLM (54–72) was found to be a better substrate than β-casein for NIMA kinase as NIMA has a tenfold higher V_{max} and threefold lower K_m when assayed with this peptide.

The identification of a peptide substrate and subsequent analysis of the structural determinants for the phosphorylation site specific to the kinase is very expensive and time-consuming, as each peptide and its analogues need to be evaluated individually based on their effects on K_m and V_{max} values. A technique for 'one-step' identification of optimal protein kinase substrate from an oriented peptide library has been recently developed and validated for several protein kinases, including NIMA kinase, with known specific phosphorylation sites (17, 18). Readers are referred to the original papers for technical details. Briefly, an oriented library of 2.5 billion distinct peptides for serine/threonine protein kinases is made using a peptide scaffold of M-A-X-X-X-X-S-X-X-X-X-A-K-K-K. For tyrosine protein kinases, serine at position 7 is replaced with tyrosine. X indicates all amino acids except tryptophan, cysteine, tyrosine, serine, or threonine.

Omission of tryptophan and cysteine is to avoid problems with subsequent peptide sequencing and potential oxidation. Omission of Y, S, and T is to ensure that position 7 is the only phosphorylation site. Purified protein kinase is used to phosphorylate this library under conditions in which only about 1% of the total peptide mixture becomes phosphorylated. It is important for a small fraction of peptides to be phosphorylated to avoid levelling effects caused by the exhaustion of the optimal substrates, as the critical step in this technique is the quantitative separation of the small fraction of phosphorylated peptide products (optimal substrate) from the vast excess of non-phosphorylated peptides. Phosphorylated peptides are separated from the rest of the library using the ferric iminodiacetic acid column and sequenced. Relative abundance of amino acids at a given cycle of sequencing indicates the preference for the amino acid at the position, and the optimal peptide substrate for the protein kinase is determined.

Protocol 5

NIMA kinase assay

Reagents

- KAB buffer: 50 mM Tris–HCl pH 7.5, 10 mM $MgCl_2$, 1 mM dithiothreitol
- Kinase substrate mix (5 μl/per reaction) β-casein as substrate: 2.5 μl of 1 mM ATP, 2 μl of β-casein (10 mg/ml, Sigma) and 0.5 μl of [γ-^{32}P]ATP (6000 Ci/mmol, Amersham)
- Kinase substrate mix (5 μl/per reaction) peptide PLM (54–72) as substrate: 2.5 μl of 1 mM ATP, 2 μl of PLM (54–72) at 4 mg/ml, and 0.5 μl of [γ-^{32}P]ATP (6000 Ci/mmol)
- Gel fixing solution: 2 parts of methanol, 2 part of distilled water, and 1 parts of acetic acid

A NIMA kinase assay using β-casein as substrate

1. Adjust NIMA kinase immunoprecipitates to 25 μl with KAB buffer.
2. Equilibrate NIMA kinase and substrate mixture to 30 °C.
3. Start kinase reaction by addition of 5 μl of β-casein substrate mixture and mix with pipette tip while pipetting in staggered intervals. We find that a 10 sec interval is adequate and comfortable to add the substrate mixture between samples.
4. Let kinase reaction proceed for 10 min in a 30 °C water-bath.
5. Terminate kinase reaction by addition of 25 μl of 2 × SDS protein sample buffer in the same sequence and the same staggered intervals as for substrate addition, and boil the samples for 5 min.
6. Separate the samples by electrophoresis on 12% (w/v) SDS–polyacrylamide gels. Stop electrophoresis while the bromophenol blue tracking dye is about an inch away from the bottom edge of the gel to prevent contamination of the electrophoresis apparatus by the unincorporated [γ-^{32}P]ATP. Disassemble the gel sandwich carefully and cut off the gel below the dye, which contains the unincorporated [γ-^{32}P]ATP.
7. Fix the gel in the gel fixing solution and dry the gels using a vacuum gel dryer.

Protocol 5 continued

8 Kinase reaction products, phosphorylated β-casein, are visualized by autoradiography and quantified with a PhosphoImager (Molecular Dynamics) or by scintillation counting of excised phosphorylated β-casein bands.

B NIMA kinase assay using PLM (54–72) peptide substrate

1 Start the kinase reactions as in part A.

2 Terminate the reactions with 1 μl of 0.5 M EGTA.

3 Briefly centrifuge the samples and transfer 5 μl of the reaction mix to phosphocellulose P81 filter discs.

4 After the discs are completely dry wash the discs several times in 75 mM phosphoric acid until blanks have very low background counts as monitored with a Geiger counter.

5 Measure the radioactivity retained in the filter by scintillation counting.

7 Ordering a signal transduction pathway

Signal transduction pathways necessarily consist of three major components: a sensing mechanism, such as a receptor, a signal transduction module, and a downstream effector. The sensing mechanism detects environmental (external or internal) changes and generates a physiological signal. The signal is transduced, propagated, and amplified through the signal transduction module, for example, through a MAP kinase cascade. Finally, the signal is transferred to the effector which elicits the proper physiological changes, such as activation of gene expression.

Genetic approaches have been used to first identify components of signalling pathways through analysis of loss of function mutants in genetically tractable model systems, like *S. cerevisiae*. Genetic analysis followed by cloning and biochemical characterization of individual gene products help establish the signalling pathway and the underlying biochemical mechanisms mediating signal transduction. These approaches have helped to define five MAK kinase signalling pathways in *S. cerevisiae*, which aid our understanding of signalling pathways in other eukaryotic systems. This classical genetic approach is very effective and continues to be used, but is very laborious and time-consuming. Our current understanding of signalling mechanisms in *S. cerevisiae* stems from the cumulative knowledge of more than two decades of painstaking research. Even for the best characterized signalling pathway (pheromone) in this model organism, research is still ongoing to understand how upstream activators act on the MAP kinase cascade and to identify new target proteins of this pathway.

The completion of the genome sequence of the first eukaryotic organism, *S. cerevisiae*, signalled that genomics had come of age. DNA microarray and proteomic technologies spawned from genome sequencing efforts are powerful tools to help rapidly elucidate gene and pathway functions. These genome-based technologies are particularly useful in identifying target genes of a signal transduction pathway, as most of the cellular responses to signal transduction are executed at the transcriptional level. The impact of these new technologies on biological research is now beginning to resonate in all research fields. Eventually, they will revolutionize the way we conduct research. However, a detailed discussion of these genome-based technologies for signal transduction studies is beyond the scope of this chapter.

In fungi, signal transduction is mainly mediated through protein-protein interactions and phosphorylation/dephosphorylation events. These mechanisms of signal transduction can thus be exploited to dissect signalling pathways with the techniques described in this chapter. As many signalling components are normally associated with each other, or undergo signalling-dependent association, co-immunoprecipitation, coupled with advanced mass spectrometry and genome database searching, can be very useful to isolate additional components of the signalling pathway. On the other hand, gel-shift assays can be applied to rapidly determine the dependency of phosphorylation, or dephosphorylation of a downstream component on the activity of upstream kinases. We have used a combination of these techniques to establish the feedback loop of NIMA and $p34^{cdc2}$ kinase and the mechanisms through which they regulate each other's mitotic function during initiation of mitosis (8, 10).

Signalling pathways have been viewed and studied in a linear fashion as important biological effects of signal transduction are induced through linear information transfer. Now it is becoming increasingly clear, however, that signalling pathways do not simply act in an isolated, linear fashion but interact with each other to form a complex signalling network. Often the final biological response is brought about in a concerted manner by several signalling pathways. In fact, several yeast signalling pathways even share common signalling components (19). Elucidation of a signal transduction pathway and the underlying signalling mechanisms is thus an arduous task and requires the application of the full arsenal of investigative tools we have, both classical and genome-based.

Acknowledgements

We thank Dr Jeffrey Radding and Dr Genshi Zhao for critical reading of the manuscript.

References

1. Hunter, T. and Plowman, G. D. (1997). *Trends Biochem. Sci.*, **22**, 18.
2. Dickman, M. B. and Yarden, O. (1999). *Fungal Genet. Biol.*, **26**, 99
3. Hunter, T. (1991). In *Methods in enzymology*, Vol. 200, p. 3, edited by T. Hunter and B. M. Sefton, Academic Press.
4. Lindberg, R. A., Quinn, A. M., and Hunter, T. (1992). *Trends Biochem. Sci.*, **17**, 114.
5. Posas, F., Wurgler-Murphy, S. M., Maeda, T., Witten, E. A., Thai, T. C., and Saito H. (1996). *Cell*, **86**, 865.

6. Hanks, S. K., Quinn, A. M., and Hunter, T. (1988). *Science*, **241**, 42.
7. Winzeler, E. A., Shoemaker, D. D., Astromott, A., Liang, H., Anderson, K., Andre, B., *et al.* (1999). *Science*, **285**, 901.
8. Ye, X. S., Xu, G., Pu, R. T., Fincher, R. R., McGuire, S. L., Osmani, A. H., *et al.* (1995). *EMBO J.*, **14**, 986.
9. Ye, X. S., Xu, G., Fincher, R. R., and Osmani, S. A. (1997). In *Methods in enzymology*, Vol. 283, p. 520. Edited by W. G. Dunphy, Academic Press.
10. Wu, L., Osmani, S. A., and Mirabito, P. M. (1998). *J. Cell Biol.*, **141**, 1575.
11. Zachariae, W., Shevchenko, A., Andrews, P. D., Ciosk, R., Galova, M., Stark, M. J., *et al.* (1998). *Science*, **279**, 1216.
12. Osmani, A. H., van Peij, N., Mischke, M., O'Connell, M. J., and Osmani, S. A. (1994). *J. Cell Sci.*, **107**, 1519.
13. Ye, X. S., Fincher, R. R., Tang, A., and Osmani, S. A (1997). *EMBO J.*, **16**, 182.
14. Osmani, A. H., O'Donnell, K., Pu, R. T., and Osmani, S. A. (1991). *EMBO J.*, **10**, 2669.
15. Lu, K. P., Osmani, S. A., and Means, A. R. (1993). *J. Biol. Chem.*, **268**, 8769.
16. Lu, K. P., Kemp, B. E., and Means, A. R. (1994). *J. Biol. Chem.*, **269**, 6603.
17. Songyang, Z., Blechner, S., Haogland, N., Hoekstra, M. F., Piwnica-Worms, H., and Cantley, L. C. (1994). *Curr. Biol.*, **4**, 973.
18. Songyang, Z., Lu, K. P., Kwon, Y. T., Tsai, L. H., Filhol, O., Cochet, C., *et al.* (1996). *Mol. Cell. Biol.*, **16**, 6486.
19. Madhani, H. D. and Fink, G. R. (1998). *Trends Genet.*, **14**, 151.

Chapter 10
Electron microscopy of filamentous fungi

Richard J. O'Connell and Raffaella Carzaniga
IACR-Long Ashton Research Station, Department of Agricultural Sciences, University of Bristol, Long Ashton, Bristol BS41 9AF, UK.

1 Introduction

Transmission electron microscopy (TEM) and scanning electron microscopy (SEM) have made major contributions to our understanding of fungal cell biology. With the application of modern cryopreparation techniques, such as freeze-substitution, and advances in instrumentation, such as field-emission SEM, electron microscopy continues to generate a wealth of new information on the structure and organization of fungal cells. In addition, cytochemical techniques such as immunogold labelling provide a link between structure and biological function by localizing cell components with exquisite specificity and precision. Despite the obvious advantages of confocal microscopy, especially for live cell imaging, optical methods can never approach the spatial resolution offered by electron microscopy. Now that complete fungal genomes are being sequenced, electron microscopy is set to play an important role in helping to ascribe functions to the vast number of genes identified, for example by revealing the morphology of gene replacement mutants and the localization of gene products.

The range of techniques now available for TEM and SEM is enormous and in this chapter we can only present a small selection of the most widely used procedures. For each of these, we have indicated the type of information they can provide to the mycologist, as well as the technical problems associated with their application to filamentous fungi.

2 Conventional preparation techniques for TEM

For critical morphological studies, conventional chemical fixation and dehydration techniques have been largely superseded by cryofixation and freeze-substitution, for the reasons outlined in Section 3. However, not all workers will have access to the equipment required for these techniques, and the use of conventional preparation is still justified for samples that are too large to cryofix adequately and where structural preservation may not be a priority, for example some cytochemical studies. Here we provide a set of basic protocols for fixation, dehydration, embedding, sectioning, and staining, which should serve as a starting point for the beginner. For further information, the reader should consult refs 1–3.

2.1 Chemical fixation and dehydration

A two-step fixation procedure is normally used for morphological studies, comprising pre-fixation with glutaraldehyde (1–5%), which crosslinks proteins, and post-fixation with osmium tetroxide (1–2%), which crosslinks both lipids and proteins. Some workers prefer to supplement the glutaraldehyde with formaldehyde (1–4%), which penetrates cells more rapidly but crosslinks proteins less strongly. The fixatives are usually diluted in sodium phosphate or sodium cacodylate buffers (0.05–0.1 M, pH 6.8–7.2). The preparation and use of these fixatives is detailed in *Protocols* 1 and 2. Following osmium fixation, samples are optionally treated with uranyl acetate (0.5–2%) to increase contrast and improve the preservation of certain structures. None of the commonly used embedding resins are soluble in water. It is therefore necessary to gradually remove water from the fixed cells by transfer through increasing concentrations of an organic solvent, usually ethanol or acetone (*Protocol* 2).

Protocol 1

Preparation of fixative solutions

Equipment and reagents

- 25% (v/v) glutaraldehyde solution (EM grade)
- Paraformaldehyde
- 0.1 M buffer (sodium phosphate or sodium cacodylate, pH 7.2)
- Deionized or distilled water (dH_2O)
- Osmium tetroxide (OsO_4)
- 10 M sodium hydroxide

A Preparation of 2.5% (v/v) glutaraldehyde in 0.05 M buffer

1 Mix 5 ml of 25% glutaraldehyde, 25 ml of 0.1 M buffer, and 20 ml of dH_2O.

B Preparation of 1% (w/v) osmium tetroxide in 0.05 M buffer

1 Remove the adhesive label from the glass ampoule and clean the surface with ethanol. To collect all the OsO_4 in the bottom of the ampoule, immerse in hot water until the crystals melt.

2 Open the ampoule containing 0.1 g OsO_4 and drop into a clean, tightly-stoppered glass flask with 5 ml dH_2O.

3 Cover the flask with aluminium foil and allow at least 24 h for the crystals to dissolve at RT, to give a 2% OsO_4 stock solution.[a]

4 Prepare the working solution by mixing equal volumes of 2% OsO_4 stock solution with 0.1 M buffer solution.

Protocol 1 continued

C Preparation of 10% (w/v) stock solution of formaldehyde

1 Add 2.5 g of *para*-formaldehyde powder to 20 ml of dH_2O and heat to 70 °C on a hot plate with continuous stirring.

2 Add a few drops of 10 M NaOH, until the solution changes from milky to clear.

3 Cool to RT, adjust to pH 7.2–7.4 with 1 M HCl, and make up to a final volume of 25 ml.

D Preparation of 4% (v/v) formaldehyde and 0.1% (v/v) glutaraldehyde in 0.05 M buffer

1 Mix 20 ml of 10% formaldehyde, 200 µl of 25% glutaraldehyde, 25 ml of 0.1 M buffer, and 4.8 ml of dH_2O.

2 Use within 2–3 h after preparation (formaldehyde solutions are unstable).

[a] Alternatively, pre-prepared aqueous solutions of OsO_4 (2% or 4%) are available commercially.

Protocol 2

Chemical fixation and dehydration

Equipment and reagents

- Glass vials with press-on lids (15 ml)
- Fixative solutions (see *Protocol 1*)
- 0.05 M buffer (sodium phosphate or sodium cacodylate, pH 7.2)
- 0.5% (w/v) aqueous uranyl acetate
- Ethanol/water series: 10%, 25%, 50%, 70%, 90% (v/v)
- Absolute ethanol (dried over molecular sieve type 4A)

A Fixation of mycelia

Exchange solutions with Pasteur pipettes and perform steps 2–7 with continuous gentle agitation (2 r.p.m.) on a rotary mixer.

1 Grow cultures on suitable agar media in Petri dishes. To obtain monolayer cultures, mycelium can be grown on sterilized cellophane membranes over nutrient agar (4).

2 Flood cultures with 2.5% buffered glutaraldehyde and fix for 1–2 h at RT. If necessary, add a drop of Tween 20 detergent to promote wetting of hydrophobic mycelia.

3 Excise small blocks (1–2 mm thick) of mycelium with a minimal amount of agar or cut cellophane into small squares (e.g. 2 × 2 mm).

4 Transfer samples to glass vials and wash in buffer (2 × 10 min).

5 Post-fix in 1% buffered OsO_4 for 2 h at RT.

6 Wash in buffer (2 × 10 min).

7 Stain the samples in 0.5% (w/v) aqueous uranyl acetate for 2–4 h at RT or overnight at 4 °C (optional step).

Protocol 2 continued

8 Dehydrate in an ethanol/water series (10%, 25%, 50%, 70%) for 30 min each step, then in 90% for 1 h, and absolute ethanol for 1 h (three changes).

B Fixation of cell suspensions and protoplasts

1 Proceed as described in *Protocol 2*, part A, but perform steps 1–4 in centrifuge tubes and exchange solutions by centrifugation/resuspension. For the primary fixation, resuspend a pellet of cells in 2.5% buffered glutaraldehyde, supplemented with an appropriate osmoticum in the case of protoplasts (osmoticum can be omitted from all subsequent steps).

2 To facilitate further processing, embed the fixed, washed cells in 2% (w/v) low gelling temperature agarose. After solidification on ice, cut the agarose into small blocks (1 mm^3) and transfer into glass vials for dehydration (see *Protocol 2*, part A, step 8).

2.2 Resin embedding

Dehydrated samples must be embedded in a resin matrix to provide support during sectioning and observation in the TEM (*Protocol 3*). The two major classes of embedding media are epoxy resins and acrylic resins. The epoxy resins generally offer better ultrastructure for morphological studies. Of these, Spurr and Quetol resins are much less viscous than Epon and Epon/Araldite, providing shorter infiltration times and greater ease of use. However, Spurr resin has fallen out of favour in some laboratories due to its greater toxicity. Acrylic resins are more suitable for immunocytochemistry because they are more hydrophilic, less reactive with tissue components, and cause less extraction than epoxy resins. Their extremely low viscosity allows good infiltration at low temperatures. Minimum working temperatures for the main types of acrylic resin are as follows: LR White (−20 °C), LR Gold (−25 °C), Unicryl (−50 °C), Lowicryl K4M (−35 °C), Lowicryl HM20 (−50 °C), Lowicryl K11M (−60 °C), and Lowicryl HM23 (−80 °C). One advantage of these resins is that they can also be polymerized at low temperature using UV or white light, whereas epoxy resins must be polymerized with heat, which denatures some antigens. Refs 5 and 6 contain detailed protocols for low temperature embedding in LR White and Lowicryl resins.

Protocol 3

Resin embedding and heat polymerization

Equipment and reagents

- Embedding oven (under a fume hood)
- Specimen moulds (flat embedding moulds and/or capsules)
- Glass microscope slides
- Spurr resin kit (medium grade)[a]
- Permanox plastic microscope slides or coverslips[b]
- Glass vials with press-on lids (15 ml)
- Propylene oxide
- 2% (v/v) dimethyldichlorosilane

Protocol 3 continued

A Resin embedding for chemical fixation samples

Exchange solutions with Pasteur pipettes and perform steps 1–4 with continuous gentle agitation (2 r.p.m.) on a rotary mixer.

1. Transfer fully dehydrated samples (see *Protocol 2*) from absolute ethanol through a 1:1 mixture of ethanol and propylene oxide (30 min) to pure propylene oxide (30 min).[c]
2. Transfer samples from propylene oxide through a series of propylene oxide/Spurr resin mixtures (3:1, 1:1, 1:3); infiltrating for 1 h at each step.
3. Infiltrate with 100% resin monomer for 1 h with the vial lid removed to allow the final traces of propylene oxide to evaporate.
4. Replace with fresh 100% resin monomer twice a day for two to four days.[d]
5. Transfer samples with a wooden cocktail stick into polyethylene capsules. If specimen orientation is important, use polyethylene flat embedding moulds or follow *Protocol 3*, part C.
6. Polymerize in a pre-warmed embedding oven at 70 °C for 8–10 h (or according to the manufacturer's instructions).

B Resin embedding for freeze-substituted samples

1. Transfer the substituted samples (see *Protocol 10*) from the Eppendorf tubes into glass vials containing anhydrous acetone.
2. Infiltrate the samples with Spurr resin[a] through the following series of resin/acetone mixtures: 10%, 20%, 30%, 40%, 50%, 60%, 70%, 80% 90%, 100% (1 h in each step).[e]
3. Continue as in *Protocol 3*, part A, steps 4–6.

C Flat embedding between microscope slides[f]

1. Dip clean glass microscope slides for several minutes in a silanizing reagent, e.g. Dow Corning 'Repelcote' (2% dimethyldichlorosilane). Rinse the silanized slides with ethanol and dH_2O and polish off any residues with a lint-free cloth.
2. Mount samples in a small drop of resin on a silanized slide. In the case of cell monolayers on membranes, ensure the cells face down (check the orientation with a light microscope).
3. Place a second silanized slide on top of the first, using strips cut from Permanox coverslips or slides as spacers. Avoid trapping air bubbles in the sandwiched layer of resin.
4. Polymerize the resin, taking care that the slide lies in a level position inside the oven.

Protocol 3 continued

5 Separate the wafer of polymerized resin from the slides using a razor blade.

[a] This protocol also works with other epoxy resins, such as Epon, Araldite (or mixtures of the two), and Quetol. We recommend the purchase of all epoxy resins as pre-measured kits to minimize exposure to their hazardous components. Mix all components thoroughly according to the manufacturer's instructions and dispense into 10 ml disposable syringes. Surplus resin can be stored frozen at −20 °C, but must be defrosted thoroughly before use to avoid contamination from condensed water, and should be used within two months since the viscosity of the stored resin gradually increases.

[b] Permanox is resistant to all solvents and resins used for microscopy and separates easily from polymerized resin. Thermanox has similar properties.

[c] Because ethanol is only slowly soluble in epoxy resins, propylene oxide is used as an intermediate solvent. Propylene oxide is highly volatile so ensure that the specimens do not dry out.

[d] Samples that are large or difficult to embed (e.g. cells with thick, melanized walls or infected plant tissues) may require infiltration for up to four days.

[e] Resin infiltration should be more gradual with freeze-substituted samples to prevent cell collapse and separation of the cell wall and cytoplasm during sectioning.

[f] Adapted from ref. 7.

2.3 Sectioning

Excellent descriptions of how to prepare ultra-thin sections are provided in refs 2 and 3. Sections through particular fungal cell types or infection sites are almost impossible to obtain by random sectioning, particularly if the desired cells are widely spaced. *Protocol 4* and refs 7 and 8 describe three different approaches that can be used to select and orient cells for sectioning by light microscopy. All these methods require that samples are polymerized in flat embedding moulds or between microscope slides (*Protocol 3*, part C). Mounting sections on Formvar-coated single-slot grids rather than mesh grids has the advantage that no part of the section is obscured by grid bars, which is especially important for viewing ribbons of serial sections. For mounting, it is convenient to use an uncoated slot grid as a loop to pick up the sections, which are then dried down onto a Formvar film stretched over a perforated aluminium platform, called a 'Domino' rack (see *Protocol 5* and ref. 9). Formvar films can be carbon-coated for greater stability in the electron beam (1).

Protocol 4

Selection and orientation of flat-embedded cells for sectioning[a]

Equipment and reagents

- Ultra-microtome and glass knives
- Glycerol
- Light microscope with calibrated eyepiece micrometer

Protocol 4 continued

Method

1 Mount the resin wafer (from *Protocol 3*, part C) in a drop of glycerol on a microscope slide and apply a coverslip to the surface of the wafer closest to the embedded cells.[b]

2 Select cells at desired stages of development by light microscopy. If prepared by freeze-substitution, select those cells without visible freezing damage using a × 100 phase-contrast objective (10). To facilitate relocation of the target cell(s), cut reference marks along the edge of the resin wafer with a razor blade and draw a map (or take a photograph).

3 Remove the glycerol with water and mount the wafer in an ultra-microtome using a vice-type specimen holder.

4 Roughly trim the wafer with a double-edged razor blade and glass knife to bring the target cell close to the block face.

5 Remount the wafer in glycerol and use an eyepiece graticule to measure the distance between the block face and target site.

6 Use a glass knife to remove the required number of 1–2 μm sections to bring the target cell precisely to the block face. Tilt the block to obtain the required orientation for sectioning (e.g. longitudinal or transverse).

7 Cut ultra-thin sections with the long axis of the block face perpendicular to the knife edge.

[a] For full details of the procedure see ref. 11.

[b] Mounting the wafers in immersion oil is better optically but glycerol is much easier to clean off. Samples polymerized in flat embedding moulds can be reduced to a wafer (1–1.5 mm thick) by grinding with abrasive papers (11).

Protocol 5

Preparation of Formvar films

Equipment and reagents

- Microscope slides
- 'Domino' rack (perforated aluminium platform)[a]
- EM grids[c]
- 0.3% (w/v) Formvar (polyvinyl formal) in anhydrous chloroform (dried over molecular sieve)[b]

A Preparation of Formvar-coated 'Domino' rack

1 Clean microscope slides with a detergent (e.g. Teepol), rinse carefully, and polish dry with a lint-free cloth.[d]

Protocol 5 continued

2 Dip the slide for 5 sec using coarse tweezers into the Formvar solution.

3 Allow the slide to drain at an angle for a few minutes until dry.

4 Score a rectangle in the coating with a diamond pen (or a single-edged razor blade), about 2 mm from the edges of the slide.

5 Breathe on the slide (scored side) and slowly immerse into a large beaker of dH_2O at an angle of approximately 20°.

6 Once the film has floated free onto the water surface, check the interference colour (a good film should be silver-grey).

7 Grasp a cleaned 'Domino' rack between thumb and forefinger and gently collect the film with a sweeping action.

8 Leave the coated 'Domino' rack to dry in a dust-free area. The Formvar film can then be stored for several months.

B Preparation of Formvar-coated grids

1 Follow part A, steps 1–6.

2 Pick up a mesh grid with fine tweezers and gently drop on to the floating Formvar film.

3 Arrange the grids in a regular pattern with approx. 5 mm spaces between them.

4 Grasp each end of a Parafilm strip and collect the film and attached grids with a sweeping action.

5 Shake off the excess dH_2O gently.

6 Dry the coated grids in a Petri dish.

[a] The 'Domino' rack is a rectangular aluminium sheet drilled with an array of 28 holes (4 mm diameter) and is available from Structure Probe Inc. For a complete description of the apparatus and its use see ref. 9.

[b] For best results, the Formvar solution should be prepared a few days in advance.

[c] Copper grids should be used for morphological studies. Nickel or gold grids should be used for all immunocytochemistry procedures and carbohydrate staining. The choice of mesh size depends upon the application. 100–200 mesh grids are suitable for routine work.

[d] If the slides are too clean the film may not float off at step 6.

2.4 Section contrasting

Ultra-thin sections are normally stained first with uranyl acetate (0.5–2%) and then with lead citrate (*Protocols* 6 and 7). Sections prepared from Spurr resin often require very long staining times with this procedure unless they are pre-treated with 1% barium permanganate for 15–30 sec (12). This pre-treatment also has the advantage of increasing the electron-opacity of fungal cell walls, which otherwise may appear highly electron-lucent and featureless.

Protocol 6

Preparation of Reynold's lead citrate[a]

Equipment and reagents

- 50 ml volumetric flask
- Lead nitrate
- Sodium citrate
- 1 M sodium hydroxide[b]
- CO_2-free dH_2O (boiled for 10 min and allowed to cool)

Method

1 Add 1.33 g lead citrate and 1.76 g sodium citrate to 30 ml boiled dH_2O in a volumetric flask.

2 Shake continuously for 1 min, and intermittently for a further 30 min, to ensure complete conversion to lead citrate.

3 Add 8 ml of 1 M NaOH and dilute to 50 ml with boiled dH_2O.

4 Dispense into 15 ml plastic screw-cap tubes (filled to the top) and tightly seal with Parafilm. The staining solution is stable for up to six months at 4°C. If a white precipitate is visible, discard the solution.

[a] After method given in ref. 13.

[b] Sodium hydroxide should be prepared fresh using boiled dH_2O and carbon dioxide pellets from a freshly opened container.

Protocol 7

Section staining with uranyl acetate and lead citrate

Equipment and reagents

- Grid storage box
- 2% (w/v) aqueous uranyl acetate[a]
- Reynold's lead citrate (*Protocol 6*)[b]
- Freshly distilled or deionized water (dH_2O)

Method

1 Float grids on 50 μl drops of uranyl acetate on a Parafilm sheet and stain for 5–30 min at RT.

2 Wash with a gentle stream of dH_2O from a Pasteur pipette.

3 Carefully remove excess dH_2O by blotting on filter paper.

4 Float grids on 50 μl drops of lead citrate inside a Petri dish containing a few NaOH pellets to absorb carbon dioxide and stain for 2–20 min at RT.[c]

5 Wash with a gentle stream of dH_2O from a Pasteur pipette.

6 Carefully dry the grids by blotting with filter paper, ensuring that all liquid is removed from between the tines of the tweezers.

Protocol 7 continued

7 Place the grids in a grid storage box and view with the TEM as soon as possible after staining (sections lose fine detail and contrast on storage).

[a] The solution is stable for one month at 4 °C. Pass through a 0.2 μm syringe filter before use.

[b] Withdraw a small amount of lead stain from below the surface of the solution using a 1 ml syringe and needle, then pass through 0.2 μm syringe filter before use.

[c] Avoid breathing carbon dioxide onto the lead citrate solution. Staining times must be optimized for particular samples. Generally, LR White and Epon/Araldite sections require 2-5 min, while Spurr sections require 10–20 min.

3 Low temperature preparation techniques for TEM

The aim of fixation should be to preserve cell structure as close as possible to the living state. However, conventional chemical fixatives take several minutes to permeate through a cell and to crosslink cell components. Immobilization of the cytoplasm is therefore slow, allowing redistribution of organelles. It is also well-known that:

(a) Fixatives induce drastic changes in membrane structure and permeability.

(b) The buffer carrier can impose osmotic stress on the cell.

(c) Solvent dehydration may cause shrinkage artefacts (14).

Over the last two decades, cryofixation followed by freeze-substitution has become the preferred method of preparing filamentous fungi for TEM (15). Cryofixation rapidly stabilizes the cytoplasm (within a few milliseconds) by freezing the cell at very low temperatures (below -180 °C). Frozen water within the cell is then replaced (substituted) at low temperatures (around -80 °C) with an organic solvent, usually containing a chemical fixative such as osmium tetroxide. When substitution is complete, the cells can be warmed to room temperature for resin embedding or processed into resin at subzero temperatures.

The more rapid stabilization of cell components and less damaging dehydration provided by freeze-substitution results in greatly superior structural preservation (see *Table 1* and *Figure 1*). Freeze-substitution also avoids some artefactual structures generated by chemical fixation, notably lomasomes. Freeze-substitution has completely revised our view of how apical vesicles and the cytoskeleton are organized within the Spitzenkörper of growing hyphal tips (16, 17). Some forms of cryofixation are even fast enough to capture dynamic events such as exocytosis and endocytosis (14). In addition to improved structural preservation, freeze-substitution results in less extraction and redistribution of water soluble components such as carbohydrates, proteins, and ions, improving the accuracy and sensitivity of localizations by cytochemistry and X-ray analysis. Some antigens that are denatured by chemical fixatives and/or dehydration are also better preserved for immunocytochemistry (see Section 4.2.1).

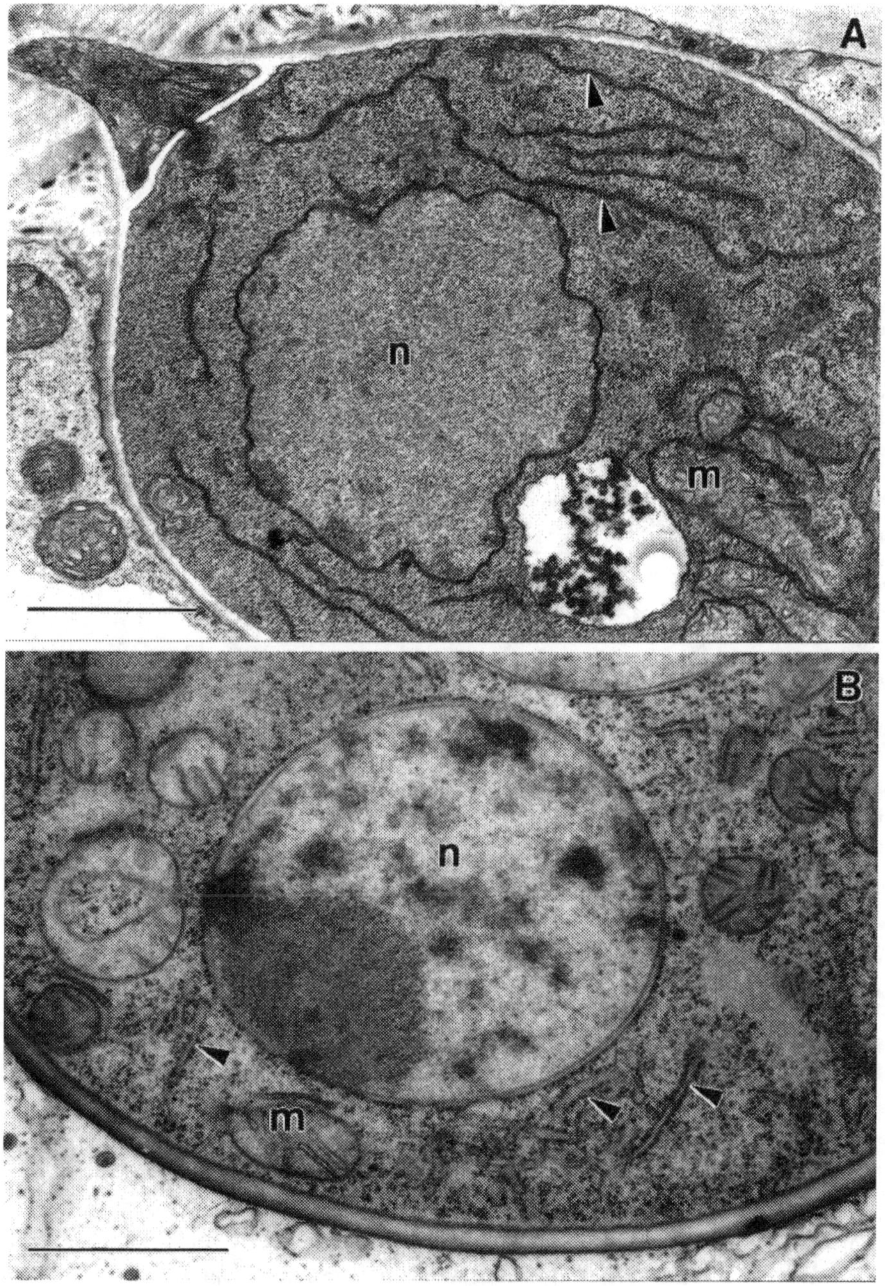

Figure 1 Transmission electron micrographs showing infection vesicles of *Colletotrichum lindemuthianum* infecting epidermal cells of *Phaseolus vulgaris* hypocotyls. Bars = 1 μm. (A) Tissue prepared by conventional fixation in 2.5% glutaraldehyde/1% OsO_4, ethanol dehydration, and embedding in Spurr's resin (*Protocols 2* and *3*). Note membranes of the nucleus (n), ER (arrowheads), and mitochondria (m) are undulating in profile. (B) Tissue prepared by high pressure freezing and freeze-substitution as described in ref. 24. Note smooth membrane profiles and turgid appearance of all organelles.

Table 1 Comparison between preservation of fungal cells by cryofixation followed by freeze-substitution and conventional fixation with glutaraldehyde/osmium followed by dehydration at ambient temperature[a]

Cell component	Cryofixation	Chemical fixation
Organelles	Rounded and 'turgid'	Shrunken
Membrane profiles	Smooth	Undulated
Plasma membrane	Smooth and tightly pressed to wall	Undulated and separated from wall
Double membranes[b]	Evenly spaced	Unevenly spaced
Vacuoles	Fine granular contents	Usually empty
Microtubules	Numerous and well preserved	Few or unrecognizable
Endoplasmic reticulum	Rough type	Smooth type
Golgi equivalents	Abundant and well preserved	Rarely if ever seen
Filasomes	Well preserved	Never seen
Multivesicular bodies	Abundant	Rarely seen
Lipid bodies	Partially or completely extracted	Not extracted
Extracellular mucilage	Retained	Often washed away

[a] Data compiled from refs 17 and 18.

[b] Double membranes of nuclear envelope, mitochondria, and endoplasmic reticulum.

3.1 Cryofixation

Successful cryofixation requires that the cell is frozen without the formation of damaging ice crystals. In practice, this means that any ice crystals should be so small that they are not detectable with the TEM (no larger than 10 nm). In order to achieve this, heat must be extracted from the sample extremely rapidly (10^4 °C/sec). Five different techniques are currently available for cryofixation.

(a) **Plunge freezing**. At its simplest, specimens are held in forceps and plunged by hand as quickly as possible into a cup containing a liquid cryogen, usually propane or a mixture of propane and ethane, cooled with liquid nitrogen to −180 °C or lower. Various commercial devices with mechanical injector systems make the depth and speed of plunging more reproducible. Although cooling rates are slow, plunge freezing is usually adequate to preserve thin samples (10–15 μm), e.g. cell monolayers. It has given good results with many different fungi, probably because their cytoplasm is relatively rich in compounds with cryoprotectant (antifreeze) properties, such as glycerol and other polyols.

(b) **Spray freezing**. An airbrush or atomizer is used to spray fine droplets (5–40 μm diameter) of a cell suspension into a cup of liquid propane. Because the small volume of the droplets optimizes heat extraction, spray freezing offers higher cooling rates than any other technique. It is only applicable to suspensions of small cells in media of low viscosity, e.g. zoospores (18).

(c) **Propane jet freezing**. Two jets of liquid propane are simultaneously ejected onto opposite sides of the specimen at high velocity, e.g. 100 m/sec. To protect the specimen from the force of the jets, it must be sandwiched between two metal plates, which are separated by a thin spacer, such as an EM grid, in order to prevent crushing. Cooling rates are high, and because heat is withdrawn from two sides, specimens up to 40 μm thick can be well-frozen (14). Propane jet freezing gives good results with monolayers and suspensions of fungal cells, and with epidermal strips taken from the leaves of infected plants (19, 20).

(d) **Cold metal block freezing**. The specimen is thrust rapidly against a polished copper block cooled with liquid nitrogen or liquid helium. Cooling rates are high because transfer of heat from the specimen to a solid metal surface is more efficient than to a liquid cryogen (14). However, cells are only well-frozen to a depth of 10–15 μm because freezing is from one side only. Crushing damage can also be a problem with fungal samples (10), although good results were obtained with haustoria and subcuticular mycelia infecting the epidermis of leaves (21).

(e) **High pressure freezing**. All of the above techniques require thin samples (10–40 μm). Fixation of thicker samples (up to 600 μm) is only possible with high pressure freezing (22). The specimen is first sandwiched between two metal holders and then pressurized to approximately 2100 bar just before (20 msec) it is frozen from both sides with two jets of liquid nitrogen, at the same high pressure. Cooling rates are slow, but ice crystal damage is restricted because high pressure reduces the freezing point of water and also slows down ice crystal nucleation and growth. High pressure freezing has allowed the examination of fungal infection structures deep inside the leaves and roots of plants (*Figure 1B*) (23, 24) and large fungal structures, such as ascocarps and teliospores (25, 26).

The choice of freezing method will be determined by the nature of the specimen and the availability of suitable apparatus. Equipment for all the above techniques can be purchased from Leica, Bal-Tec, and RMC. However, the expense and complexity of devices for propane jet freezing, cold metal block freezing and, in particular, high pressure freezing has so far limited the use of these techniques in fungal cell biology. Propane plunge freezing is by far the simplest and cheapest method, requiring equipment that can be easily constructed in a laboratory workshop, and it has been used for the vast majority of freeze-substitution studies on fungi. The basic procedure is outlined in *Protocols 8* and *9*, but the reader is referred to several excellent reviews which provide more detailed protocols, designs for the freezing apparatus, and advice on safety issues (7, 15).

Protocol 8

Sample preparation for propane plunge freezing

Equipment and reagents

- Cellophane (e.g. dialysis tubing)[a]
- Polycarbonate membrane filters (20 μm thick)[b]
- Low gelling-temperature agarose (2%, w/v)
- Platinum or gold-palladium wire (0.25 mm thick)
- Petri dishes with nutrient agar or water agar

A Mycelia growing on nutrient agar[c]

1 Cut cellophane membrane or polycarbonate filters into small pieces (5 × 5 mm) and sterilize by autoclaving between sheets of filter paper.

2 Place the sterile membrane pieces adjacent to the margin of a colony growing on nutrient agar in a Petri dish.

3 Allow the mycelium to grow three-quarters of the way across the membrane.

4 Cut around the membrane with a de-greased double-edged razor blade.

5 Transfer the membrane to a fresh agar plate and leave undisturbed for 30–60 min to recover prior to freezing.

B Germlings and infection structures growing without nutrients[c]

1 Place cellophane or polycarbonate filters[d] (prepared as in part A, step 1) on the surface of 2% water agar in a Petri dish or onto glass slides supported in a humid chamber.

2 Deposit spores on the membrane, either by dusting with dry spores or by applying droplets of spore suspension, and incubate as required.

3 Blot away excess water or medium immediately before freezing to leave the cells just slightly moist.[e]

C Cell suspensions[f]

1 Prepare small wire loops (3 mm diameter with a 2 cm handle) and coat with Formvar (approx. 200 nm film thickness).

2 Dip the coated loops into the cell suspension.[g]

3 Blot off excess liquid with filter paper to leave a thin film over the Formvar and freeze immediately.

D Ungerminated, unimbibed spores

1 Pick up ungerminated spores directly from a sporulating colony by touching the surface with a piece of hydrated cellophane (5 × 5 mm) and freeze immediately.

[a] Dialysis tubing contains sulfide preservatives which should be removed by boiling for 10 min in 2% (w/v) sodium bicarbonate and 1 mM EDTA pH 8.0. Rinse thoroughly in dH_2O before autoclaving.

Protocol 8 continued

[b] Polycarbonate membrane filters (e.g. Nuclepore or Osmonic filters) are available with varying sizes and densities of pores. Select types that are not coated with polyvinyl pyrrolidone.

[c] Adapted from ref. 15.

[d] Other support membranes can also be used, e.g. Mylar, Melinex, Teflon, polyethylene, or aluminium foil (de-greased with chloroform). However, these tend to separate from the resin during sectioning.

[e] This step is critical: water films covering the cells will result in poor freezing, but serious artefacts will be produced if the cells dry out. Monitor drying of the sample with a stereomicroscope.

[f] Adapted from ref. 27.

[g] To minimize possible loss of cells during subsequent processing, they can be embedded in an agar matrix by mixing the cell suspension with an equal volume of 2% (w/v) low gelling temperature agarose (e.g. Sigma Product N° A5030 is molten down to 17 °C). The cells can be concentrated by centrifugation but this may induce artefacts.

Protocol 9

Plunge freezing in liquid propane

Equipment and reagents

- Freezing apparatus (Dewar flask, freezing cup, and compressed air-driven stirrer)[a]
- Fine tweezers (five pairs)
- Cylinder of propane gas (commercial grade)
- Liquid nitrogen (LN_2)
- Small Dewar flask for specimen transfer
- Thermocouple-based digital thermometer (for monitoring propane temperature)

Method

Propane is highly flammable and should be handled and disposed of with extreme caution (see ref. 7). It should be used in a high efficiency fume hood where there is no source of sparks, e.g. electrical devices. Liquid oxygen will gradually condense on the cold propane, producing a potential explosion hazard. Therefore, aim to complete freezing of all samples within 1 h from the start of step 2.

1. Cool the freezing cup with LN_2.
2. Condense liquid propane inside the freezing cup and stir vigorously.[b]
3. Cool the propane to between −185 °C and −191 °C.[c]
4. Pick up the sample (prepared as described in *Protocol 8*) with a pair of tweezers[d] and plunge rapidly (without hesitation and as deeply as possible), into the liquid propane, and hold submerged for a few seconds.[e]
5. Transfer the frozen sample under LN_2 in a small Dewar directly into the freeze-substitution medium (see *Protocol 10*) or into a storage Dewar for long-term storage under LN_2.[f]

Protocol 9 continued

6 Dispose of the liquid propane by tipping into a container inside the fume hood.

[a] Two alternative designs for the freezing apparatus are given in refs 7 and 15.
[b] Continuous stirring is important to minimize temperature gradients, to prevent solidification of the propane, and to allow the propane to be cooled below its freezing point.
[c] Lower temperatures can be achieved with propane–ethane mixtures.
[d] To avoid pre-cooling of the sample, ensure tweezers have returned to ambient temperature before reusing.
[e] It is vital to plunge the sample rapidly through the greatest possible depth (at least 3 cm) of propane to maximize the cooling rate. Any hesitation can result in slow freezing (and hence ice crystal damage) in the cold air above the liquid propane.
[f] During transfer, minimize movement of the frozen sample through the air to prevent warming up.

3.2 Freeze-substitution

During freeze-substitution, frozen water inside the cells is replaced by an organic solvent at a temperature low enough to avoid ice re-crystallization (less than -80 °C). The solvent is usually either acetone or methanol, to which one or more chemical fixatives are added to stabilize the cytoplasm, reduce extraction of tissue components, and improve membrane contrast (10). Osmium tetroxide (1–2%) and uranyl acetate (0.01–0.2%) is the combination most widely used for freeze-substitution of fungi (*Protocol 10*). After substitution, the specimen is gradually warmed to ambient temperature for resin embedding. It is essential that substitution is complete before warming the specimen otherwise it will be damaged by the re-growth of ice crystals. The time required for substitution will depend on the solvent and temperature used and the size of the specimen.

Protocol 10

Freeze-substitution

Equipment and reagents

- Freezers (-80 °C and -20 °C)[a]
- Aluminium block[b]
- Uranyl acetate
- Anhydrous acetone (dried over calcium sulfate or freshly opened HPLC-grade acetone)
- Anhydrous absolute ethanol (dried over molecular sieve type 4A)
- 25 ml Pyrex conical flask with a rubber bung
- Osmium tetroxide
- LN_2

A Preparation of the substitution fluid (1% (w/v) OsO_4 and 0.1% (w/v) uranyl acetate in acetone)

1 Dissolve 0.02 g uranyl acetate in 20 ml acetone in a conical flask on a magnetic stirrer at RT.

Protocol 10 continued

2 Cool the uranyl acetate solution by dipping the conical flask into LN_2 until the acetone just starts to freeze.

3 Open a vial of OsO_4 crystals (0.2 g), add to the cold uranyl acetate solution, and gently agitate the flask until the crystals dissolve.

4 Dispense aliquots (1 ml) of substitution fluid into 1.5 ml Eppendorf tubes and quickly close the lids to avoid condensation of moisture.

5 Equilibrate the tubes for 30 min in the aluminium block, pre-cooled to -80 °C in a freezer.

B Substitution

1 Quickly transfer frozen samples (see *Protocol 9*) from LN_2 into the substitution fluid[c] using a pair of pre-cooled tweezers, keeping the tubes open for the shortest possible time to avoid condensation of moisture. Each tube can accommodate two samples.

2 Substitute the samples for 48–96 h at -80 °C (longer times may be necessary for large specimens).

3 Remove the substitution fluid with a Pasteur pipette and replace with fresh substitution fluid, pre-cooled to -80 °C.

4 Gradually warm the samples by transferring the aluminium block to a -20 °C freezer for 6 h. Then remove the tubes from the block and hold at 4 °C for 1 h, followed by RT for 1 h.

5 Remove the substitution fluid and replace with two changes of anhydrous acetone (15 min each) and process for resin embedding (see *Protocol 3*). For low temperature embedding in acrylic resins, interrupt the warming (step 4) and continue with the solvent rinses and resin infiltration (e.g. LR White or Lowicryl) at the required working temperature. Because acetone inhibits polymerization of these resins, rinse the samples in a 1:1 mixture of acetone and ethanol (15 min) followed by two changes of pure anhydrous ethanol (15 min each). Polymerize at low temperature with UV light as described in *Protocol 12*.

[a] If a -80 °C freezer is not available, use a Dewar flask filled with a mixture of dry ice (solid CO_2) and acetone, pre-cooled with liquid nitrogen. The mixture should maintain a temperature of approximately -79 °C. Automated freeze-substitution devices available from Leica Bal-Tec and RMC offer more flexible programming of the substitution temperature and warming process but are expensive.

[b] An aluminium block drilled with holes to accommodate Eppendorf tubes, e.g. a dry block heater insert.

[c] It is vital that samples do not warm up above the substitution temperature to avoid ice crystal damage.

3.3 Artefacts of cryofixation and freeze-substitution

Although freeze-substitution has many advantages, it is associated with certain artefacts. The most serious of these is ice crystal damage, resulting from slow initial freezing or subsequent re-crystallization of the ice. Solutes become concentrated into regions between the ice crystals (eutectic boundaries), producing a characteristic honeycomb pattern within the cytoplasm (see examples in ref. 10). Mitochondria and the nucleoplasm are the most sensitive damage indicators. It is not unusual for a large proportion of the cells to show freezing damage, and for this reason it is important to select only well-frozen cells for sectioning by light microscopy of flat-embedded samples (*Protocol 4*). Sometimes the cytoplasm separates from the cell wall during sectioning, probably due to poor resin infil-

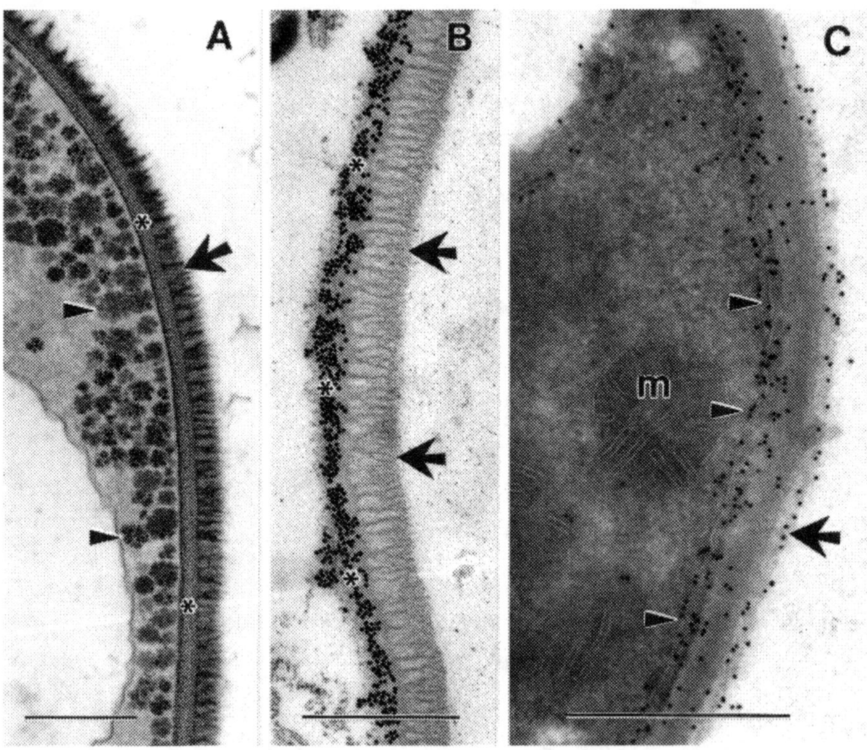

Figure 2 Transmission electron micrographs showing conidia of *Colletotrichum lindemuthianum*. Bars = 0.5 μm. (A) Ungerminated conidium prepared by freeze-substitution (*Protocol 10*), embedded in Epon resin, and stained for carbohydrates (*Protocol 11*). The cell wall (asterisks), glycogen granules (arrowheads), and spore coat (arrow), all stain strongly. The latter is never preserved by chemical fixation. Reproduced with permission from ref. 60. (B) Empty, germinated conidium prepared as in (A) but labelled with wheat germ lectin conjugated to colloidal gold (*Protocol 14*). Chitin in the cell wall (asterisk) is intensely labelled, while the spore coat (arrows) is not labelled. (C) Ultra-thin frozen section through an ungerminated conidium, showing immunogold labelling of glycoproteins recognized by mAb UB20 (28). Labelling is concentrated on the spore coat (arrow) and along the plasma membrane/wall interface (arrowheads). Membranes, e.g. those of mitochondria (m), appear in negative contrast in frozen sections.

tration. Another problem can be excessive loss of lipids during the substitution step, resulting in poor membrane contrast and complete extraction of lipid bodies, which appear as electron-lucent regions.

4 Cytochemical techniques

4.1 PATAg staining of carbohydrates

The periodic acid–thiocarbohydrazide–silver proteinate (PATAg) test (*Protocol 11*) is equivalent to the periodic acid–Schiff stain for polysaccharides used in light microscopy. The test is based on three main steps:

(a) Oxidation of free hydroxyl groups on sugars to form aldehyde groups.

(b) Condensation of the aldehyde groups with thiocarbohydrazide.

(c) Detection of the thiocarbohydrazide binding sites with an electron-opaque silver complex.

Cell walls, plasma membranes, Golgi equivalents, and glycogen usually stain strongly, and this procedure is particularly useful for revealing the microstructure of extracellular matrices, especially in samples prepared by freeze-substitution (*Figure 2A*, ref. 28). Staining indicates the presence of polysaccharides, glycoproteins, or glycolipids containing vicinal hydroxyl groups (1, 2). However, two major components of fungal cell walls, β-1,3-linked glucans and chitin, do not have this structure and so are not stained (29). Thus, a negative reaction does not necessarily indicate an absence of polysaccharide. Further information can be obtained by combining PATAg staining with selective extraction of specific polysaccharides by enzyme digestion (2).

Protocol 11

Staining carbohydrates with PATAg

Equipment and reagents

- Gold grids
- Aqueous solutions of acetic acid: 1%, 5%, 15%, and 20% (v/v)
- 10% (v/v) hydrogen peroxide
- 1% (w/v) periodic acid[a]
- 0.2% (w/v) thiocarbohydrazide (TCH) in 20% (v/v) acetic acid[b]
- 1% (w/v) silver proteinate[c] (TAAB Laboratories Ltd.)

Method

Grids are transferred between drops of reagent using fine tweezers or a platinum loop. All steps are performed in a humid chamber.

1 Collect sections on gold grids.

2 Float the grids on drops of 1% periodic acid for 30 min at RT.[d]

Protocol 11 continued

3 Wash thoroughly on drops of dH_2O (3 × 10 min).

4 Float grids on 0.2% TCH in 20% acetic acid for 30 min up to 24 h at RT.[e]

5 Wash grids in decreasing concentrations of acetic acid, i.e. 20%, 15%, 10%, 5%, and 1% (10 min in each) and then dH_2O (2 × 15 min).

6 Stain with 1% silver proteinate for 30 min at RT in a humid chamber in the dark.

7 Wash grids with dH_2O (2 × 15 min), dry carefully, and observe in the TEM. A fine electron-opaque deposit indicates positive staining (2).

[a] Stable for one month at 4 °C. Corrosive.

[b] Stable for several days at 4 °C. Highly toxic.

[c] Prepare a 1% aqueous silver proteinate solution at least 12 h in advance by sprinkling the powder onto dH_2O in a beaker and leaving to stand for 30 min (avoid stirring). Briefly microcentrifuge before use (2 min at 11 600 g). Protect the solution from light and store for up to one week at 4 °C.

[d] Treat negative control sections with 10% hydrogen peroxide in place of 1% periodic acid. This bleaches sections but does not form aldehyde groups, so that positive staining should be absent.

[e] The duration of this step depends on the carbohydrate to be visualized. Glycogen may require only 30 min treatment, whereas glycoproteins may require up to 24 h.

4.2 Immunogold labelling

Immunogold labelling (IGL) allows cell components such as proteins, glycoproteins, and polysaccharides to be localized with very high spatial resolution. Applications in fungal cell biology are far too numerous to mention here but have been reviewed recently (30). Normally an indirect labelling approach is used, in which binding of an unconjugated primary antibody to the target antigen is detected with a gold-conjugated secondary antibody specific for IgG of the animal species providing the primary antibody (*Protocol 13*). Primary antibodies can also be detected using colloidal gold conjugated with protein A or protein G (31). Post-embedding labelling is by far the most widely used technique, in which antigens are detected on the surface of resin sections prepared from fixed, embedded tissues.

4.2.1 Specimen preparation

Specimen preparation for immunocytochemistry is always a compromise between preserving adequate cell structure and retaining antigenicity. The optimum method must be found empirically for each antibody/antigen combination, although Western blots, dot blots, or ELISA assays can help predict the sensitivity of an antigen to particular reagents (32). The antigenicity of proteins is usually inversely proportional to the level of aldehyde crosslinking, so low concentrations of glutaraldehyde (0.1–1%), possibly supplemented with formaldehyde (1–4%), are preferred for immunocytochemistry (*Protocol 12*). On the other hand, crosslinking must be sufficient to avoid artefactual diffusion or extraction of the antigen. Osmium post-fixation is usually not recommended but is worth trying because a surprising number of antigens survive exposure to osmium, and membrane preservation and contrast are poor without it. In some cases, the antigenicity of osmium-fixed tissues can be restored by 'etching' the thin sections with sodium metaperiodate just prior to immunolabelling (5).

Dehydration at room temperature in organic solvents can extract soluble antigens and causes major conformational changes in proteins by removing water 'hydration shells' from their surface. These damaging effects can be mitigated by progressively lowering the temperature as the concentration of organic solvent increases (the so-called PLT method) or by freeze-substitution (*Protocol 10*). With both techniques, the samples are maintained at low temperature (−35 to −80 °C) during subsequent infiltration with Unicryl or Lowicryl resins and polymerization with UV light (see Section 2.2). For antigens that are sensitive to even low concentrations of aldehyde or osmium, samples can be freeze-substituted in pure solvent (6). Although acrylic resins are usually preferred for immunocytochemistry (see Section 2.2), labelling efficiency for certain antigens, e.g. cell wall polysaccharides, may be higher in Epon-embedded samples (K. Mendgen, personal communication).

Cryosectioning avoids all exposure of antigens to solvents, resins, and heat and provides better antibody access because the antigens are not encased in resin. Following mild aldehyde fixation, tissues are cryoprotected with sucrose (1.6–2.3 M), rapidly frozen, and sectioned at low temperature (3, 32). The sections are then thawed prior to immunogold labelling. It is not easy to cut frozen sections through the cell walls of fungi and the technique requires expensive attachments for the ultra-microtome but it may be the only way to detect certain labile antigens (*Figure 2C*) (28). Pre-embedding labelling has the advantage that labelling is completed before fixation, and ultrastructure is excellent because glutaraldehyde, osmium, and epoxy resins can be used for subsequent processing. Unfortunately, the technique is only applicable to cell surface antigens (32). Immunogold labelling can also be combined with negative staining for labelling fibrillar surface components such as fungal fimbriae (33).

Protocol 12

Fixation and embedding for immunocytochemistry

Equipment and reagents

- Glass vials with press-on lids (15 ml)
- Buffered 4% formaldehyde/0.1% glutaraldehyde (see *Protocol 1*, part D)
- 0.05 M buffer (sodium phosphate or sodium cacodylate, pH 7.2)
- Ethanol/water series: 10%, 25%, 50%, 70%, 90% (v/v)
- Absolute ethanol
- LR White (hard grade)[a]
- Benzoin methyl ether

Protocol 12 continued

Method

Exchange solutions with Pasteur pipettes and perform steps 1–5 with continuous gentle agitation (2 r.p.m.) on a rotary mixer.

1. Fix samples (prepared as in *Protocol 2*, part A or B) using buffered formaldehyde/glutaraldehyde for 1–2 h at RT.
2. Wash in buffer (2 × 10 min).
3. Dehydrate in a pre-cooled ethanol/water series for 30 min each step (10% and 25% at 4 °C; 70%, 90%, 100% at −20 °C).
4. Infiltrate with LR White resin (containing 0.5% (w/v) benzoin methyl ether) through an ethanol/resin series (3:1, 1:1, 1:3) pre-cooled to −20 °C (1 h in each step).
5. Replace with fresh 100% resin, twice a day for two to four days, at −20 °C.
6. Transfer samples into gelatin capsules or polyethylene flat embedding moulds at −20 °C.[b]
7. Polymerize with UV light (350 nm) at −20 °C (see ref. 6 for details).

[a] LR White resin requires no mixing and can be stored for up to six months at 4 °C. For UV polymerization, dissolve benzoin methyl ether in the resin with gentle stirring.

[b] Oxygen must be excluded during the polymerization of LR White, either by using sealed capsules or by covering flat embedding moulds with cling film or a layer of paraffin oil. Alternatively, polymerization can be performed in a chamber flushed with a continuous gentle flow of dry nitrogen gas.

4.2.2 Primary antibodies

Methods for the production and purification of polyclonal antibodies (pAbs) and monoclonal antibodies (mAbs) are given in Chapter 12. Although the majority of commercial antibodies are raised to mammalian antigens, those recognizing highly conserved molecules, such as actin and tubulin, may be suitable for labelling homologous proteins in fungi (34). Commercial pAb and mAb probes specific for β-1,3-linked glucans (available from Sigma-Genosys and Biosupplies Australia, respectively) are particularly useful in studies of cell wall organization (28). For immunocytochemistry, an antibody should be highly specific for its target antigen and have both high affinity and high titre. mAbs can give fewer problems with undesirable cross-reactions because, unlike pAbs, they are not contaminated with non-specific antibodies and they recognize just one epitope. However, mAbs can still give cross-reactions if the same epitope occurs on other unrelated molecules. To obtain meaningful localization data, it is therefore essential to establish the specificity of mAbs and pAbs with Western blots of total protein extracts or other methods (see Chapter 12). After fixation and processing for TEM, pAbs are more likely to recognize their target antigen because they bind multiple epitopes, whereas the single epitope recognized by a mAb could be denatured. It may therefore be necessary to use fixed antigen when screening mAbs for use in immunocytochemistry (35).

Protocol 13

Indirect immunogold labelling

Equipment and reagents

- Nickel or gold grids (optionally Formvar-coated, see *Protocol 5*)
- Primary antibodies
- PBSA: phosphate-buffered saline containing 1% (w/v) BSA and 20 mM sodium azide[a]
- Blocking buffer: PBSA containing 5% (v/v) normal goat serum[a]
- Secondary antibodies conjugated with 10 nm colloidal gold particles[b]
- Uranyl acetate and lead citrate (see *Protocol 7*)

Method

Grids are floated on drops (40 μl) of reagents arranged on a sheet of Parafilm in a humid chamber. Transfer grids between drops using fine tweezers or a wire loop, removing liquid from the previous drop by carefully touching the edge of the grid on a piece of filter paper. Do not allow grids to dry out at any stage of the procedure.

1. Collect ultra-thin resin sections on nickel or gold grids.[c]
2. Incubate the grids on drops of blocking buffer for 30 min at RT.
3. Blot away excess blocking buffer (without rinsing).
4. Transfer the grids to drops of the primary antibody diluted in PBSA and incubate at 4 °C overnight or at RT for 1 h.[d, e]
5. Wash on drops of PBSA (5 × 5 min).
6. Repeat steps 2 and 3 (only if non-specific binding of the secondary antibody is a problem).
7. Transfer the grids to a drop of the gold-conjugated secondary antibody diluted 1:50 in PBSA (or see manufacturer's instructions) and incubate at RT for 1 h.
8. Wash on drops of PBSA (5 × 5 min) and then dH_2O (2 × 5 min).
9. Stain with uranyl acetate and lead citrate (see *Protocol 7*) and observe with TEM. Shorter staining times may be required to image the gold particles with good contrast.

[a] PBSA and blocking buffer should be passed through a 0.2 μm membrane filter before use.

[b] Antibody–gold conjugates are available from many suppliers, including Amersham-Pharmacia, British BioCell International, Aurion, Nanoprobes Inc., and are stable for 12 months at 4 °C.

[c] Always use freshly cut sections. To increase labelling intensity, mount the sections on uncoated grids and immerse the grids in the reagent drops, giving labelling on both sides of the section. To 'unmask' antigens that are poorly detected after osmium fixation, pre-treat the sections with saturated aqueous sodium metaperiodate for 10–30 min and rinse with dH_2O.

[d] Test a series of dilutions of the primary antibody to maximize the signal and minimize background labelling (e.g. 1:10 to 1:10 000 for pAbs, neat to 1:100 for mAb tissue culture supernatants).

[e] Replace the primary antibody with PBSA to identify non-specific binding of the gold conjugate. To test for non-specific binding of the primary antibody, replace it with an antibody to an antigen known to be absent from the tissue, or (for pAbs) replace the primary with pre-immune serum from the same animal. Non-immune serum from a different animal of the same species is less informative because it could display very different immune responses. Other negative controls include pre-adsorption of the primary antibody with excess antigen and labelling tissues known to lack the target antigen (e.g. non-expressing cell types or mutants, and cells from which the antigen has been removed using enzymes). As a positive control to test the entire labelling procedure, use an antibody known to label the tissue.

4.2.3 Gold conjugates

Colloidal gold has become the marker of choice for EM immunocytochemistry. Gold particles are easily distinguished by TEM due to their high electron-opacity and their clearly defined spherical shape. Because the gold particles can be produced in a wide range of non-overlapping sizes (3–150 nm) they can be used to co-localize multiple antigens on the same tissue section (see ref. 32 for a description of double labelling methods). Gold particles also lend themselves to quantification of labelling intensity, i.e. counts of particles per unit area, although it is difficult to relate this to antigen concentration (32). Greatest labelling intensity is obtained with small gold particles (1–5 nm), due to lower steric hindrance, but larger particles (10–20 nm) are much easier to visualize with TEM, especially at low magnifications. However, small particles can be enlarged to improve their visibility by silver enhancement. This simple, one-step procedure involves deposition of metallic silver onto the gold after immunolabelling (36). For conventional SEM, large gold particles (> 30 nm) or silver-enhanced gold particles can be visualized in either secondary electron or back-scattered electron imaging modes (37). Smaller gold particles (5 nm) can be imaged with field-emission SEM (see Section 6). Although colloidal gold can be produced and conjugated to antibodies in the laboratory (2), most workers prefer to use commercial conjugates, which can be purchased with a range of sizes and secondary antibody specificities.

4.2.4 Background suppression

The most frequently encountered problem in immunocytochemistry is unwanted background labelling (32). Non-specific protein-binding sites in tissue sections can usually be blocked by the judicious use of blocking agents, such as BSA (1–3%), fetal calf serum (1–5%), normal goat serum (1–5%), or non-fat dried milk (1–5%). However, blocking agents should only be used when necessary, because many of them also reduce specific binding to the target antigen. Nuclei are a common site of background labelling due to binding of the negatively charged gold particles to positively charged histone proteins. This type of electrostatic binding can be reduced by raising the pH of the buffer from 7.2 to 8.2, increasing the concentration of sodium chloride from 0.15 M to 0.5 M, and adding proteins with a net positive charge, e.g. fish skin gelatin (0.1–1%), or net negative charge, e.g. acetylated BSA (0.1–0.2%, available from Aurion). Background due to hydrophobic interactions can be blocked with Tween 20 detergent (0.05–1%) or hydrophobic proteins, such as casein or acetylated BSA.

Background can be caused by unwanted 'specific' binding when antisera are contaminated with natural antibodies recognizing non-target antigens in the tissue. For example, binding of pAbs to fungal cell walls is a very common problem, probably due to immune responses to fungi encountered by the animal through infections or in food (38, 39). If the contaminating antibodies are of low affinity and low titre, simply increasing the dilution should solve the problem. However, it may be necessary to purify the desired antibodies by antigen affinity chromatography (see Chapter 12) or by depleting the contaminating antibodies through adsorption against tissue powders, whole fungal cells, or isolated cell walls (38).

4.3 Lectin–gold cytochemistry

Lectins are a diverse group of carbohydrate-binding proteins that are of non-immune origin and non-enzymic in function. Several hundred lectins have been identified in all types of organism, and many are available commercially. Lectins offer five main types of sugar-binding specificity: mannose/glucose, *N*-acetyl glucosamine, galactose/*N*-acetyl galactosamine, fucose, and sialic acid. Within these major groupings, lectins vary in their affinity according to the anomeric configuration of the target sugar, the type of linkage involved, and the nature of adjoining sugars (40). Most lectins only recognize the non-reducing terminal sugar residues of polysaccharides and glycoproteins, but some can also recognize internal residues, e.g. Concanavalin A (ConA) and wheat germ agglutinin (WGA).

Lectins have been widely used for both light and electron microscopy in studies on fungal morphogenesis, secretion, and adhesion. They are especially useful for localization of glycoproteins and polysaccharides in cell walls, extracellular mucilage, the endomembrane system, and vacuoles. For example, chitin can be detected with WGA (*Figure 2B*) (29) and mannosylated glycoproteins are labelled by ConA (41). Lectin labelling can be performed either before embedding (pre- or post-fixation) or after resin embedding, on sections. It is compatible with osmium fixation and works well with both epoxy and acrylic resins. Colloidal gold is used to visualize lectin binding sites with both TEM and SEM (42).

A direct labelling approach is normally used, where a lectin adsorbed onto the surface of the gold particles binds directly to the target sugar in the section (*Protocol 14*). More intense labelling is provided by indirect procedures, in which unconjugated lectin bound to the target sugar is detected with colloidal gold conjugated to a glycoprotein for which the lectin has high affinity, e.g. horseradish peroxidase for ConA and ovomucoid for WGA (41, 43). Alternatively, binding of the unconjugated lectin can be detected with a lectin-specific primary antibody followed by a gold-conjugated secondary antibody.

Results must be interpreted with caution. Lectin specificity is usually described in terms of the monosaccharide that most strongly inhibits binding but the lectin may bind to a different sugar residue within a complex carbohydrate on a tissue section (40). Due to their ability to distinguish subtle differences in carbohydrate structure, it is not unusual for a set of lectins with the same nominal monosaccharide specificity to give very different patterns of labelling. A further problem is that some lectins recognize more than one sugar, e.g. ConA binds preferentially to mannose but also binds glucose to a lesser extent.

Protocol 14

Direct lectin–gold labelling

Equipment and reagents

- Nickel or gold grids (optionally Formvar-coated, see *Protocol 5*)
- Lectin–colloidal gold conjugates[a]
- Buffers[b]
- Uranyl acetate and lead citrate (see *Protocol 7*)

Method

Grids are floated on drops (40 μl) of reagents arranged on a sheet of Parafilm in a humid chamber. Transfer grids between drops using fine tweezers or a wire loop, removing liquid from the previous drop by carefully touching the edge of the grid on a piece of filter paper. Do not allow grids to dry out at any stage of the procedure.

1. Collect ultra-thin resin sections on nickel or gold grids.
2. Incubate the grids on drops of lectin–gold conjugate diluted in appropriate buffer for 1 h at RT.[c, d]
3. Wash the grids on drops of buffer (4 × 5 min) and then dH_2O (2 × 5 min).
4. Counterstain the sections with uranyl acetate and lead citrate (see *Protocol 7*) and observe with TEM. Shorter staining times may be required to image the gold particles with good contrast.

[a] A limited range of lectin-gold conjugates can be purchased from Vector Laboratories, EY Laboratories, and Sigma. Alternatively, they can be easily prepared in the laboratory (44).

[b] The choice of buffer for grid washing and dilution of the lectin–gold conjugates will depend on the binding requirements of the lectin (40). If non-specific binding is a problem, supplement the buffer with 1% (w/v) BSA and/or 0.1% (v/v) Tween 20. Pass all buffers through a 0.2 μm syringe filter before use.

[c] Test a series of dilutions of the lectin-gold conjugates (e.g. 100 μg, 10 μg, 1μg, and 0.1μg) protein/ml to determine the optimum concentration. To increase labelling intensity, mount the section on uncoated grids and immerse the grids in the reagent drops, giving labelling on both sides of the section.

[d] As a negative control to check that the labelling is sugar-specific, pre-incubate the lectin conjugate with an excess of its corresponding sugar inhibitor (0.1–0.2 M) for 30 min before use. Because most lectins have a greater affinity for complex carbohydrates, monosaccharides may only reduce, rather than completely inhibit, labelling.

4.4 Enzyme–gold cytochemistry

Enzyme-gold labelling exploits the specificity of an enzyme for its corresponding substrate to localize the latter in tissue sections (45). Many different enzymes have been successfully conjugated with colloidal gold, but those degrading wall polymers are most often used for labelling fungal cells (29, 42). These include chitinase, chitosanase, β-1,3-glucanase, and exo-β-1,4-glucanase (cellobiohydrolase). The latter is probably the best cytochemical probe for the localization of cellulose (43). The enzyme–gold conjugates must be prepared in the laboratory because none are available commercially. The labelling procedure is similar to direct labelling with lectins, although indirect labelling is also possible using anti-enzyme antibodies. Clearly, the specificity of this technique depends on the use of highly purified enzymes.

5 Shadowing and replication techniques

5.1 Freeze-fracture

Freeze-fracture is a powerful technique for imaging the ultrastructure of plasma membranes and intracellular membranes, which can be applied to cell monolayers, cell suspensions, and tissues (46, 47). In brief, cells are cryofixed, fractured at low temperature and high vacuum, and optionally 'etched', which involves sublimation of water from the fractured surface. A replica of the fractured surface is then made by coating with platinum and carbon films. After thawing, the cells are removed from the replica, which can then be examined by TEM (*Protocol 15*, *Figure 3*).

The fracture plane preferentially follows the hydrophobic interior of biological membranes, splitting the lipid bilayer into two halves. Freeze-fracture therefore reveals internal fracture faces rather than true membrane surfaces, although the latter can be exposed by etching. Transmembrane proteins appear as small

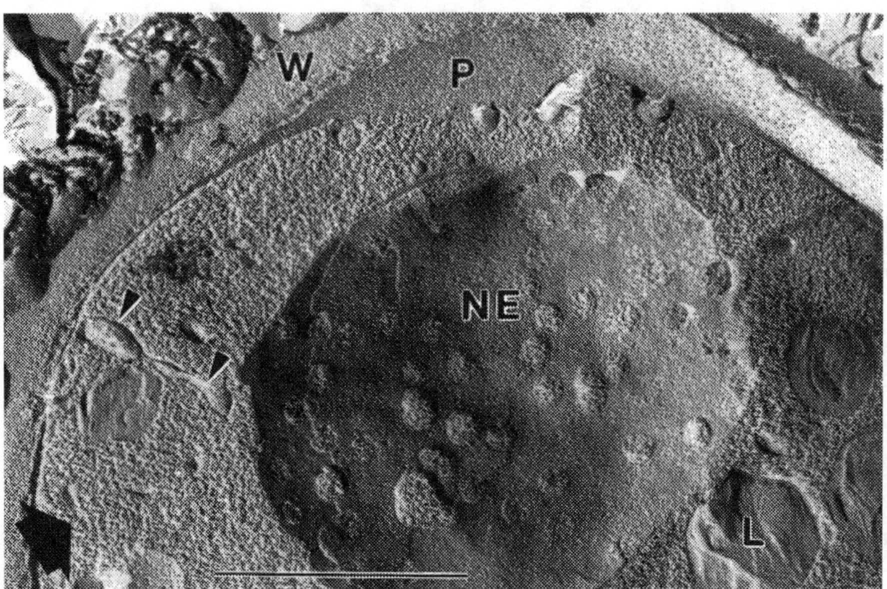

Figure 3 Transmission electron micrograph showing conidium of *Colletotrichum lindemuthianum* prepared by freeze-etching (*Protocol 15*). Note pores in nuclear envelope (NE), lipid globules (L), ER (arrowheads), plasma membrane (P), and cell wall (W). Bar = 1 μm. Arrow indicates direction of shadowing.

bumps (intramembrane particles) protruding from the fracture face. Deep etching of the fracture surface followed by rotary shadowing can reveal details of filamentous structures, e.g. the cytoskeleton and wall microfibrils. Freeze-fracture can also be combined with lectin–gold and immunogold labelling for the localization of glycoproteins and other membrane components (46). In fungal cell biology, freeze-fracture has been used to demonstrate specialized membrane domains at plant–fungal interfaces and in fungicide-treated cells (24, 48). The technique has also been used to visualize hydrophobic rodlet layers on the spores of many different fungi (49). We describe here a widely-used procedure involving fixation with glutaraldehyde and cryoprotection with glycerol, which reduces the risk of ice crystal damage during cryofixation (*Protocol 15*). However, where possible these pre-treatments should be avoided since they can cause changes in the distribution of intramembrane particles, membrane fusions, and phase separations (46).

Protocol 15

Freeze-fracture

Equipment and reagents

- Freeze-fracture apparatus and gold–nickel alloy specimen holders
- Platinum wire loop
- 40% (v/v) chromic acid
- White porcelain cavity plate
- 2.5% buffered glutaraldehyde (see *Protocol 1*)
- Glycerol series: 5%, 10%, 15%, and 20% (v/v) in 0.05 M phosphate buffer pH 7.2

Method

1. Fix samples (optimum size 1 mm^3) in buffered 2.5% glutaraldehyde for 2 h at RT (see *Protocol 2*).
2. Infiltrate samples through a buffered glycerol series (5%, 10%, 15%, and 20%), with gentle agitation on a rotary mixer for 30 min in each step.
3. Mount cryoprotected samples in specimen holders.
4. Freeze the mounted samples by plunging in liquid propane (see *Protocol 9*).[a]
5. Transfer a frozen specimen to the cold stage of the freeze-fracture apparatus.
6. Fracture the specimen with a cold knife (–150 °C) under high vacuum.
7. Optionally, etch the fractured surface for 30 sec to 3 min at a specimen temperature of between -95 °C and -100 °C.
8. Shadow the fractured surface by evaporating a layer of platinum/carbon (1.5–2 nm) at an angle of 45°.
9. Evaporate a supporting layer of carbon at an angle of 90° to the fracture plane (5–20 nm film thickness).
10. Remove the specimen holder from the freeze-fracture apparatus, thaw, and place into a 20% glycerol solution in a porcelain cavity plate.

Protocol 15 continued

11 Clean the replica in 40% chromic acid at RT (one change every 24 h for a total of 72 h).[b] This digestion treatment removes adhering cells from the replica. Carefully transfer the replica between solutions using a platinum wire loop.

12 Transfer the replica to a saturated KCl solution.[c]

13 Rinse the replica several times in dH_2O.[d]

14 Pick up the replica on coated or uncoated 200-mesh copper grids. Allow to dry and observe with TEM.

[a] Alternatively, cryoprotected samples can be frozen in sub-cooled nitrogen (-210 °C). This is prepared by placing a small insulated container of LN_2 into a vacuum chamber evacuated by a rotary pump. When the nitrogen solidifies, air is admitted to the chamber, producing a mixture of liquid and solid nitrogen ('slush'). Frozen samples can be stored under LN_2 until use.

[b] Heating the acid to 50–60 °C can accelerate cleaning.

[c] Transfer through the KCl solution is less damaging to the replica than direct transfer from chromic acid to water.

[d] If required, replicas may be further cleaned by transfer from water to 6% sodium hypochlorite solution for 1–24 h.

5.2 Surface replication

This technique resembles freeze-fracture except that a platinum–carbon replica is made from the surface of dry samples at ambient temperature, without fracturing (*Protocol 16*). It is particularly useful for imaging the hydrophobin rodlet layer on fungal spores (49, 50). Surface replication is technically simpler than freeze-fracture and the equipment required is likely to be available in most EM laboratories. It is also less disruptive to the rodlet layers of some fungi, which can be partially or completely removed when the spores are mounted in aqueous media for freeze-fracture (50). However, freeze-fracture may be preferred in those fungi where outer cell wall layers must be removed in order to reveal an underlying rodlet layer.

Protocol 16
Surface replication

Equipment and reagents

- Vacuum coating unit
- Mica sheets
- 40% (v/v) chromic acid
- Glass Petri dishes (6 cm)

Method

1 Scatter dry conidia (or air dry a cell suspension in dH_2O) onto a sheet of freshly cleaved mica (25 mm × 25 mm).

Protocol 16 continued

2 Place the mica sheet in a vacuum coating unit (or a freeze-fracture machine).

3 Shadow the cell surface by evaporating platinum/carbon at an angle of 45°.

4 Evaporate a supporting layer of carbon at an angle of 90°.

5 Gently lower the coated mica sheets into 40% chromic acid in a glass Petri dish to clean the replica and separate it from the mica.

6 Wash and mount the replica as described in *Protocol 15*, steps 11–13.

6 Scanning electron microscopy

Scanning electron microscopy (SEM) allows the surface topography of specimens to be examined with great depth of field over a wide range of magnifications, e.g. $\times$ 10 up to $\times$ 300 000. The morphology of internal structures can also be examined after fracturing the specimen. In addition, SEM can be combined with analytical techniques, such as immunogold labelling (see Section 4.2) and X-ray microanalysis for the identification and quantification of elements (51). For high magnification work, an instrument equipped with a field-emission electron gun is required. Field-emission SEM offers much higher resolution (1–2 nm) than conventional SEM (6–10 nm), sufficient to resolve intramembrane particles in freeze-fractured membranes (52). Combined with cryopreparation techniques and low accelerating voltages, field-emission SEM can provide images of extraordinary quality and detail (53). With variable pressure SEM and environmental SEM, samples can be imaged at chamber pressures up to 3×10^2 Pa and 3×10^3 Pa, respectively. This allows completely untreated, hydrated specimens to be examined at low accelerating voltage without any conductive coatings (54). These instruments have great potential for live cell imaging, cytochemical localization of extracellular components and observation of spores attached to leaf surfaces, which frequently become dislodged during preparation for conventional and low temperature SEM.

6.1 Ambient temperature SEM

In conventional SEM, samples are examined at ambient temperatures and low pressure ($<10^{-2}$ Pa). Specimen preparation involves chemical fixation, either by immersion in fixative solutions (*Protocol 2*) or by exposure to fixative vapour, followed by dehydration and drying. For critical point-drying, specimens are dehydrated through an ethanol or acetone series and dried in a critical point-drying apparatus using liquid CO_2 as a transition fluid (55). Freeze-substituted material (*Protocol 10*) can also be dried in this way. For freeze-drying, specimens are frozen in sub-cooled nitrogen and slowly dried by vacuum sublimation at -60 to -65 °C in a freeze-drying apparatus or vacuum-coating unit (55). Before examination, the dried specimens are sputter-coated with a thin film (10–12 nm) of gold to reduce electrical charging and to conduct away heat during exposure to the electron beam. For high resolution field-emission SEM, specimens are coated with finer grained films of platinum or carbon, or viewed uncoated at low accelerating voltages (51). A serious disadvantage of ambient temperature SEM is that the removal of water by freeze-drying or critical point-drying causes major artefactual changes, including shrinkage and collapse of cells and surface wrinkling. In addition, solvent extraction during critical point-drying can strip away extracellular mucilage from the surface of fungal cells (56).

6.1.1 Osmium maceration

Internal structures, such as mitochondria and endomembranes, can be viewed by SEM after removal of the cytoplasmic matrix. Cells are fixed with osmium tetroxide, freeze-fractured, macerated in a dilute osmium tetroxide solution, and critical point-dried. This technique has been combined with field-emission SEM to image septal pore caps in *Rhizoctonia* (57).

6.2 Low temperature SEM

Specimens for low temperature SEM are cryofixed and viewed in a fully hydrated or partially freeze-dried condition at temperatures of between −100 °C and −175 °C (51, 56). By avoiding chemical fixation, solvent dehydration, and drying, low temperature SEM eliminates most of the artefacts associated with critical point-drying and freeze-drying. Fungal cells have a turgid appearance and surface structures, such as extracellular matrices, are very well preserved. The main steps of the procedure are outlined in *Protocol 17*.

Protocol 17

Low temperature SEM

Equipment and reagents

- Cryopreparation device[a]
- Cryo-stubs[b]
- LN_2
- Low temperature adhesive, e.g. Tissue-Tek O.C.T. compound

Method

1. Mount the specimen on a cryo-stub with low temperature adhesive mixed with colloidal graphite to improve conductivity. Cell suspensions can be collected on Millipore membrane filters and mounted in the same way (56).
2. Cryofix the specimen by plunging into sub-cooled nitrogen (see *Protocol 15*).[c]
3. Freeze-fracture the specimen on the cold stage of the cryopreparation device using a pre-cooled needle or knife in order to expose internal structures[d] (optional).
4. Transfer the stub to the microscope cold stage (below −150 °C) and check the uncoated specimen at low accelerating voltage for ice contamination.

Protocol 17 continued

5 Partially freeze-dry the specimen by increasing the stage temperature to between −60 °C and −90 °C (optional).[e]

6 Sputter-coat the specimen with a gold.

7 Observe the specimen in the SEM below −150 °C.[f]

[a] Cryopreparation devices are manufactured by Oxford Instruments, Bal-Tec, VG Microtech, and Emitech, and are either directly mounted onto the SEM specimen chamber or are free-standing units (56).

[b] Side views of cell–substratum interfaces can be obtained using angled stubs (45°–90°) (58, 59).

[c] Alternatively, the stub can be placed directly onto the cold stage of the cryopreparation device pre-cooled below −150 °C. This gives much slower freezing, with the attendant problems of ice crystal damage, freeze-drying, and removal of volatile surface films. However, it may be the only way to preserve delicate structures, e.g. the conidial chains of powdery mildews, which are frequently dislodged by plunging into liquid cryogens (59).

[d] Freeze-fracture is particularly useful for imaging fungal hyphae and haustoria within infected leaf tissue (56).

[e] This step may be necessary to remove surface water vapour contamination, or to reveal internal structures after freeze-fracture. The sublimation process can be monitored in real time with the SEM. Note that highly hydrated mucilages around fungal cells may collapse during partial freeze-drying to produce a reticulum of fine threads.

[f] Frozen hydrated specimens are particularly susceptible to beam damage at high magnifications, causing fine cracks to appear in the specimen surface (56).

7 Safety

Many of the substances used in TEM and SEM procedures are extremely hazardous to health and workers should use appropriate protective clothing and containment facilities. Ref. 1 contains an excellent discussion of risk assessment and safety precautions for EM laboratories. Formaldehyde, glutaraldehyde, osmium tetroxide, and sodium cacodylate are all volatile and harmful by inhalation, skin contact, and ingestion. In addition, osmium vapour can seriously damage the eyes and formaldehyde may cause cancer. Lead citrate and uranyl acetate are toxic by ingestion and the latter is also mildly radioactive. Propylene oxide is extremely volatile and flammable, toxic by inhalation and skin contact, and may cause cancer. Acrylic resins can cause sensitization by skin contact and inhalation, and some components of epoxy resins are carcinogenic.

Acknowledgements

We would like to thank Prof. Kurt Mendgen and Dr Alan Beckett for reading the manuscript, and Dr Richard Howard and Mr Barry Thomas for helpful discussions. IACR-Long Ashton receives grant-aided support from the Biotechnology and Biological Sciences Research Council of the UK.

References

1. Robards, A. W. and Wilson, A. J. (ed.) (1993). *Procedures in electron microscopy*, Vols 1 and 2. Wiley, Chichester.
2. Roland, J. C. and Vian, B. (1991). In *Electron microscopy of plant cells* (ed. J. L. Hall and C. Hawes), p. 1. Academic Press, London.
3. Reid, N. and Beesley, J. E. (ed.) (1991). *Practical methods in electron microscopy*, Vol. 13. Elsevier, Amsterdam.
4. Grove, S. N. and Bracker, C. E. (1970). *J. Bacteriol.*, **104** (2), 989.
5. Newman, G. R. and Hobot, J. A. (ed.) (1993). *Resin microscopy and on-section immuno-cytochemistry*. Springer-Verlag, Berlin.
6. Hobot, J. A. (1989). In *Colloidal gold: principles, methods and applications* (ed. M. A. Hayat), Vol. 2, p. 75. Academic Press Inc., San Diego.
7. Howard, R. J. and O'Donnell, K. L. (1987). *Exp. Mycol.*, **11**, 250.
8. Mims, C. W., Richardson, E. A., and Taylor, J. (1988). *Mycologia*, **80**, 586.
9. Moran, D. T. and Rowley, J. C. (1987). In *Correlative microscopy in biology. Instrumentation and methods* (ed. M. A. Hayat), p. 1. Academic Press Inc., NY.
10. Hoch, H. C. (1991). In *Electron microscopy of plant pathogens* (ed. K. Mendgen and D.-E. Lesemann), p. 1. Springer-Verlag, Berlin.
11. Zeyen, R. J. and Bushnell, W. R. (1981). *Can. J. Bot.*, **59**, 397.
12. Hoch, H. C. (1977). *Mycologia*, **69**, 1209.
13. Reynolds, E. S. (1963). *J. Cell Biol.*, **17**, 208.
14. Gilkey, J. C. and Staehelin, A. L. (1986). *J. Electron Microsc. Tech.*, **3**, 177.
15. Hoch, H. C. (1986). In *Ultrastructure techniques for micro-organisms* (ed. H. C. Aldrich and W. J. Todd), p. 183. Plenum, NY.
16. Roberson, R. W. and Fuller, M. S. (1988). *Protoplasma*, **146**, 143.
17. Howard, R. J. and Aist, J. R. (1979). *J. Ultrastruct. Res.*, **66**, 224.
18. Shields, J. P. and Fuller, M. S. (1996). *Protoplasma*, **191**, 84.
19. Hippe, S. (1985). *Protoplasma*, **129**, 52.
20. O'Connell, R. J., Pain, N. A., Hutchison, K. A., Jones, G. L., and Green, J. R. (1996). *J. Microsc.*, **181**, 204.
21. Dahmen, H. and Hobot, J. A. (1986). *Protoplasma*, **131**, 92.
22. Studer, D., Michel, M., and Müller, M. (1989). *Scanning Microsc. Suppl.*, **3**, 253.
23. Rodríguez-Gálvez, E. and Mendgen, K. (1995). *Protoplasma*, **189**, 61.
24. Knauf, G. M., Welter, K., Müller, M., and Mendgen, K. (1989). *Physiol. Mol. Plant Pathol.*, **34**, 519.
25. Czymmek, K. J. and Klomparens, K. L. (1992). *Can. J. Bot.*, **70**, 1669.
26. Roberson, R. W. (1993). *Mycol. Res.*, **97**, 195.
27. Lancelle, S. A., Callaham, D. A., and Hepler, P. K. (1986). *Protoplasma*, **131**, 153.
28. Hughes, B. H., Carzaniga, R., Rawlings, S. L., Green, J. R., and O'Connell, R. J. (1999). *Microbiology*, **145**, 1927.
29. Benhamou, N. (1991). In *Fungal cell wall and immune response* (ed. J. P. Latgé and D. Boucias), p. 205. Springer-Verlag, Berlin.
30. Hardham, A. R. and Mitchell, H. (1998). *Fungal Genet. Biol.*, **24**, 252.
31. Bendayan, M. (1989). In *Colloidal gold: principles methods and applications* (ed. M. A. Hayat), Vol. 1, p. 34. Academic Press Inc., San Diego.
32. Griffiths, G. (ed.) (1993). *Fine structure immuno-cytochemistry*. Springer-Verlag, Berlin.
33. Svircev, A., Smith, R., Gardiner, R. B., Racki, I. M., and Day, A. W. (1986). *Exp. Mycol.*, **10**, 19.
34. Bourett, T. M. and Howard, R. J. (1991). *Protoplasma*, **163**, 199.
35. Hardham, A. R., Gubler, F., and Elliot, J. (1991). *J. Microsc.*, **162**, 305.

36. Hayat, M. A. (ed.) (1995). *Immunogold-silver staining*, p. 2. CRC Press, Boca Raton.
37. Horisberger, M. (1989). In *Colloidal gold: principles, methods and applications* (ed. M. A. Hayat), Vol. 1, p. 217. Academic Press Inc., San Diego.
38. Heller, A. (1995). *Micron*, **26**, 15.
39. Srebotnik, E., Messner, K., Foisner, R., and Pettersson, B. (1988). *Curr. Microbiol.*, **16**, 221.
40. Goldstein, I. J. and Poretz, R. (1986). In *The lectins: properties, functions, and applications in biology and medicine* (ed. I. E. Liener and I. J. Goldstein), p. 33. Academic Press, New York.
41. Bourett, T. M. and Howard, R. J. (1994). *Mycol Res.*, **98**, 769.
42. Latgé, J. P. (1991). In *Electron microscopy of plant pathogens* (ed. K. Mendgen and D. E. Lesemann), p. 43. Springer–Verlag, Berlin.
43. Chérif, M., Benhamou, N., and Bélanger, R. (1992). *Can. J. Microbiol.*, **39**, 213.
44. Benhamou, N. (1989). In *Colloidal gold: principles, methods and applications* (ed. M. A. Hayat), Vol. 1, p. 95. Academic Press, NY.
45. Bendayan, M. (1989). In *Colloidal gold: principles, methods and applications* (ed. M. A. Hayat), Vol. 2, p. 117. Academic Press Inc., San Diego.
46. Severs, N. J. and Shotton, D. M. (ed.) (1995). In *Techniques in modern biomedical microscopy*. Wiley-Liss, NY.
47. Fujikawa, S. (1991). In *Electron microscopy in biology: a practical approach* (ed. J. R. Harris), p. 173. IRL Press, Oxford.
48. Hippe, S. (1985). *Pest. Biochem. Physiol.*, **24**, 161.
49. Talbot, N. J., Kershaw, M. J., Wakley, G. E., de Vries, O. M. H., Wessels, J. G. H., and Hamer, J. E. (1996). *Plant Cell*, **8**, 985.
50. Dempsey, G. P. and Beever, R. E. (1979). *J. Bacteriol.*, **140**, 1050.
51. Jeffree, C. E. and Read, N. D. (1991). In *Electron microscopy of plant cells* (ed. J. L. Hall and C. Hawes), p. 313. Academic Press, London.
52. Walther, P., Hentschel, J., Herter, P., Müller, T., and Zierold, K. (1990). *Scanning*, **12**, 300.
53. Braun, E. J. and Howard, R. J. (1994). *Protoplasma*, **181**, 202.
54. Collins, S. P., Pope, R. K., Scheetz, R. W., Ray, R. I., Wagner, P. A., and Little B. J. (1993). *Microsc. Res. Tech.*, **25**, 398.
55. Cole, G. T. (1986). In *Ultrastructure techniques for micro-organisms* (ed. H. C. Aldrich and W. J. Todd), p. 1. Plenum, NY.
56. Beckett, A. and Read, N. D. (1986). In *Ultrastructure techniques for micro-organisms* (ed. H. C. Aldrich and W. J. Todd), p. 45. Plenum, NY.
57. Müller, W. H., Stalpers, J. A., van Aelst, A. C., van der Krift, T. P., and Boekhout, T. (1998). *Mycologia*, **90**, 170.
58. Clement, J. A., Porter, R., Butt, T. M., and Beckett, A. (1994). *Mycol. Res.*, **98**, 1217.
59. Carver, T. L. W., Thomas, B. J., and Ingerson-Morris, S. M. (1994). *Can. J. Bot.*, **73**, 272.
60. O'Connell, R. J., Pain, N. A., Bailey, J. A., Mendgen, K., and Green, J. R. (1996). In *Histology, ultrastructure and molecular cytology of plant-microorganism interactions* (ed. M. Nicole and V. Gianinazzi-Pearson), p. 79. Kluwer Academic Publishers, Dordrecht.

Chapter 11
Biochemical methods for filamentous fungi

Hans von Döhren
Technische Universität Berlin, Franklinstr. 29, 10587 Berlin, Germany.

1 Introduction

Although biochemical approaches in the study of filamentous fungi may be considered more complex than bacterial or yeast studies, this is not obvious from current protocols. Fairly little attention has so far been given to morphological specialities (except for mycelium and spore structures) and ultrastructural compartmentation. This will hopefully improve in the future with rising interest in cell biological questions. Biochemistry of fungi includes the description of constituents, and here focus is given to proteins and metabolites. At this time, the genomes of filamentous fungi have not been fully sequenced, and/or no genome sequence is as yet available in the public domain. Still, proteomics is entering the field, and will be involved in problem solution in the near future.

Current emphasis is on the functional identification of genes and the characterization of gene products. Still dominating these investigations is the expression of fungal genes in hosts like *Escherichia coli* or yeasts, but fungal hosts will be preferred in the future, due to processing and modification. Protein production in fungi is closely linked to their ultrastructure, and the respective events in post-translational modification, transport, and export. These events are still largely unexplored. In this context, focus will be given to the preparation of proteins from fungal biomass. Protein isolation procedures are essentially identical to other sources, so initial steps will be emphasized. As a class of specific fungal proteins some attention is given to hydrophobins.

Fungal metabolites are of significant industrial importance, and thus basic research in this area is clearly most advanced. Not only penicillins and cephalosporins, but cyclosporins, new antifungals such as pneumocandin and aureobasidin, and antihelminthic depsipeptides are established as emerging products. The class of multienzymes responsible for their biosynthesis represent a unique class of fungal enzymes of unusual structural complexity. It is thus obvious that some protocols included have been taken from these extensive studies.

Increasing attention is given to more complex mathematical approaches of metabolic control analysis. These rely on the accurate determination of the respective parameters, and new approaches including nuclear magnetic resonance and advanced mass spectrometric methods will be of significant help.

2 Preparation of extracts

2.1 Preparation of mycelium

To obtain reproducible results conditions of growth and time of harvesting need to be controlled carefully. Depending on scale or equipment, shake flask cultures or fermentation is usually applied. Therefore, usually submerged cultures are prepared, which may cause some limitations with respect to the expression of metabolic activities. Some activities may require specific growth rates (antibiotics), limitations (siderophores), supplements such as solid materials (corn steep solids, plant cell walls, insect cuticles), or need to be induced (e.g. polymer degrading enzymes). The handling of mycelium prior to extraction is described in *Protocol 1*. Note that extracellular proteins are lost. It may also be suitable to inactivate exoproteases by the addition of proteinase inhibitors, as indicated. Additional comments are found in protocols in the sections.

Protocol 1

Preparation of mycelium from shake flask cultures

Equipment and reagents

- Shake flasks, possibly baffled flasks or metal spirals included for better aeration
- Miracloth (Calbiochem) or handi-wipes for filtration
- Buffer A: 10 mM Tris-HCl, 1 M sorbitol, 1 mM PMSF

Method

1 Using appropriate medium, inoculate conidia to a concentration of 2×10^6 conidia/ml.

2 Incubate at appropriate temperature (e.g. 25, 30, or 37 °C/250 r.p.m.) for 16–20 h.

3 Harvest mycelium through Miracloth and wash with cold ddH_2O; an alternative procedure is filtration through two layers of handi-wipes followed by a wash with cold buffer A.

4 Blot mycelium pad dry and determine the mass. The usual yield is 3–5 g of mycelium per 400 ml culture.

5 Freeze immediately in liquid nitrogen. The mycelium can be stored at −80 °C if desired. Alternatively use directly as fresh material, or freeze with dry-ice, at −80 °C, or least recommended in a −20 °C freezer.

2.2 Extraction of mycelium

Extraction of mycelium presents no problems in most cases. Various procedures have been employed starting from fresh, frozen, shock frozen (liquid nitrogen), or lyophilized material. A convenient method for protein preparation described for *A. nidulans* directly extracts shock-frozen mycelium which has been ground to a fine powder in a stainless steel Waring blender (1). Extraction is performed by stirring mycelium powder in aqueous buffer in a 1 g per 3 ml ratio for 1 h at 4 °C. This procedure has been used to purify the nuclear ribonuclease P complex, a 580 kDa complex of seven polypeptides and one RNA molecule. Lyophilized mycelium also permits storage of large amounts of cells from a defined fermentation; homogenization by grinding with sand is recommended, followed by extraction with 20 volumes of buffer/g dry cell mass (2). Such methods have been used for the preparation of the largest multienzymes known, catalysing the multistep formation of peptides like penicillin, enniatin, and cyclosporin (3).

Extracts from spores have been made far less frequently, and acetone powder methods are described in *Protocol 6* in Section 4. Initial studies on the water repellent rodlet layer of macroconidia, which comprises about 2% of the dry weight of *N. crassa* conidia, have been hampered by solubility problems (4). This field is now progressing with the evaluation of the hydrophobin family of fungal proteins (5, 6). *Protocol 2* provides a procedure for hydrophobin isolation (7).

Protocol 2

Isolation of hydrophobins[a]

Equipment and reagents

- Mortar and pestle
- Centrifuge
- Buffer A: 50 mM sodium citrate pH 6.0, 2% SDS, 1 mM PMSF

Method

1 Grind 1 g of frozen or freeze-dried mycelium in liquid nitrogen.

2 Extract with 4 ml buffer A for 30 min at room temperature.

3 Centrifuge for 10 min at about 10 000 g to obtain mycelium debris.

4 Extract with 0.1% trifluoroacetic acid in 20% acetonitrile for 5 min at room temperature.

[a] The procedure has been adapted from ref. 7.

The protein samples obtained may contain pigments and can be further purified by reverse-phase HPLC. It may be useful to include a washing step with 4 ml hot 4% SDS solution before the final extraction.

3 Preparation of cellular fractions

Whole cells and, in particular, intact mycelium, are inadequate for the study of metabolic processes usually involving various transport steps between compartments. Morphology and thus compartmentation of the mycelium may differ for various strains, especially for industrially significant over-producers. Proteins involved in transport processes can be studied employing membrane fractions. In many cases the localization of proteins can be determined by the isolation of the respective compartments.

3.1 Plasma membranes

Only few experiences are so far available for the isolation of plasma membranes from filamentous fungi. Plasma membrane vesicles devoid of metabolic activities are suitable for the study of transport processes. By fusion with liposomes containing cytochrome *c* oxidase hybrid membranes with a high proton motive force can be prepared, which permits the study of , for example, amino acid uptake systems (8) (*Protocol 3*).

Protocol 3

Isolation of plasma membranes (example *Penicillium chrysogenum*)[a]

Equipment and reagents

- Centrifuge
- Buffer: 25 mM MOPS/KOH pH 7.2, 0.25 M sucrose, 5% glycerol (w/v), 1 mM $MgCl_2$, 2 mM dithiothreitol, 1 mM EDTA, 1 mM PMSF, 0.15 mM tetracaine, 1 μg/ml pepstatin, 1 μg/ml leupeptin
- Cell homogenizer employing glass beads (e.g. Braun MSK or Biospec products, Bartlesville, OK)
- Percoll
- 50% Nycodenz

Method

1. Harvest mycelium and wash with an equal volume of 0.9% NaCl.
2. Suspend cells at about 10 mg/ml in cold (4 °C) buffer.
3. Homogenize with 1 mm diameter glass beads: fill a 75 ml flask containing about 50 g of glass beads completely with the cell suspension, and homogenize using a Braun MSK homogenizer for 2 min at full speed with cooling by liquid CO_2 expansion.
4. Remove cells and debris by 3000 g centrifugation for 5 min in a swing-out rotor.
5. Add to the supernatant Percoll to a final concentration of 24%.
6. Centrifuge at 45 000 g for 30 min.
7. Remove upper band containing sealed plasma membranes, and dilute fourfold with buffer.
8. Remove Percoll by centrifugation at 100 000 g for 2 h.

Protocol 3 continued

9 Collect membrane pellet on top of the Percoll pellet and suspend in buffer to a final volume of about 5% of the homogenate used (membrane suspension).
10 Mix 4.8 ml of the membrane suspension with 3.2 ml of a 50% Nycodenz solution in buffer.
11 Pour the mixture into a thick-walled tube and add carefully on top of each mixture 6 ml of 15% Nycodenz in buffer and 4 ml buffer (total volume 18 ml).
12 Centrifuge in a swing-out rotor at 90 000 g for 90 min.
13 Recover plasma membranes at the interphase of 15% Nycodenz and buffer as a white band and dilute tenfold with buffer.
14 Centrifuge the diluted membrane fraction at 25 000 g for 30 min.
15 Collect pellet by suspending in a small volume of buffer and store in liquid nitrogen.

[a] Method adapted from ref. 8.

3.2 Organelles

Many cell constituents, especially proteins, are found in specific compartments, like vacuoles, mitochondria, peroxisomes or microbodies, and nuclei. Besides verifying their localization, preparation from cell organelles may prove advantageous, especially with respect to protein degradation.

3.2.1 Vacuoles

Vacuoles of fungi are similar in many respects to vacuoles of higher plants or lysosomes of mammalian cells. They contain proteases and nucleases as well as a number of basic amino acids including arginine (9). The vacuolar membrane thus controls the flux of this reserve pool by regulated transport processes involving specific carrier proteins. It has recently been shown that treatment of *N. crassa* mycelium with cupric ions permeabilized both plasma and mitochondrial membranes, permitting uptake studies into vacuoles (10). For the isolation of vacuoles 1 M sorbitol is used as osmoticum, reducing the volume of the vacuoles, making them more resistant to mechanical and osmotic lysis. For the isolation of mitochondria (see below) this procedure is inadequate.

Protocol 4

Isolation of vacuoles from *N. crassa*[a]

Equipment and reagents

- Bead beater (Biospec products)
- Teflon glass homogenizer
- Centrifuge, possibly ultracentrifuge, swing-out rotor
- Buffer A: 10 mM MES/Tris, 1 M sorbitol, 2 mM $MgCl_2$, 1 mM *o*-phenanthroline, 2 mM ATP, 1 mM PMSF pH 6.9

Protocol 4 continued

Method

1 Fill 340 ml chamber to half with 0.3–0.5 mm diameter glass beads wetted with buffer A and add mycelium (10–50 g) in sufficient buffer to fill the chamber.

2 Homogenize with 4 × 30 sec bursts and 1 min rests.

3 Decant content of the chamber and wash beads once with buffer A.

4 Centrifuge combined homogenate and wash for 10 min at 4000 g, discard pellet.

5 Centrifuge supernatant for 20 min at 10 000 g.

6 Suspend pellet in buffer A using two or three strokes of a Teflon glass homogenizer.

7 Layer the suspension on top of an 18 ml gradient of 1.0 M sorbitol to 1.8 M sorbitol in buffer A.

8 Centrifuge for 45 min at 100 000 g in a swing-out rotor. Alternatively if no ultracentrifuge is available, gradients can be made in 15 ml centrifuge tubes by layering the sample on top of 4 ml of 1.8 M sucrose. A continuous 1.8 M sucrose to 1.0 M sorbitol 6 ml gradient can be run for 2 h at 43 000 g in a high speed centrifuge.

9 Collect pellets, resuspend in buffer A, and store at −80 °C.

[a] Adapted from refs 11 and 12.

3.2.2 Mitochondria

Work on mitochondria has led to new insights into the translocation of pre-proteins across biological membranes with the characterization of components of the translocation machinery. The pre-protein translocase of the outer membrane of mitochondria facilitates the recognition, insertion, and translocation of nuclear-encoded mitochondrial pre-proteins., and contains a cation-selective high-conductance channel (13). Certain pathways may be localized within mitochondria, and the presence of proteins within mitochondria may be confirmed employing the green fluorescent protein technology.

Protocol 5

Isolation of mitochondria[a]

Reagents

- Buffer A: 10 mM TES–NaOH pH 7.5, 1 mM EDTA
- Buffer A with 1.2 M sucrose
- 1.2 M sucrose in ddH_2O

Method

1 Follow *Protocol 4*, steps 1–5, obtaining a 10 000 g pellet.

2 Resuspend in buffer A.

Protocol 5 continued

3 Layer 10 ml suspension on top of a gradient of:
   - 10 ml of 1.2 M sucrose
   - 10 ml of 1.2 M sucrose in buffer A

4 Centrifuge for 2 h at 43 000 g.

5 Vacuoles will form a pellet, while mitochondria are collected at the interphase; remove mitochondria with a Pasteur pipette.

6 Dilute with an equal amount of buffer A.

7 Centrifuge at 19 000 g for 20 min.

[a] Adapted from ref. 11.

3.2.3 Nuclei

The preparation of nuclei is of specific interest to isolate DNA-binding proteins, including histones, transcription factors, and DNA-dependent RNA polymerases. *Protocols* 6 and 7 describe the isolation and extraction of nuclei.

Protocol 6

Isolation of nuclei[a]

The following volumes are appropriate for 5–10 g of mycelium mass. Mycelium can be disrupted by glass beads in a bead beater or ground in a mortar and pestle.

Equipment and reagents

- Bead beater or mortar and pestle.
- Centrifuge, ultracentrifuge.
- Buffer A: 50 mM Tris–HCl pH 7.5, 5 mM $MgOAc_2$, 20% glycerol, 5 mM EGTA, 3 mM $CaCl_2$, 1 M sorbitol, 7% Ficoll. Sterilize by autoclaving and store at 4 °C. Immediately prior to use add 1 M DTT to 5 mM and 0.01 volumes each of 100 × protease inhibitors A, B, and C. Store on ice and discard any excess after use.
- Buffer B: 25 mM Tris–HCl pH 7.5, 5 mM $MgOAc_2$, 10% glycerol, 5 mM EGTA. Sterilize by autoclaving and store at 4 °C. Immediately prior to use add 1 M DTT to 5 mM and 0.01 volumes of 100 × protease inhibitors C. Store on ice and discard any excess after use.
- Buffer C: 25 mM Tris–HCl pH 7.5, 5 mM $MgOAc_2$, 10% glycerol, 1 M sucrose. Sterilize by autoclaving and store at 4 °C. Immediately prior to use add 1 M DTT to 5 mM. Store on ice and discard any excess after use.
- Buffer D: 25 mM Tris–HCl pH 7.5, 5 mM $MgOAc_2$, 25% glycerol, 0.1 mM Na_2EDTA. Sterilize by autoclaving and store at 4 °C. Immediately prior to use add 1 M DTT to 3 mM. Store on ice and discard any excess after use.
- 100 × protease inhibitors A: 200 μg/ml aprotinin, 50 μg/ml leupeptin, 250 μg/ml antipain, dissolved in H_2O, store at −20 °C. These protease inhibitors are stable for about 12 h at room temperature.
- 100 × protease inhibitors B: 100 μg/ml pepstatin, dissolved in methanol, store at −20 °C. This protease inhibitor is stable for about 6 h at room temperature.
- 100 × protease inhibitors C: 100 mM PMSF, dissolved in absolute ethanol, store at −20 °C. This protease inhibitor is unstable in aqueous solutions and should be added immediately before use.

Protocol 6 continued

A Bead beater method

1. Break up frozen mycelium pad and place in bead beater with 20 ml of buffer A.
2. Disrupt mycelium with six 30 sec pulses with 60 sec delay between the pulses.
3. Pour contents into beaker on ice and stir gently with a magnetic stirrer.
4. Slowly add 40 ml buffer B.

B Mortar and pestle method

1. Grind mycelium to a very fine powder under liquid nitrogen in a mortar and pestle for 5 min.
2. Place into 150 ml beaker on ice and immediately add 20 ml buffer A. The entire sample will freeze at this stage due to the liquid nitrogen. Allow it to thaw for about 10 min.
3. Once thawed stir gently but thoroughly on ice using a magnetic stirrer. Once slurry is homogeneous slowly add 40 ml buffer B.

C Method continuation for both A and B

1. Transfer to 30 ml Corex tube and centrifuge at 1500 g in a swinging bucket rotor at 2 °C for 7 min.
2. Remove the supernatant and gentle layer over 5 ml pads of buffer C in 30 ml Corex tubes.
3. Pellet nuclei at 30 000 g in a swinging bucket rotor at 2 °C for 25 min.
4. Discard supernatant and resuspend appropriately. If nuclei are to be stored resuspend in buffer D and store at −80 °C. If nuclei will be extracted, proceed as in *Protocol 7*.

[a] Provided by A. Brakhage, see ref. 31.

Protocol 7

Extraction of nuclei[a]

Equipment and reagents

- Buffer E: 15mM HEPES-KOH pH 7.6, 500 mM NaCl, 5 mM $MgCl_2$, 1 mM Na_2 EDTA, 5% glycerol. Add 0.01 volumes of 100 × protease inhibitors B and C.
- 5 × buffer F: 125 mM Hepes-KOH pH 7.6, 200 mM KCl, 5 mM Na_2EDTA, 50% glycerol. Sterilize by autoclaving and store at 4 °C. Immediately prior to use make up 1 × BIND buffer and add 1 M DTT to 5 mM and 100 × protease inhibitor C to 0.1 ×. Store on ice and discard any excess after use.
- Dialysis tube or ultrafiltration equipment (e.g. Amicon Centriprep 10 unit)
- Ultracentrifuge

Protocol 7 continued

Method

1. Resuspend the crude nuclear pellet in 4 ml buffer E. Thorough resuspension is best achieved by aspirating up and down with a micropipettor.
2. Incubate on ice for 30 min with gentle agitation.
3. Pellet the extracted nuclei at 100 000 r.p.m. for 30 min at 4 °C.
4. Dialyse supernatant against 2 × 1 litre of buffer F at 4 °C for 4 h.
5. Concentrate sample to 0.5–1 ml using vacuum dialysis or ultrafiltration through a Centriprep 10 unit.
6. Determine the protein concentration of the sample. Aliquot the sample in 50–100 μl lots and store at −80 °C

[a] Provided by A. Brakhage, see ref. 12.

3.2.4 Microbodies

Peroxisomes, glyoxysomes, and glycosomes have each been classified as microbodies; subcellular organelles with an electron-dense matrix that is bound by a *single* membrane. As a targeting signal the terminal tripeptide sequence SKL-COOH has been identified. A recent example of microbody-involvement is the last step in penicillin biosynthesis, where the acyl side chain exchange takes place within this compartment (14). *Protocol 8* has been taken from this work, providing a mild lytic method to obtain organelles via protoplasts, which have been studied by electron microscopy.

Protocol 8

Preparation of microbodies[a]

Equipment and reagents

- Centrifuge
- Potter homogenizer
- Buffer A: 2 g/litre citric acid pH 6.4, 60 g/litre KCl
- Buffer B: 0.1 M MES pH 7.5, 60 g/litre KCl
- Buffer C: 0.1 M MES pH 7.5, 5 mM DTT, 1 mM EDTA, 25 g/litre NaCl
- Novozyme 234 (Novo Biolabs, Bagsvaerd, Denmark)

Method

1. Harvest mycelium 48–72 h after inoculation, and wash once with 0.9% NaCl.
2. Resuspend 20 g wet mycelium in 100 ml buffer A.
3. Add 100 mg Novozyme 234.
4. Shake suspension for 4 h at 25 °C at 100 r.p.m. for protoplast generation.

Protocol 8 continued

5 Wash protoplasts three times with buffer B.
6 Resuspend protoplasts in buffer C.
7 Lyse cells further by five passages with a Potter homogenizer.
8 Centrifuge at 1000 g for 15 min at 4°C and discard pellet.
9 Centrifuge supernatant at 14000 g for 30 min at 4°C. The pellet contains mitochondria, vacuoles, endoplasmic reticulum, and microbodies. The supernatant may be further fractionated by centrifugation at 100000 g, 1 h, 4°C to obtain soluble proteins (supernatant) and smaller organelles and membrane vesicles (pellet).

[a] Adapted from a procedure developed for *Penicillium chrysogenum* from ref. 14.

4 Protein analysis

4.1 Protein survey

For surveys of total proteins standard procedures of two-dimensional gel electrophoresis will be followed. For detailed techniques see ref. 15, including the mass spectrometric identification of protein spots from the gel. Standard procedures miss large proteins above 200 kDa as well as hydrophobic proteins and peptides. For these, *Protocol 1* can be followed.

4.2 Detection of specific proteins

Defined proteins can be detected with specified antibodies by standard blotting techniques. Wet-harvested conidiospores (5) can be suspended in buffer and lysed by passage through a French Press cell at 20000 psi. The extracts are centrifuged, and both supernatant and pellet subjected to standard SDS electrophoretic techniques. An alternative procedure employing about 50 mg of spores or 1 g of dry mycelium is described in *Protocol 9* (16).

Protocol 9

Small scale protein extraction of conidiospores (A) or mycelium (B)[a]

Equipment and reagents

- Centrifuge
- Cooling bath
- Magnetic stirrer
- Vacuum pump
- Mortar and pestle
- Acetone (−60 °C), ice, ethanol
- Buffer: 50 mM BES pH 7.3 with KOH

Method A

1 Place 50 mg dry spores in a polypropylene centrifuge tube and cool to −60 °C.

Protocol 9 continued

2 Add 10 ml acetone and stir at −60 °C for 5 min.
3 Centrifuge for 5 min at −15 °C at 13 000 g.
4 Place tube in an ice-ethanol bath and remove acetone with a pipette.
5 Place tube with the pelleted spores quickly under vacuum to remove excess acetone.
6 Extract dry acetone powder with 5 ml of buffer by stirring for 30 min at 0 °C.

Method B

1 Grind 1 g mycelium with 2 g of acid washed sand and 5 ml buffer (20 mM potassium phosphate pH 7.2) for several minutes.
2 Transfer the suspension into a polypropylene centrifuge tube at −60 °C.
3 Extract with 40 ml acetone at −60 °C by stirring for 5 min.
4 Centrifuge for 5 min at −15 °C at 13 000 g.
5 Remove excess acetone in vacuum.
6 Extract by resuspending the pellet in 20 ml buffer as above.

[a] Adapted from ref. 16.

4.3 Protein isolation and purification

The following protocols focus on the initial isolation steps, while standard purification protocols should be consulted for further details. These can be found in other volumes of this series (17, 18) as well as in other similar compilations (19–22).

4.3.1 Isolation of extracellular proteins

Proteins may either be excreted into the medium or conidia, grown on solid medium, or may be leached to obtain extracellular enzymes (*Protocol 10*). Proteins excreted in submerged cultures may be precipitated by salting out with ammonium sulfate. It may be helpful to remove pigments by treatment with activated charcoal, as these may interfere with enzyme assays.

Protocol 10

Isolation of an extracellular enzyme from solid cultures (lipase from *N. crassa*)[a]

Equipment and reagents

- Centrifuge
- Freeze-dry equipment
- 0.9% NaCl
- Activated charcoal
- Buffer: 50 mM potassium phosphate pH 7.0, 1 mM DTE, 1 mM PMSF (suggested)

Protocol 10 continued

Method

1 Conidia were harvested from solid cultures and suspended in 0.9% NaCl, filtered, and kept in 0.9% NaCl (0 °C) for 60 min.
2 The suspension is centrifuged for 15 min at 20 000 g, and an orange supernatant obtained.
3 Treat with 50 mg/ml of activated charcoal for 30 min at 4 °C to remove pigments.
4 Centrifuge 10 min at 20 000 g.
5 Lyophilize colourless supernatant.
6 To prepare proteins, dissolve and dialyse against buffer.

[a] Adapted from ref. 23.

4.3.2 Isolation of intracellular soluble proteins

In many cases limited proteolysis presents a major problem in purification attempts. A procedure designed especially for sensitive proteins is given in *Protocol 11*, which has been used for the isolation of the three nuclear RNA polymerases from *N. crassa* (24). In this protocol mycelium is resuspended in fresh medium containing PMSF, and the initial steps are carried out at −15 °C, which helps to avoid proteolytic artefacts of RNA polymerase II. *Protocol 12* provides this approach for less sensitive proteins, which has been successful in the isolation of a peptide synthetase from *Claviceps purpurea* involved in synthesis of ergot peptide alkaloids (25). A convenient method starting from shock-frozen mycelium has been outlined in Section 2. A convenient method starting from lyophilized mycelium is also given in *Protocol 13*. This method has been used routinely in the isolation of peptide synthetases, representing multienzymes with sizes of 400–1700 kDa. This has been useful for a variety of organisms including *Aspergillus nidulans*, *Acremonium chrysogenum*, several strains of Fusaria, *Tolypocladium niveum*, *Beauveria bassiana*, and others. As indicated above, a variety of methods have been employed in the extraction of mycelium, including glass beads in a bead beater or Dyno mill, a French Press cell, sonication, and grinding with quartz sand. Lyophilized mycelium is often conveniently disintegrated by grinding or blending, with or without buffer. The inclusion of glycerol in buffers ranging in concentration from 10–50% has often been found to be useful in extraction and stabilization. Proteolytic degradation is a major problem, and various approaches have been introduced. What will be obvious to the reader from looking at the above protocols, is that no specific method has been favoured. It is generally agreed that addition of PMSF is beneficial, although it does react quite unspecifically with seryl and cysteinyl side chains. So some proteins may be irreversibly modified. Many proteins require thiol reducing agents, such as DTT, DTE, or 2-mercaptoethanol, which will also activate cysteinyl proteinases, indicating a balance between conservation and destruction. Metal proteinases can be inactivated by chelators such as EDTA or EGTA, but the protein of interest might likewise be metal ion-dependent for activity. The effects of more specific proteinase inhibitors such as leupeptin and pepstatin has not been clearly documented. If this problem is not obvious, any of the methods could well work without proteinase inhibitors, or simply a commercial proteinase inhibitor kit like Pefabloc an be employed as a simple solution.

Protocol 11

Isolation of intracellular proteins from fresh mycelium, method for proteins highly susceptible to proteolytic degradation, initial steps at −15 °C[a]

Equipment and reagents

- Incubator shaker
- Vacuum filtration unit
- Blender
- Sonicator
- Centrifuge
- Freeze-dry equipment
- Paper towels
- Extraction buffer: 220 mM Tris–HCl pH 8 (20 °C), 11% glycerol, 22% DMSO, 11 mM $(NH_4)_2SO_4$, 5.5 mM $MgCl_2$, 0.55 mM EDTA, 4.4 mM PMSF, 30 μg pepstatin A (1 mg/ml in MeOH) is added for every 10 g of wet weight mycelium
- 5% polymin P
- 400 mM $(NH_4)_2SO_4$ in 10% glycerol

Method

1. Harvest mycelium by filtration with very gentle suction to avoid drying.
2. Rinse with growth medium at 34 °C.
3. Transfer into one-half the culture volume of growth medium containing 5 mM PMSF (34 °C).
4. Incubate with aeration for 20 min.
5. Filtrate and rinse with ice-cold water under strong suction to cool to 0 °C.
6. Cool to −15 °C with extraction buffer (one to two volumes of wet mycelium).
7. Blot mycelium between paper towels and weigh.
8. Transfer immediately into ten times the wet weight of extraction buffer at −15 °C.
9. Homogenize the hard mat with a blender for 1–2 min at the highest speed.
10. Sonicate (for example in 400 ml portions at 100 watts, 20 kHz, in four to eight 1 min bursts) with stirring (high viscosity) and cooling.
11. Remove cell debris by centrifugation at 10 000 g for 15 min at −15 °C.
12. Bring supernatant to 0 °C and add 0.075 vol. of a 5% polyimin P solution, stir for 5 min. All steps are now carried out at 0–4 °C.
13. Centrifuge at 10 000 g for 20 min and collect pellet.

Protocol 11 continued

14 Resuspend pellet by gentle homogenization, avoiding foaming, in 400 mM $(NH_4)_2SO_4$, 10% glycerol buffer (half the volume of the preceding step).

15 Centrifuge at 16 000 g for 30 min and collect the supernatant.

16 Add 35 g $(NH_4)_2SO_4$/100 ml to the supernatant, and allow to precipitate (1 h to overnight).

17 Centrifuge for 1 h at 16 000 g and collect pellet.

18 Freeze and store pellet at −80 °C for further chromatographic steps.

[a] Adapted from ref. 24.

Protocol 12

Isolation of intracellular proteins from fresh mycelium[a]

Equipment and reagents

- French Press cell
- Centrifuge
- Buffer C: 0.2 M Tris–HCl pH 8.0, 40% glycerol, 10 mM DTE, 0.3 M NaCl

Method

1 Suspend 50 g mycelium in 150 ml buffer.

2 Pass through French Press at 10 000 psi.

3 Centrifuge for 30 min at 15 000 r.p.m. in a Sorvall SS34 or equivalent.

4 Fractionate supernatant by polyimin precipitation and ammonium sulfate fractionation.

[a] Adapted from ref. 25.

Protocol 13

Isolation of intracellular protein from freeze-dried mycelium

Equipment and reagents

- Mortar and pestle
- Alternative: blender and cheesecloth for filtration
- Centrifuge
- Polyimin
- Buffer A: 50 mM potassium phosphate pH 7.0, 1 mM DTE, 1 mM PMSF
- Buffer B: 20 mM Tris–HCl pH 7.5, 5 mM $MgCl_2$, 3 mM EGTA, 10 mM benzamidine, 0.25 μg/ml pepstatin, 0.2 μM aprotinin, 0.5 μg/ml leupeptin, 10 mM 2-mercaptoethanol/g mycelium
- Ammonium sulfate

Protocol 13 continued

Method A

1 Grind dry mycelium with 2% PMSF at room temperature for 5–10 min.
2 Stir the resultant fine powderous material into 20 vol. of buffer A (ice-cold).
3 Centrifuge at 10 000 g for 15 min.
4 Process the supernatant by polyimin precipitation and ammonium sulfate fractionation.

Alternative method

1 Mix dry mycelium with buffer B in a 1:15 ratio.
2 Disrupt in a Waring blender.
3 Filter through cheesecloth.
4 Clear by centrifugation for 30 min at 30 000 g.
5 Process supernatant by column chromatography, e.g. ion exchange resin.

4.3.3 Isolation of membrane proteins

For the preparation of membrane proteins, especially of integral membrane proteins, the use of purified membranes is highly recommended. Detailed protocols can be found in standard references (e.g. 17, 20). As an example the preparation of plasma membranes has been described in *Protocol 3*. Likewise membranes of organelles are prepared. The proteins are solubilized from membrane vesicles. In a typical protocol for the preparation of the mitochondrial protein translocase complex the membrane preparation is treated with 10 mM MOPS buffer pH 7, containing 50 mM KCl and 1 mM PMSF, with 1% digitonin as solubilizer (13). After centrifugation the supernatant can be further fractionated by column chromatography.

5 Metabolite identification

Due to the complex morphological structures of filamentous fungi, different pools of metabolites will be encountered. This fact is generally neglected, but any careful analysis should consider the isolation of organelles, especially vacuoles for local estimates. While usually whole mycelium are extracted, procedures for the isolation of conidiospores are available. Wet-harvested conidiospores of, for example, *N. crassa* are prepared by washing cultures with 0.2% Tween 80, filtering through cheesecloth, and pelleting by centrifugation. The obtained spores are washed with distilled water and lyophilized. Dry harvested spores are obtained by inverting 10 ml slant cultures grown in 18 × 150 mm test-tubes, tapping the sides (26).

5.1 Direct determination of metabolites

A recently introduced procedure makes use of matrix-assisted laser desorption time-of-flight mass spectrometry (MALDI-TOF MS) to directly determine the presence of metabolites in fungal mycelium (27). The procedure has been widely tested with bacterial, fungal, and plant samples, and permits the analysis of microgram amounts of material without extraction, or microlitre samples of culture filtrates. Extractions are performed directly on the sample metal plate of the mass spectrometer employing about 1 μl of a cell suspension in a mixture of water, methanol, and acetonitrile (equal volumes). This solvent composition may be altered for specific kinds of metabolites. Co-crystallized sample with matrix (usually dihydroxybenzoic acid) is energized by laser shots leading to the release of the dissolved metabolites into the gas phase by yet unknown mechanisms (15). Fragmentation of compounds is only rarely observed, and usually molecular ions are detected. Often the procedure yields comparable results to HPLC, but has the additional mass information. Further structural information is obtained if the mass spectrometer is equipped with post-source decay facilities. Thus MS-MS information may be obtained as well.

5.2 Metabolite extraction

Procedures for the extraction of metabolites depend on the method of sample processing. If extracts are to be subjected to enzymatic assay procedures, mycelium or conidiospores should be extracted with aqueous buffers. The example protocol supplied concerns the determination of 5-formyltetrahydropteroylpolyglutamates in *N. crassa* (16) (*Protocol 14*).

Protocol 14

Metabolite extraction for enzymatic assays

Equipment and reagents

- Extraction buffer: 50 mM Hepes pH 7.6 with KOH, 0.2 mM EDTA, 10 mM $(NH_4)_2SO_4$, 50 mM 2-mercaptoethanol, 2% sodium ascorbate
- Sonicator
- Centrifuge

Method

1. Place about 5 mg dry mycelium or spores in a 1.5. ml Eppendorf tube.
2. Add 0.5 ml extraction buffer.
3. Boil for 5 min.
4. Cool on ice.
5. Sonicate.
6. Centrifuge.
7. Dilute supernatant accordingly.

5.3 Isotope labelling of metabolites

For pathway elucidation labelling with stable (^{13}C, ^{15}N, or ^{2}H) or radioisotopes (^{3}H, ^{14}C) is commonly applied. These well established procedures are illustrated with labelling studies of the peptide cyclosporin in *Tolypocladium inflatum* (28, 29). To establish the origin of the constituent amino acids Senn *et al.* grew a producer strain on ^{13}C-labelled glucose. After one week of cultivation the culture was transferred into a defined medium containing a 1:1 mixture of labelled and unlabelled D-glucose. After two weeks the complete broth was mixed with an equal volume of methanol, the mycelium removed by filtration, the filtrate concentrated *in vacuo*, extracted with ethyl acetate, and the hydrophobic peptide chromatographically purified for NMR analysis. The labelling pattern provides information on the biosynthetic pathway leading to the direct amino acid precursors of cyclosporin (28).

To directly study the enzyme systems producing cyclosporin, Lawen and Zocher (29) employed purified protein fractions, which were incubated with the constituent amino acids, ATP, $MgCl_2$, and [^{14}C]S-adenosyl-methionine. As the undecapeptide is methylated in seven positions, [^{14}C]methyl labelled product is obtained. In a typical experiment 100 μl of extract are incubated with the same volume of a mix containing all components required at 25 °C for 15 min. The total mixture is then extracted with ethyl acetate, the extracts concentrated and analysed by TLC followed by autoradiography. This procedure has been applied to detect the presence of a single multienzyme, the cyclosporin synthetase, which produces this peptide.

6 Enzyme analysis

In various applications the direct detection of enzyme activities is required, without the need for purification and characterization. As an example the determination of intracellular neutral trehalases in *A. nidulans* and *N. crassa* is considered (30). Conidia were collected by centrifugation or filtration, washed with distilled water, and suspended in 50 mM potassium phosphate pH 7. Disruption was done with glass beads, and the lysate obtained assayed directly after an activation step including incubation with cAMP. This signal-dependent phosphorylation requires the presence of membranes, and can also be performed in permeabilized cells. The assay measures the release of glucose from trehalose.

Generally enzyme assays not requiring an activation step would be performed after clearing the lysate by centrifugation.

References

1. Han, S. J., Lee, B. J., and Kang, H. S. (1998). *Eur. J. Biochem.*, **251**, 244.
2. Zocher, R., Nihira, T., Paul, E., Madry, N., Peeters, H., Kleinkauf, H., *et al.* (1986). *Biochemistry*, **25**, 550.
3. von Döhren, H., Keller, U., Vater, J., and Zocher, R. (1997). *Chem. Rev.*, **97**, 2675.
4. Beever, R. E., Redgwell, R. J., and Dempsey, G. P. (1979). *J. Bacteriol.*, **140**, 1063.

5. Wessels, J. G. (1997). *Adv. Microb. Physiol.*, **38**, 1.
6. Kershaw, M. J. and Talbot, N. J. (1998). *Fungal Genet. Biol.*, **23**, 18.
7. Nakari-Setala, T., Aro, N., Kalkkinen, N., Alatalo, E., and Penttilä, M. (1996). *Eur. J. Biochem.*, **235**, 248.
8. Hillenga, D. J., Versantvoort, H. J. M., Driessen, A. J. M., and Konings, W. N. (1994). *Eur. J. Biochem.*, **224**, 581.
9. Klionsky, D. J., Herman, P. K., and Emr, S. D. (1990). *Microbiol. Rev.*, **54**, 266.
10. Keenan, K. A. and Weiss, R. L. (1997). *Fungal Genet. Biol.*, **22**, 177.
11. Paek, Y. L. and Weiss, R. L. (1989). *J. Biol. Chem.*, **264**, 7285.
12. Cramer, C. L., Ristow, J. L., Paulus, T. J., and Davis, R. H. (1983). *Anal. Biochem.*, **128**, 384.
13. Kunkele, K. P., Hein, S., Dembowski, M., Nargang, F. E., Benz, R., Thieffry, M., *et al.* (1998). *Cell*, **93**, 1009.
14. Müller, W. H., van der Krift, T. P., Krouwer, A. J., Wosten, H. A., van der Voort, L. H., Smaal, E. B., *et al.* (1991). *EMBO J.*, **10**, 489.
15. Kellner, R., Lottspeich, F., and Meyer, H. E. (1999). *Microcharacterization of proteins*, 2nd edn. Verlag Chemie, Weinheim.
16. Kruschwitz, H. L., McDonald, D., Cossins, E. A., and Schrich, V. (1994). *J. Biol. Chem.*, **269**, 28757.
17. Harris, E. L. V. and Angal, S. (ed.) (1990). *Protein purification applications.* IRL Press, Oxford.
18. Harris, E. L. V. and Angal, S. (ed.) (1989). *Protein purification methods.* IRL Press, Oxford.
19. Coligan, J. E., Dunn, B. M., Ploegh, H. L., Speicher, D. W., and Wingfiled, P. T. (ed.) (1999). *Current protocols in protein science*, updated collection. Wiley, New York.
20. Scopes, R. K. (1994). *Protein purification—principles and practice.* Springer-Verlag, New York, Heidelberg, Berlin.
21. Deutscher, M. P., Simon, M. I., and Abelson, J. N. (ed.) (1997). *Guide to protein purification, Methods in enzymology*, Vol. 182. Academic Press, New York.
22. Rosenberg, I. M. (1996). *Protein analysis and purification: benchtop techniques.* Springer-Verlag, New York, Heidelberg, Berlin.
23. Kundu, M., Basu, J., Guchhait, M., and Chakrabarti, P. (1987). *J. Gen. Microbiol.*, **133**, 149.
24. Armaleo, D. and Gross, S. R. (1985). *J. Biol. Chem.*, **260**, 16169; 16174.
25. Riederer, B., Han, M., and Keller, U. (1996). *J. Biol. Chem.*, **271**, 27524.
26. Guignard, R. and Brody, S. (1983). *Exp. Mycol.*, **7**, 133.
27. Erhard, M., von Döhren, H., and Jungblut, P. (1997). *Nat. Biotechnol.*, **15**, 906.
28. Senn, H., Weber, C., Kobel, H., and Traber, R. (1991). *Eur. J. Biochem.*, **199**, 653.
29. Lawen, A. and Zocher, R. (1990). *J. Biol. Chem.*, **265**, 11355.
30. D'Enfert, C., Bonini, B. M., Zapella, P. D. A., Fontaine, T., da Silva, A. M., and Terenzi, H. F. (1999). *Mol. Microbiol.*, **32**, 471.
31. Heeswijck, R. and Hynes, M. J. (1991). *Nucl. Acids Res.*, **19**, 2655.

Chapter 12
Immunological methods for fungi

Christopher R. Thornton
University of Exeter, School of Biological Sciences, Washington Singer Laboratories, Perry Road, Exeter EX4 4QG, UK.

1 Introduction

Immunological approaches to the study of fungi are well documented, beginning with the studies of Flor, DeVay, and others (1, 2) in the 1960s and earlier of the common antigenic relationships between fungal plant pathogens and their hosts using rabbit polyclonal antisera (pAbs). However, it was not until the advent of hybridoma technology and the production of monoclonal antibodies (mAbs) pioneered by Köhler and Milstein in the mid-1970s (3), that immunological methods began to provide the degree of specificity and reproducibility required by modern day fungal biologists. The complex structure of fungal cell walls and their extracellular glycoproteins, compared to organisms such as viruses, and the complexity of the soil and plant matrices in which they grow and reproduce, meant that traditional approaches employing pAbs found only limited applicability in studies of fungal biology. Hybridoma technology and the production of mAbs overcame many of these problems and revolutionized immunological studies of fungi, allowing detailed investigations of morphology, biochemistry, and physiology in pure culture through to the detection, quantification, and visualization of the components of fungal biomass in naturally infected plant and infested soil samples.

With hybridoma technology, antibody-producing B cells, of mouse or rat origin, are isolated from an immunized animal and fused *in vitro* with a lymphoid tumour cell (myeloma). The resulting hybrid cell (hybridoma) contains characteristics of both parents, that is, the production of specific mAbs and the ability to multiply *in vitro*. Using this technique the humoral immune response, which acts against complex molecules containing many antigenic sites, each of which gives rise to an individual antibody, can be separated into component antibodies. The identification and culture of hybridomas secreting identical antibodies with predefined specificity, allows the production of limitless quantities of highly reproducible antibodies directed against epitopes that are unique to individual genera, species, and even isolates of fungi.

The production of pAbs and mAbs requires an animal house, thereby limiting antibody production to individuals or groups who have access to such a facility.

While the production of human monoclonal antibodies via phage display technology holds enormous potential for immunological studies of fungi, the current restricted access to antibody gene libraries means that, for the foreseeable future, immunoglobulins derived from immunized animals remain the only viable source of antibodies for the fungal biologist.

The broad aim of this chapter is to provide greater accessibility to antibody-based methodology for the fungal biologist. Greater emphasis is given to the routine culture and maintenance of hybridoma cell lines than to detailed immunization regimes and fusion protocols. Standard protocols for a number of routine immunological assays, that have found broad applicability to the study of fungi, are also provided. This chapter will therefore be of greatest relevance to the majority of the readership who do not have routine access to animal house facilities, but who may, during the course of their studies, develop pAbs or mAbs via commercial organizations (Section 7). The reader is referred to Harlow and Lane (4) and other texts throughout the chapter, where further details may be required with regard to specific procedures and applications.

2 Properties of a molecule that make it immunogenic

The first step in the production of polyclonal or monoclonal antibodies is immunization of the host animal with an appropriate antigen. Prerequisites of a good immunogen, i.e. one that elicits strong primary and secondary immune responses, are:

(a) It must have an epitope that can be recognized by the cell surface antibody located on B cells.

(b) It must be degradable.

(c) It must have at least one site that can be recognized simultaneously by a class II protein.

Class II protein are receptors that bind antigen fragments that migrate to the surface of B cells once immunogens have been partially degraded. The antigen fragment–class II protein complex is then bound by a T cell receptor, an essential step in the differentiation of a B cell into a plasma cell (a terminally differentiated B lymphocyte that synthesizes and secretes immunoglobulin). The necessity for both an epitope and a class II–T-cell receptor binding site means that molecules of less than approximately 3000–5000 Daltons are poor immunogens. Such molecules may bind to B cell surface antibodies, but in doing so do not present suitable sites for simultaneous binding of a class II protein and a T-cell receptor. Coupling of small molecules such as synthetic peptides or mycotoxins to larger immunogenic molecules such as bovine serum albumin (BSA) or keyhole limpet haemacyanin (KLH) (4) overcomes this by providing the missing class II–T-cell receptor binding sites and allows the induction of a potent antibody response against the small molecule.

2.1 Selection of fungal material for immunization

Several approaches have been employed for the production of pAbs and mAbs to fungi. It is worth noting that none of these approaches have provided definitive procedures for the production of genus-, species-, or isolate-specific antibodies. However, certain trends in specificity have become apparent. For example, when pAbs are raised to mycelial fragments, extracts from lyophilized mycelia, surface washings of solid cultures, or culture filtrates, they commonly cross-react with both related and unrelated fungi and extracts of host plant tissues. The reason for this is that, unlike viruses that typically possess single unique coat proteins, fungal cell walls are considerably more complex, and, perhaps more importantly, are shrouded in extracellular matrices containing glycoprotein molecules. Glycoproteins are much more immunogenic than simple compounds. It is believed that the repeating carbohydrate epitopes on these immunogens make better targets for binding to B cells. Unfortunately, many fungal glycoproteins contain epitopes that are conserved amongst related and unrelated genera making production of mono-specific pAbs problematic. Indeed, no pAb has yet been raised to a fungus that is unequivocally species-specific and/or non cross-reactive with unrelated fungal genera or plant extracts, but recent developments in recombinant DNA technology may provide a novel means of improving pAb specificity, rather than by cross-absorption which drastically reduces antibody titres or by other less conventional means such as tolerization (5). For example, complementary DNA of the obligate plant parasitic fungus *Polymyxa betae* was selectively isolated from infected sugar beet roots, expressed *in vitro*, and the purified protein used to raise pAbs. Polyclonal antiserum developed using a gene product highly expressed by the parasite *in planta* resulted in the production of a highly specific pAb that did not cross-react with plant tissues or with other micro-organisms found in association with sugar beet, barley, cabbage, and potato roots (6).

Unlike pAbs, mAbs can be developed that are highly specific, recognizing single genera, species, or even isolates of fungi. Nevertheless, this varies considerably from fungus to fungus. Apparent improvements in the likelihood of raising highly specific mAbs to individual fungi have been made using co-immunization techniques (7, 8). Further studies are needed to absolutely confirm this. In the meantime, careful thought and planning prior to immunization remains the most successful strategy for developing specific antibodies and maximizing their usefulness. Careful consideration should be given to the intended final use of the antibodies and, as important, the biology of the fungus in question as this can have a profound influence on the most appropriate immunogen for antibody production. Key areas for consideration might be:

(a) Complexity of the matrix in which the antibodies are to be used. Are the antibodies intended for studies *in vitro*, *in planta*, or in soils or composts?

(b) Is the target antigen expressed in the natural environment and, if so, is its expression determined by the presence of inducers or is it produced constitutively.

(c) How stable is the target antigen and how easily is it extracted?

(d) Antigen distribution. Is it intracellular or extracellular? Is it found in the spores or mycelium or both. Where is it secreted—at the growing hyphal tip or along the length of young and mature hyphae? Are stage-specific antibodies required and do you need to differentiate between hyphae and other differentiated cell types (appressoria, spores, fruit bodies)?

(e) Are the antibodies intended for detection, quantification, or visualization?

(f) Is the antigen produced by the target species only or is it produced by a range of other related and unrelated fungi? Do you need to differentiate between genera or species?

2.2 Preparation of immunogens

With the exception of a minority of studies, most antibodies to fungi have been raised using a 'black box' approach. Simple saline extracts prepared from lyophilized mycelium, surface washings of solid slant cultures, or whole organelles, are injected into an animal. Such heterogeneous populations of fungal antigens typically elicit a very strong immune response over a short period of time due to the presence of highly immunogenic polysaccharides and glycoproteins. A major drawback of this approach is that variability in the immunogenicity of the different antigens often results in some of the molecules dominating the antibody response. Therefore, while this approach to antibody production is extremely effective it is never likely to result in the generation of specific polyclonal antisera. Therefore, a pure antigen preparation, such as a protein band excised from an SDS–PAGE gel, represents the best opportunity for raising relatively specific pAbs. However, not all pure antigens elicit good immune responses and there is no means of testing whether an antigen is immunogenic, other than the obvious empirical test of injecting the antigen and monitoring antibody production by test bleeds. Furthermore, because the antigen is denatured by this method, the resulting antibodies are often denaturation-specific and may not bind to the native antigen.

The magnitude and speed of the immune response to heterogeneous fungal antigen preparations and the ability to produce large volumes of antigen extracts for the purpose of hybridoma screening, continues to make them the most attractive candidates for the production of monoclonal antibodies. Nevertheless, identification and recovery of species-specific mAbs can still only be achieved using highly efficient and stringent selection procedures during the initial stages of hybridoma screening (Section 4.4).

Whole fungal organelles such as conidia, chlamydospores, germ tubes, haustoria, or appressoria are more tangible as antigens as they are visible under the microscope. Particulate antigens such as these can be much more immunogenic than soluble molecules. However, it is important to remember that an animal's immune system reacts to an epitope on the surface of the organelle or to an intracellular component released during antigen processing and not to the organelle as a whole. Consequently, an antigenic determinant on the surface of a conidium may also be present on the surface of a germ tube and the shared epitope may also be conserved on similar structures in other species or genera. Notwithstanding this, mAbs and pAbs have been raised that are able to discriminate between components of fungal biomass within individual genera or species (9, 10). *Protocol 1* describes a number of methods for the preparation of soluble antigens from filamentous fungi. The preparation of whole organelles for immunization is more complex and depends to a large extent on the fungus in question. Procedures for the isolation of specialized organelles (appressoria, infection vesicles, intracellular hyphae, haustoria) of facultative biotrophes such as *Erysiphe* and *Colletotrichum* are described elsewhere (8, 11–13). Similarly, procedures for the preparation of zoospores and cysts of the pythiaceous fungi *Phytophthora* and *Pythium* and resting spores *of Plasmodiophora brassicae* are well documented (10, 14, 15).

Protocol 1

Preparation of fungal antigens for immunization

Equipment and reagents

- Phosphate-buffered saline (PBS): 80 g NaCl, 2 g KCl, 11.5 g Na_2HPO_4, and 2 g KH_2PO_4 per litre dH_2O. This automatically buffers to pH 7.2. It is a 10 × concentrate and should be diluted in a wash bottle before use.
- 1 ml pipette and tips
- Inoculating loop
- Glass wool or Miracloth (Calbiochem)
- Microcentrifuge

A Surface washings from mycelium (16, 17)

1. Grow the fungus on slopes of an appropriate agar.
2. Into slopes of young, actively growing, mycelial cultures pipette 2 ml of PBS. Gently stroke the surface of the culture with a 1 ml pipette tip.
3. Transfer the suspension containing solubilized antigens to a Eppendorf tube and centrifuge at 12 000 g for 10 min to pellet cellular debris.
4. For immunization, transfer saline extract to fresh Eppendorf tubes and add adjuvant if desired (see Sections 3.2 and 4.2).

B Lyophilized mycelium and culture filtrates (18, 19)

1. Grow fungus in appropriate liquid medium to obtain mycelial cultures.
2. Filter cultures through glass wool, snap-freeze mycelium and culture filtrates in liquid nitrogen, and lyophilize.
3. For long-term storage place in sealed plastic bottles in freezer.
4. Prior to immunization, reconstitute with PBS at 1 mg solid/ml PBS (rabbits) or 10 mg solid/ml PBS (mice) and add adjuvant if desired.

C Acetone precipitates (18, 20)

1. Prepare surface washings and extracts from lyophilized mycelium as described and mix with acetone to give a final acetone concentration of 70% (v/v).

Protocol 1 continued

2 Incubate overnight at 4°C and centrifuge for 5 min at 2000 g. Discard the supernatant, resuspend the white precipitate in dH_2O, and freeze-dry.

3 For long-term storage place in sealed plastic bottles in freezer.

4 For immunization of mice prepare a suspension of 0.5 mg solid/ml PBS and add adjuvant if desired.

D Antigens purified by SDS–PAGE or transferred electrophoretically to membranes (20)

1 Separate sample on a polyacrylamide gel. After electrophoresis, locate the protein band either by staining a replicate lane cut from the gel or by weakly staining the gel itself with Coomassie. Excessive fixation of the protein in the gel by over-staining should be avoided.

2 Excise the gel fragment and rinse the gel slice in several changes of deionized H_2O.

3 For rabbits, the gel fragment can be injected with the protein intact, by passing it repeatedly through a syringe needle in a small volume of PBS. Alternatively lyophilize the whole gel slice, grind into a powder, and add a volume of deionized H_2O equal to one-half of the original gel volume. The samples are now ready for injection.

4 For smaller animals such as mice, the protein should be electroeluted or transferred electrophoretically to a membrane (4).

E Conidia (8)

1 Grow the fungus in Petri dishes under conditions necessary to induce sporulation.

2 To each culture add approx. 10–20 ml of PBS. Stroke the culture with a sterile inoculating loop to remove the spores. Filter the spore suspension through glass wool or Miracloth to remove mycelium and conidiophores etc.

3 Transfer to Eppendorf tubes and centrifuge at 12 000 g for 5 min.

4 Remove supernatant and resuspend pellet in fresh PBS. Mix thoroughly and repeat procedure twice. Thorough washing of the spores is imperative to remove any mycelium-associated antigens that may interfere with the production of spore-specific antibodies.

5 Adjust the spore concentration to 10^6 spores/ml PBS and add adjuvant if desired (see Sections 3.2 and 4.2).

3 Raising polyclonal antisera

The production of polyclonal antisera is a simple and humane procedure when carried out by suitably trained personnel. It should not, and cannot, be performed by unskilled operators. The use of animals for the production of antisera or hybridomas is governed in most countries by strict legal requirements. In the UK, the Animals (Scientific Procedures) Act 1986 stipulates that an individual must hold both a personal licence and also a project licence endorsed by the Home Office. These regulations are designed not only to protect the welfare of the animals, but also to ensure that the operator is skilled in animal manipulations and that the raising of antiserum is justified. Consequently, only a relatively small number of people have the necessary credentials and most investigators must turn to other sources for antibody production. Many companies now provide antibody production as a chargeable service. A list of these are given in Section 7.

3.1 Which species/strain and how many animals?

While a wide range of vertebrate species can be used for polyclonal antisera production, the most commonly used animal for fungal studies is the rabbit. There are a number of reasons for this. Rabbits are easy to keep and handle, can be safely and repeatedly bled, and the antibodies they produce are well characterized and easily purified. Rabbits do not require large amounts of immunogen per injection (*cf.* sheep and goats) but often yield sizeable volumes of high quality serum over the course of an immunization regime (*cf.* rats, mice, hamsters, and guinea pigs). Rabbits used for antiserum production are outbred. They exhibit a broader spectrum of class II proteins and other immune response proteins than inbred animals such as mice. Even in genetically identical animals such as BALB/c mice (the host of choice for mAb production), a single antigen preparation will elicit different antibodies. Thus, in a rabbit, these differences are amplified. Therefore, if large amounts of antigen are available, several rabbits should be used for a single immunization regime, and individual rabbits should be monitored for antiserum production separately. Two rabbits should be used as a minimum, with three to four being preferable. The strain most commonly used for the production of antisera to fungi is the male New Zealand white rabbit.

3.2 Immunization procedures, determination of immune response by test bleeds, and boosters

While a molecule may fill the three requirements of a good immunogen outlined above, an animal's response to an antigen relies on a number of additional factors. These include the dosage, the route of injection, and the use of adjuvants. For pAb production in rabbits, between 50–1000 μg protein should be administered per injection. This figure applies to soluble proteins (enzymes), particulate proteins (structural proteins), insoluble proteins (immunopurified proteins bound to beads), carbohydrates (glycoproteins), and nucleic acids conjugated to carrier proteins. For pAb production to fungal antigens the most appropriate route of injection is subcutaneously (sc) as injections can include both particulate immunogens and/or adjuvants. Adjuvants are non-specific stimulators of the immune response. Their use results in more persistent antibody responses and can often be the deciding factor in whether an immune response is elicited or not. Consequently, adjuvants should, if possible, always be used where soluble antigens are to be administered or where only small amounts of immunogen are available. They are not always required for particulate or whole cell antigens such as spores. The most commonly used adjuvant was Freund's, administered either as complete Freund's adjuvant (CFA) or incomplete Freund's adjuvant (IFA). With the exception of primary injections, it is now illegal in the UK to use CFA but most Home Office inspectors will still allow the use of IFA. If large amounts of immunogen are available or the antigen is known to be highly immunogenic, then less harsh adjuvants should be used such as the saponin QuilA (Superfos a/s, Denmark) (18, 21) at 25 μg per dose.

Rabbits should be given multiple sc injections on the back. For each immunization, 800 μl doses of immunogen should be administered at each of ten sites with or without adjuvant. Polyclonal antibodies were traditionally raised by hyperimmunization in which the rabbit was repeatedly boosted with the same antigen. This is now frowned upon. For fungal antigens, immunization regimes now typically consist of two injections at 4–6 week intervals followed by a terminal bleed. After an injection, samples of serum can be taken for testing. Test bleeds should be performed 7–14 days after an injection as this will coincide with a peak in antibody titres for sc injections. Test bleeds taken from the marginal ear vein should be tested using an assay that most closely resembles the technique for which the antibodies are being raised such as ELISA (*Protocol 4*) against the immunogen itself. Sera from different rabbits should be compared to other test bleeds in titrations to determine the strength of the antibody responses (Section 3.4).

3.3 Serum preparation, purification, storage

After collection, blood should be allowed to clot and contract at 4 °C, during which time the clot should be discouraged from sticking to the surface of the collection tube by inserting glass capillary tubes. The serum should be carefully removed from the tubes and any insoluble matter removed by centrifugation at 3000 g for 5 min at 4 °C. The serum should then stored neat in aliquots of say 100 μl in Eppendorf tubes at −20 °C. Once frozen the serum will keep for many years. If the serum is frozen straight away, no preservative needs to be added. However, if the serum is to be stored at 4 °C or if an aliquot is thawed, a preservative such as 0.02% (w/v) sodium azide (NaN_3) should be added. Sera do no tolerate repeated freezing and thawing. Therefore, once thawed, samples should be kept at 4 °C. Should the sodium azide need to be removed, samples should be dialysed against an appropriate buffer such as PBS.

One of the most commonly used steps in the purification of polyclonal antiserum is ammonium sulfate precipitation. However, this alone will not yield pure antibodies as they will be contaminated with other high molecular weight proteins. In contrast, the addition of short-chain fatty acids such as caprylic acid (*n*-octanoic acid) under mildly acidic conditions, precipitates most serum proteins with the exception of the IgG molecules. Caprylic acid precipitation followed by ammonium sulfate precipitation provides a simple and efficient means of preparing relatively pure antibody. A combined precipitation procedure is given in *Protocol 2*.

Protocol 2

Antibody purification using caprylic acid and ammonium sulfate precipitation

Equipment and reagents

- Dialysis tubing: MWCO 12–14 kDa (Medicell International Ltd.)
- 60 mM sodium acetate buffer pH 4.0: 18.0 ml of 60 mM NaOAc, 82.0 ml of 60 mM HOAc
- PBS (see *Protocol 1*)
- Caprylic acid (*n*-octanoic acid): free acid solution (Sigma C2875)
- Saturated (4.1 M) ammonium sulfate solution: 761 g $(NH_4)_2SO_4$ in 1 litre dH_2O; adjust to pH 7.0 with NaOH (aq)

Method

1. Measure the volume of serum (original volume). Do not use all of the serum in one go. 10 ml of serum is a good volume to start with, but it depends on the total volume available. Transfer to a clean beaker, add a stir bar, and place on a magnetic stirrer at room temperature.
2. Add two volumes of 60 mM sodium acetate buffer pH 4.0. Check the pH and adjust to pH 4.8, if necessary.
3. Add the caprylic acid slowly drop-wise and continue stirring for 30 min at room temperature. The amount of caprylic acid added depends on the source of the antiserum. For rabbit antiserum add 0.75 ml per 10 ml of original volume, for sheep add 0.7 ml per 10 ml, and for mouse add 0.4 ml. A thickish off-white precipitate will form.
4. Centrifuge at 5000 g for 10 min. Carefully decant and save the supernatant. If the supernatant contains suspended matter after centrifugation, filter through glass wool or Miracloth (see *Protocol 1*).
5. Transfer the supernatant to dialysis tubing and dialyse versus PBS overnight at 4 °C.
6. Transfer the dialysed supernatant to a clean beaker, add a stir bar, and return to a magnetic stirrer at 4 °C. While the antibody solution is stirring, slowly add enough saturated ammonium sulfate solution to bring the final concentration to 50% saturation (i.e. add an equal volume of the volume after step 5). Continue stirring at 4 °C for 6 h or overnight. A fine white precipitate will form.
7. Centrifuge the precipitate at 3000 g for 30 min.
8. Carefully remove and discard the supernatant. Drain the pellet well and resuspend in 0.3–0.5 volumes of the original volume in PBS. Transfer the antibody solution to dialysis tubing and dialyse versus three changes of PBS overnight, allowing enough room for expansion.
9. Remove the antibody solution from the tubing and centrifuge to remove any remaining debris. Store as for neat antiserum, in the presence of 0.02% (w/v) sodium azide, if appropriate.

3.4 Titre determination

3.4.1 Microtitre plates, coating buffers, and detergents

Both the type and pH of buffers, the make of microtitre well, and the use of detergents can have a profound effect on the outcome of solid support immunoassays. If little or nothing is known about the antigen, the most appropriate buffer for immobilization to microtitre wells for assay by plate-trapped-antigen-enzyme-linked immunosorbent assay (PTA-ELISA) is PBS (pH 7.0) (see *Protocol 1*). This might be the case where the animal was immunized with surface washings, culture filtrates, or extracts of lyophilized mycelium that contain heterogeneous mixtures of antigens. The high content of immunogenic glycoproteins in such preparations demands the use in ELISA of microtitre wells that have a high capacity for glycoprotein binding. Certain makes of microtitre wells, such as those arranged in strips, have proved invaluable for this purpose (see *Protocol 3*). For purified peptides or bands cut from PAGE gels, the most suitable buffer has often proved to be bicarbonate (pH 9.6) (see *Protocol 3*). As with PBS, immobilization of fungal peptides in bicarbonate buffers for assay by PTA-ELISA is best achieved using microtitre plates activated for high protein binding (see *Protocol 3*). During the immobilization procedure, the presence of detergents in antigen extracts inhibits or prevents binding of fungal antigens to the plastic well. This is the only stage of a PTA-ELISA where the absence of detergents is of paramount importance. In contrast, the inclusion of detergents in washing and diluent buffers is imperative to prevent non-specific binding of primary antibodies and secondary antibody–enzyme conjugates to plastic surfaces.

For any given antigen–antibody interaction, there will be optimum ratios of antigen:antibody concentrations where maximum absorbance values are achieved in enzyme-linked immunoassays. This applies both to the PTA-ELISA described in *Protocol 3* but also to the double-antibody-sandwich-ELISA (DAS-ELISA) described in *Protocol 10*. In both assays we can assume that we are dealing with an excess of antibody, equating say to a 1 in 1000 dilution of rabbit pAb or neat hybridoma supernatant (Section 4.6). Under such circumstances an excess of antigen might result in suboptimal conditions for antibody–antigen interaction. For example, in the case of large decavalent IgM immunoglobulins, the geometric arrangement of high concentrations of antigen molecules bound to a microtitre well might prevent a strong antibody–antigen complexes from forming since not all of the potential ten antigen binding sites on each of the IgM molecules are able to interact with an epitope. This steric hindrance is alleviated by reducing the antigen concentration, thereby allowing a greater number of multivalent interactions to form which in turn greatly stabilizes the immune complexes. Similar needs for assay optimization occur where contaminant molecules in plant or soil extracts compete for antigen binding to microtitre wells (22). By diluting the extract an optimum signal-to-noise ratio is achieved which in turn translates to elevated absorbance values.

The need to optimize antigen–antibody interactions highlights the need for standardization of immunoassays. This is most easily achieved through the construction of calibration curves, enabling not only a simple and rapid means of quantifying fungal biomass, but also preventing bias in estimations due to day-to-day variability and incomplete extraction of fungal antigens from plant and soil samples (21, 23). Choice of antigen source for construction of calibration curves and techniques for relating amounts of antigen derived from standard curves to units of fungal biomass are described amply elsewhere (7, 21–23).

Calibration curves also form the basis of antibody titre determinations. The titre is an estimate of the amount of antibody activity. The antibody containing fluid is diluted serially until an end-point (for example no colour production in ELISA relative to a control) is reached. *Protocol 3* describes a PTA-ELISA for the determination of antisera titres, but is equally applicable to the analysis of mAbs in hybridoma supernatants (Section 4.4). In the example described, microtitre wells are coated with 50 μl of a 1 in 10 dilution of surface washings (see *Protocol 1*) prepared either in PBS or bicarbonate buffer. 100 μl volumes can be used which increase the sensitivity of the assay but is more costly. Antibody titre is determined by twofold dilution of the antiserum into saline buffer containing detergent. The last well in the dilution contains diluent only. This acts as a negative control and provides a threshold value for titre determination. It is important to note that the titre of an antiserum will change according to the source and concentration of the antigen used to coat the microtitre wells and so titres should always be quoted against a named antigen preparation. Titres are often expressed as the reciprocal of the end-point dilution.

The results from an experiment using this procedure are shown in *Figure 1*.

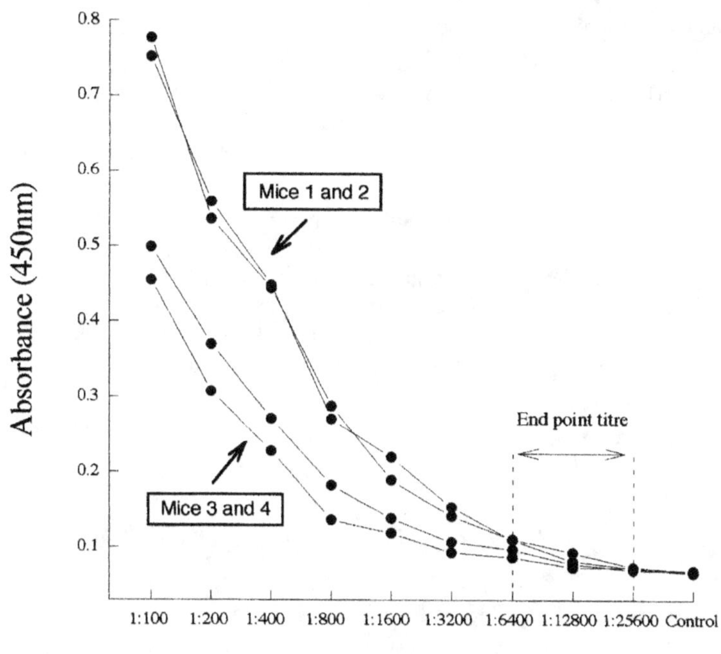

Figure 1 Titre determination by ELISA of antisera taken from the tail veins of four mice immunized with a hydrolytic enzyme preparation from *Trichoderma harzianum*.

Four BALB/c female mice were each immunized by intraperitoneal (ip) injection (see Section 4.2) with an extract from *Trichoderma harzianum* containing a cocktail of enzymes including chitinase, cellulase, protease, and xylanase. Each mouse was bled from the tail vein after the third injection and respective titres determined against the immunogen immobilized to flat-bottomed microtitre plates using bicarbonate buffer (pH 9.6). The titres of all four antisera were very similar (approximately 1/6400 to 1/12 800) and yet the dilution profiles of two of the mice (1 and 2) differed considerably from the other pair (mice 3 and 4). On the basis of their dilution profiles, mice 1 and 2 were selected for hybridoma production (Section 4.3) and resulted in the production of cells secreting IgM immunoglobulins specific to *T. harzianum* exo-1,4-β-glucanases and xylanases.

Protocol 3

ELISA for pAb titre determination and assay of mAb supernatants

Equipment and reagents

- Microtitre wells arranged in 12 well strips (Life Sciences Ltd., 9502037) and multiframes (Life Sciences Ltd., 9503060) or flat-bottomed plates such as Immulon 2 HB plates (Dynex 3455)
- Variable volume (50–100 μl) multichannel pipette (Anachem)
- PBST: 10 × PBS (see *Protocol 1*) containing 0.05% (v/v) Tween 20 (polyoxyethylene(20)sorbitan monolaurate; Sigma P1379). This is a 10 × concentrate and should be diluted in a wash bottle before use.
- Bicarbonate buffer: 1.59 g Na_2CO_3 and 2.93 g $NaHCO_3$ per litre dH_2O; this automatically buffers to pH 9.6
- Blocking solution: PBS containing bovine serum albumin (BSA, Sigma A7030) or casein (BDH, 440164R)
- Microplate reader (MRX-II, Dynex Technologies) fitted with 450 nm filter
- Goat anti-rabbit IgG (whole molecule) or goat anti-mouse polyvalent (IgG, IgA, IgM) peroxidase conjugates (Sigma A6154 and A0412 respectively)
- Peroxidase substrate buffer: 5 ml dH_2O, 5 ml of 0.2 M solution of sodium acetate (5.4 g in 200 ml dH_2O), 195 μl of 0.2 M solution of citric acid (8.4 g in 200 ml dH_2O), and 5 μl of 30% H_2O_2
- Peroxidase substrate solution: 10 mg of 3,3′,5,5′-tetramethyl benzidine (TMB, Sigma T2885) dissolved in 1 ml of dimethyl sulfoxide (DMSO, Sigma D5879). This should be kept in the fridge as 100 μl aliquots. DMSO solidifies at 4 °C. Thaw and add 100 μl to substrate buffer.
- Substrate reaction stop solution: 3 M H_2SO_4

Method

1 Grow slants of appropriate fungi and prepare surface washings in PBS or bicarbonate buffer (see *Protocol 1*).

2 Dilute the surface washings tenfold in PBS or bicarbonate buffer and transfer 50 μl to wells of microtitre strips held in frames or plates. Tap gently to ensure even distribution over well surface. Place in a sealed plastic bag and incubate overnight at 4 °C or 1 h at 37 °C to allow immobilization of antigens.

Protocol 3 continued

3 Remove extract from wells by inverting wells and gently flicking wrist. Fill each well to the rim with PBST dispensed from a wash bottle. Leave for 5 min and repeat twice more (5 min each time) with PBST, once with PBS, and once with dH_2O. Remove dH_2O as described and bang dry on paper towel. Air dry the wells in a flow bench at room temperature (RT; 23 ± 2 °C) or proceed immediately with ELISA. Dried plates can be stored at 4 °C in sealed plastic bags for several months with no deleterious effect on the antigens.

4 Carry out the following primary antibody incubation step, once the wells have been blocked[a] with 100 μl of blocking solution if needed:

(a) For pAb titre determination add 50 μl of PBST to wells 2 to 12. Into well 1 place 100 μl of serum diluted into PBST. Start at a 1 in 100 dilution. Control wells should contain a similar dilution of pre-immune serum or commercial normal rabbit serum. Remove 50 μl of the 1 in 100 dilution from well 1 and transfer to well 2. Mix thoroughly by sucking up and expelling 50 μl of solution through the pipette tip, avoiding bubbles. Repeat this three times and transfer 50 μl of the 1 in 200 dilution from well 2 to well 3. Mix as described and repeat the procedure in strict sequence as far as well 11. After mixing in well 11, remove 50 μl of what is now a 1 in 102 400 dilution of pAb and discard, leaving well 12 containing PBST only.

(b) For testing of mAbs, place 50 μl of neat hybridoma supernatant into each well. Control wells should contain an equal volume of mAb-free tissue culture medium or mAb supernatant known not to react with the fungus in question.

Place microtitre wells in a sealed plastic bag and incubate for 1 h at RT.

5 Remove pAb or mAb by inversion and flicking and wash four times with PBST (3 min each time) as described in step 3.

6 Prepare a 1 in 1000 dilution of secondary anti-mouse or anti-goat peroxidase[b] conjugates by diluting in PBST (with blocking agent if needed). Add 50 μl to each well and incubate for 1 h at RT in a sealed plastic bag.

Note: the concentrations of secondary enzyme conjugates used should be optimized for each antibody–antigen interaction.

7 Wash wells three times with PBST and once with PBS as described. Add 50 μl substrate solution to each well and incubate in a sealed plastic bag for 30 min at RT. Stop the blue coloured reaction by the addition of 50 μl stop solution to each well. Solutions turn yellow.

8 Read absorbance values at 450 nm.

[a] An important component of any immunoassay is blocking of wells prior to the addition of the primary antibody and the addition of a blocking agent in the washing and secondary antibody conjugate diluent buffers. This should not be necessary under normal working conditions, but may occasionally be required because of non-specific reactivity of reagents, particularly secondary antibody enzyme conjugates. After antigen immobilization to the solid support (microtitre well or membrane), unbound sites can be blocked to prevent non-specific binding of the

Protocol 3 continued

primary antibody. Blocking agent can also be added to the mAb supernatant (18) or the supernatant diluted with PBST (17). Different blocking solutions should be tested and adapted to suit each antigen–antibody interaction, as no one buffer or blocking agent has universal applicability. However, buffers containing milk proteins such as casein often prove to be the most effective in reducing or eliminating non-specific binding. Casein is a mixture of milk proteins which differ in size, amino acid sequence, and degree of glycosylation and sulfation, and also surface charge, which contrasts with the homogeneity of bovine serum albumin (BSA) and gelatin, two commonly used alternative blocking agents. Products such as Marvel and Five Pints bought from the supermarket provide a cheap source of milk proteins for blocking solutions. In addition to casein, BSA, and gelatin, detergents such as Tween 20 can be added to blocking solutions. The optimum concentration of blocking agent and buffer type should be determined for each assay. Whether blocking agents are used or not, polyclonal antisera and secondary enzyme conjugates should always be diluted in buffer containing a detergent such as Tween 20, to prevent non-specific binding to microtitre wells.

[b] Enzyme conjugates other than peroxidase can be used. Alkaline phosphatase-based assays (25) are appropriate where endogenous peroxidases are likely to interfere with ELISAs, such as with fungal antigens extracted from plant material. The advantage of peroxidase-mediated assays is that the colour reaction step is much faster and often more sensitive. Sensitivity and speed of solid phase ELISAs can be improved by incorporating enzyme amplification systems. For peroxidase-based assays, biotin–extravidin peroxidase conjugates can be used (26). The speed and sensitivity of alkaline phosphatase-based assays can be significantly improved using the AMPAK amplification system (DAKO, K6200). Using the AMPAK system, the substrate step is reduced to 30 min or less.

4 Raising monoclonal antibodies

4.1 Setting up a 'tissue culture facility'

The only requirement for establishment of a successful tissue culture facility is cleanliness. The facility and the equipment contained therein should be used solely for tissue culture. Several basic pieces of equipment are needed:

(a) A 37 °C CO_2 incubator.

(b) A Class II flow cabinet fitted with Hepa filter and preferably with UV for sterilization.

(c) Inverse microscope.

(d) A liquid nitrogen storage vessel.

(e) Pipetteboy.

(f) Benchtop centrifuge with variable volume buckets.

4.2 Immunization procedures, the immune response, and titre determination

The dose of immunogen and routes and frequencies of immunization for hybridoma production differ considerably to those used for antiserum production in rabbits (Section 3.2). However, the procedures used for monitoring the immune response by PTA-ELISA (see *Protocol 3*) remain the same. The species most commonly used for fungal hybridoma production is the female mouse strain BALB/c although rats have also found applicability under certain circumstances. For example, interacting populations of fungi have been studied using mixtures of species-specific mouse and rat mAbs with binding of the mAbs detected by ELISA or visualized by IF (see *Protocol 5*) using host-specific enzyme or fluorochrome conjugates (24). Injection into the peritoneum (ip) is the most commonly used method of introducing antigens into mice. Because of the large volume of the peritoneal cavity, the volume of the immunogen can be larger for ip injections than for other sites. Furthermore, because the injections do not deliver the antigen directly into the blood system, particulate antigens such as spores or mycelial fragments can be used. The antigen should be dispersed in approximately 250–300 μl of PBS and contain between 50 μg (minimum) and 200 μg (maximum) of protein. As with pAb production in rabbits, the primary injection should contain an adjuvant such as Quil A at 10 μg per dose (18). Mice should be immunized fortnightly (preferably with adjuvant) and after the third or fourth injection test bleeds should be taken from the tail veins (4) and their titres determined by ELISA (see *Protocol 3*). If the mice have responded satisfactorily, they should be given a final booster injection and a fusion performed three days later (Section 4.3). Inherent differences in the antigenicity of fungal preparations means that the timetable for hybridoma production will vary. However, in most instances the time taken from primary immunization to fusion is approximately two months. Hybridoma production including screening, expansion, and freezing of selected, uncloned, cell lines takes roughly the same amount of time.

4.3 Production of murine hybridomas

Protocol 4 describes the production of hybridomas from an immunized mouse. This method of fusing myeloma cells and splenocytes (B cells) is based on the modification of Kennet *et al.* (27) of the original method pioneered by Köhler and Milstein (3). It has been used extensively and with consistent success for the production of mAbs to a wide range of fungal antigens. For a comprehensive review of antibody production by hybridomas the reader is referred to Goding (28).

Protocol 4

Production of murine hybridomas by B cell–myeloma fusion

Equipment and reagents

- Sterile 10 ml pipettes (Becton Dickinson, 7551)
- Sterile 30 ml plastic Universal tubes (Greiner, 201170)
- Sterile 10 ml round-bottomed centrifugation tubes
- Sterile 96-well multiwell plates (Becton Dickinson, 3072)
- 1 ml and 100 μl pipettes (Gilson) and sterile tips
- Variable volume (50–100 μl) multichannel pipette (Anachem) and sterile tips

Protocol 4 continued

- Sterile scalpel and forceps
- Penicillin–streptomycin (pen-strep) stock antibiotic solution: 0.5 g streptomycin sulfate (Sigma, S9137) and 0.3 g penicillin G (Sigma, PEN-NA) dissolved in 5 ml distilled water, sterile filtered (0.2 μm sterile filter units, e.g. Sartorius), and stored frozen in 100 μl aliquots in sterile 1.5 ml Eppendorf tubes
- L-glutamine stock solution: sterile 200 mM L-glutamine solution (Sigma, G7513), stored frozen in 1 ml aliquots in sterile 1.5 ml Eppendorf tubes
- 20% tissue culture medium (20% TCM): RPMI 1640 medium (Sigma, R0883) containing 20% fetal bovine serum (FBS)[a] (Sigma, F7524), supplemented with 100 μl of pen-strep solution and 1 ml of 200 mM L-glutamine per 100 ml of culture medium. This should be warmed to 37 °C prior to use.
- Centrifuge with variable volume buckets
- SP2/0-Ag-14 myeloma cells
- RPMI 1640 serum-free medium warmed to 37 °C
- Polyethylene glycol solution: 33% PEG (Av. Mol. Wt. 1000, Sigma, P3515) solution prepared by melting sterilized PEG at 65 °C and then mixing 0.2 ml with 0.4 ml RPMI 1640 serum-free TCM. Kept at 37 °C until needed.
- HAZA selective medium: 95 ml of 20% TCM containing HAZA supplement. This can be obtained as a γ-irradiated lyophilized powder (Sigma, A9666). When reconstituted in 10 ml sterile water each vial contains 2.85×10^{-5} M azaserine and 5×10^{-3} M hypoxanthine. This is a 50 × concentrate and should be diluted accordingly in sterile 20% TCM to give a working solution.

Method

1. Determine myeloma cell concentration using a haemocytometer. Use cells from a culture that is in the exponential phase of growth and calculate volume needed to obtain a total of 10^7 cells. Return cells to CO_2 incubator.
2. Take a 30 ml Universal bottle containing 10 ml warm 20% TCM to the animal house.
3. Kill the immunized mouse with CO_2. Douse the mouse with 70% (v/v) ethanol. Locate the spleen by opening the mouse on the left side and pull back skin. Using fresh sterile forceps and scissors make an incision and remove the spleen aseptically and place in Universal. Return immediately to tissue culture facility.
4. Place the spleen in a 9 cm Petri dish containing 15 ml of fresh 20% TCM. Using sterile hypodermic needles tease the spleen open; it disintegrates easily as it is little more than a bag of cells. Transfer the cell suspension to a sterile 50 ml Falcon tube and leave the contents to stand momentarily while myeloma cells are transferred.
5. Transfer myeloma cells containing a total of 10^7 cells to a sterile 50 ml centrifuge tube. Transfer spleen cells to a fresh sterile 50 ml centrifuge tube; the connective tissue should have settled out leaving the supernatant containing a cell suspension. Spin both myeloma and spleen cells at 250 g for 5 min.
6. Discard supernatants and resuspend cells in 2.5 ml of RPMI serum-free medium. Combine the samples, mix, and centrifuge in a sterile round-bottomed 10 ml tube at 250 g for 5 min.

Protocol 4 continued

7 Aspirate to dry the pellet and add 0.3 ml of 33% PEG solution allowing exposure of the cells for 7 min including 1 min centrifugation at 500 g.

8 Gently suck off the PEG and resuspend the pellet in 5 ml of serum-free medium. Add the medium slowly so as not to disrupt the cells from fusing. Dilute the 5 ml of fusion products directly into 95 ml HAZA[b] selective medium and dispense the cell suspension at 100 μl per well of ten 96-well multiwell plates. Transfer the plates to the incubator.

9 Examine the cells regularly. Cell selection will not occur for 3–4 days:
   (a) Day 1 fusion.
   (b) Days 2 and 3 lymphocytes die, SP2 cells bright.
   (c) Days 4 and 5 SP2 cells die.
   (d) Days 6 and 7 (approximately) hybridomas begin to appear. At this stage the cells should be fed with an equal volume (100 μl) of 20% TCM per well.
   (e) Day 9 check and ring wells containing cells. Assay supernatants by ELISA, IF (see Section 4.4 and *Protocols* 3 and 5 respectively), or appropriate assay from clones that cover one-third of the field of vision under × 10 objective of inverse microscope.

[a] Many companies supply FBS from a large number of countries and sources. Before purchasing large volumes of serum, it is advisable to test small volumes of larger FBS batches for suitability and quality. Companies such as Sigma routinely supply 100 ml samples for this purpose free of charge. Once selected, larger volumes can be reserved from single batches. By quoting a personal reserve number with each order, clients are supplied with serum from their own verified stock, that is held in cold storage until expiry (typically one to two years).

[b] The HAZA selective medium used here contains hypoxanthine and azaserine in place of aminopterin and thymidine. It is the method used by Foung *et al.* (29).

4.4 Screening procedure for selection of hybridomas

It has long been suggested that mAbs of the IgG class or subclass of immunoglobulins are the most ideal antibodies to aim for. There are a number of reasons for this, the most important being that IgG mAbs bind strongly to protein A and can therefore be more easily purified for direct conjugation to enzymes and fluorochromes. It has also long been perceived that IgG mAbs are the most specific, but studies by a number of groups have clearly shown that, in the case of fungal antibody production, this is simply not the case. A number of mAbs of the immunoglobulin class IgM and IgA have been raised against fungal antigens that not only are highly specific (18) but which also provide, under certain circumstances, an increased measure of assay sensitivity due to their increased valency compared to IgG mAbs. Immunoglobulin M antibodies are pentamers and therefore have ten antigen binding sites compared to the two of monomeric IgGs. Undoubtedly, the best rule-of-thumb is to aim for a panel of mAbs that bind to the same antigen but which are of different Ig classes or subclasses. Antibodies recognizing different epitopes on the same molecule are an added bonus and, in the case of a fungal glycoprotein, these might bind to carbohydrate or protein moieties on the same structure. The availability of a range of Ig classes binding to different epitopes on the same antigen, gives the operator much greater flexibility in the design of different assay formats. It follows that the usefulness of an antibody increases significantly if it performs well in a range of assay formats.

In any antibody selection procedure, the first priority is to identify hybridomas secreting mAbs to the target antigen. After fusion, the hybridoma cells are dispersed in 100 μl volumes in the wells of 96-well plates (see *Protocol* 4). Approximately one week later the cells are fed with 100 μl of fresh 20% TCM. From day 9 onwards cells will be ready for screening. Allowing for evaporation and the need to leave the cells bathed in at least 50 μl of TCM (on no account should the cells be allowed to dry out), the amount of antibody supernatant available for a primary screen is 50 μl. The huge number of hybridomas that can be created from a single fusion (typically 300 but in some cases near to 1000) means that, realistically, a plate-trapped-antigen (PTA)-enzyme-linked immunosorbent assay (ELISA) is the most appropriate method for the primary screen. However, it cannot be assumed that because a mAb recognizes an antigen strongly by ELISA when immobilized to a microtitre well, it will do so in whole cell immunofluorescence (IF). Such variation among different techniques is often due to multivalent binding. Similarly, a cross-reaction that is undetectable in one technique may dominate the results of another assay. Consequently, if the intended final use of an antibody is for IF, then a secondary screen should be based on IF, once potential candidates have been identified in a primary screen by ELISA. For screening by PTA-ELISA, an assay such as that described in *Protocol 3* can be used. A method for screening of mAbs by IF is given in *Protocol 5*.

Protocol 5

Immunofluorescence

Equipment and reagents

- Photographic film (e.g. Kodak TMAX monochromatic print film or Kodak Ektachrome 320 Tungsten colour reversal film)
- Fixative: ethanol/chloroform/3% paraformaldehyde solution (6:3:1, by vol.). For the paraformaldehyde solution, prepare a 10% solution by dissolving 2.0 g of paraformaldehyde powder (Lancaster, 4737) in 20 ml dH_2O and heating to 60–65 °C in a fume cupboard. Add a few drops of 1 M NaOH until the solution clears. Allow the solution to cool and mix with PBS to give a 3% solution.
- Teflon-coated multitest slides (ICN biomedicals) and coverslips
- Epifluorescence UV microscope fitted with 495 nm (excitation) and 525 nm (emission) filters (Leitz Axiophot or Nikon)
- 95% methanol
- Mounting medium (Sigma, 1000-4)
- Goat anti-mouse polyvalent (IgG, IgA, IgM) fluorescein isothiocyanate (FITC) conjugate (Sigma, F1010) or goat anti-rabbit (IgG) FITC conjugate (Sigma, F6005)
- PBS and PBST (see *Protocols* 1 and 3)

Protocol 5 continued

Method

1 Sterilize the Teflon-coated slides by autoclaving and embed the slides in an appropriate agar medium in 9 cm Petri dishes. Remove the agar above each well using a sterile cork borer and scalpel. Inoculate the agar with the fungus and incubate under appropriate conditions. Once the glass in each well has been overgrown, carefully excise the remaining agar and thoroughly air dry the slide at room temperature.

2 Fix the air dried slide by immersing in fixative for 3 min, 95% methanol for 4 min, and a brief rinse in dH_2O. Air dry the slides at room temperature.

3 Place the slides in a plastic tray lined with moist tissue paper. Into each well pipette mAb supernatant or pAb diluted into PBS. The dilution of a pAb should be optimized for use in IF but start with a 1 in 40 dilution.[a] Incubate control wells with TCM or pre-immune serum. Seal the tray to prevent evaporation and incubate at room temperature for 30 min to 1 h.

4 Wash the slides three times in PBS (5 min each wash) and carefully remove as much PBS from the Teflon coating surrounding each well as possible using tissue paper. Do not dry the glass wells themselves. Pipette anti-mouse or anti-rabbit conjugate into each well. The conjugates should be diluted 1 in 40 in PBS. Return to the sealed tray and incubate in the dark for a further 30 min at room temperature.

5 Wash the slides in PBS and dry as described in step 4. Place a drop of mounting medium on the slide and cover with a coverslip. Place the mounted slides in clean Petri dishes and wrap the dishes in tin foil to prevent bleaching of the FITC. Store at 4 °C. Examine under UV.

Note: in IF, a sample would be judged as positive if it fluoresced and would be negative if it didn't. Clearly, though, there are many degrees of fluorescence in between, and so a positive response can only be determined by comparison with a negative control.

[a] Unlike ELISA, pAbs and secondary fluorescent conjugates work best in IF when diluted into PBS rather than PBST. They are also used at higher concentration (1 in 40 in IF compared to say 1 in 1000 in ELISA).

4.5 Subcloning, expansion, and routine culture of hybridomas and long-term storage of mAb supernatants

Because a hybridoma contains characteristics of both of its parents, it will continue to multiply indefinitely *in vitro*. Consequently, cells contained in the wells of fusion plates may be derived from more than one cell line each secreting a different mAb. It is therefore essential that the cells are subcloned at the earliest opportunity in order to separate mixed cell lines. Subcloning will also eliminate any non-producing cells that can outgrow antibody-producing cells.

Protocol 6

Subcloning by limiting dilution

Equipment and reagents

- Sterile 30 ml plastic Universal tubes (Greiner, 201170)
- Sterile 96-well multiwell plates (Becton Dickinson, 3072)
- 1 ml and 100 μl pipettes (Gilson) and sterile tips
- Sterile pipettes (Becton Dickinson, 7551)
- Variable volume (50–100 μl) multichannel pipette (Anachem)
- Vortex mixer
- 20% TCM at 37 °C (see *Protocol 3*)

Method

1. Into four Universal tubes pipette 5 ml, 10 ml, 4 ml, and 4.5 ml pre-warmed 20% TCM and label each B, C, D, and E dilution respectively.
2. Transfer 50 μl of hybridoma cell suspension to tube B and mix gently by vortexing. **Note**: many hybridoma cell lines grow on the surface of multiwell plates or Petri dishes. To ensure that the suspension contains an adequate concentration of cells, the hybridomas should first be dislodged by gently circulating the bathing supernatant over the cells using a pipette. In doing so, care should be taken not to blow oxygen into the culture.
3. Transfer 100 μl of suspension from tube B to tube C and vortex.
4. To tubes D and E add 1 ml and 0.5 ml respectively of suspension from tube C. Vortex both D and E dilutions.
5. Using a multichannel pipette transfer 100 μl volumes of B dilution cell suspension to the wells of a 96-well multiwell plate. This will fill half of the 96-well plate (approx. 5 ml). Into the remaining half transfer 100 μl volumes of the C dilution cell suspension (the remaining 5 ml is superfluous). Into the wells of a second plate, disperse the cells from the D and E dilutions. Label each half B, C, D, and E respectively. Examine the wells under the microscope.
6. Transfer the two plates to the CO_2 incubator. After five days, feed the cells with 100 μl of fresh 20% TCM per well. After a further 24 h, check the wells for colonies of cells and mark each well with a cross. Continue to check the well over the following few days for slower growing cells.
7. When ready for screening (4), supernatant should be tested by immunoassay such as ELISA (see *Protocol 3*).

The inherent ability of hybridomas to multiply indefinitely means that they need to be routinely subcultured into fresh TCM to prevent death by overcrowding, depletion of nutrients, and increases in acidity indicated by a yellowing of the TCM. Two characteristics make culturing of hybridomas an easy task. The first is that hybridomas do not stick to plastic and the second is that they do not produce toxic by-products. Consequently, cultures can simply be replenished by pipetting cell suspensions into fresh medium.

The first stage in subculturing is the expansion of selected hybridoma cells from 96-well multiplates after fusion and preliminary screening. Clones are expanded by transfer into ever increasing volumes of TCM until the cells can be routinely cultured in Petri dishes for freezing or large scale antibody supernatant production in culture flasks. During the expansion process the percentage volume of FBS in the TCM is gradually decreased from 20%, through 10% to 5% and in some cases lower. In addition, the proportions of cell suspensions subcultured into fresh TCM can be adjusted according to the growth rate of each cell line and to suit the demands of the operator. Knowing when to expand or subculture hybridomas is simply a matter of practice. A novice should ideally spend some time with an experienced tissue culturist before embarking on the culture of a prized cell line, but the procedures are simple and are quickly learnt.

Once the cells in wells of 96-well multiplates are confluent (4), they should be taken up to the 24-well stage. Most hybridoma cells will not tolerate a sudden large jump in volume, so transfer from a 96-well multiplate to a Petri dish is not advised. Again, once the cells are confluent at the 24-well stage, they should be transferred to a 6-well plate and then on to a Petri dish. Some hybridomas adhere lightly to the plastic surface, but are easily removed by carefully sucking up and expelling the culture fluid from the well with a sterile pipette tip. Once a cell line is growing well in a Petri dish, dilute the cells 1:1 with fresh medium and monitor how quickly the culture grows. Average doubling time is usually between 12–24 h. Once the properties of the cell line are known, begin to divide the cells into larger volumes of fresh TCM. Typical dilutions are 1:5 or 1:10. Larger dilutions may be possible, but this will depend on the individual cell line. Occasionally the growth of hybridomas is retarded for no apparent reason. The most common reason for this is bacterial contamination. In such cases, the surfaces of plastic culture vessels have a grey, mottled appearance. Examination under high magnification will reveal the presence of large numbers of bacterial cells. Most bacterial contamination is checked by the penicillin and streptomycin incorporated in TCM. However, additional antibiotics will be needed in such cases. First choice is gentamycin. This can be obtained in sterile solution (Sigma, G1379) and should be added at 100 μl per 100 ml TCM. Once bacterial growth is eliminated (commonly after two rounds of subculturing), the additional antibiotic should be omitted.

For bulk production of mAb supernatants, hybridomas should be cultured in flasks. Cells should be allowed to die prior to harvesting of mAbs to permit antibody saturation of supernatants. Cells should be removed by centrifugation for 5 min, the supernatants decanted and, after the addition of NaN_3 at 0.02% (v/v), stored in glass bottles at 4°C. Hybridoma supernatants stored in this way will keep for many years. Supernatants should not be frozen.

4.6 Long-term storage of hybridoma cell lines

Once stable, hybridoma cells can be slowly frozen in a suitable medium and stored indefinitely in liquid nitrogen (N_2). After storage, cells are thawed quickly and transferred to TCM. Provided the cells are viable, they will continue to divide and to secrete their specific mAb. Freezing of hybridomas is a quick and straightforward procedure. However, three critical factors determine whether the cells will survive the freezing process:

(a) That the cells are actively growing at the time of freezing.

(b) The speed of centrifugation (350–400 g) calculated from the formula:

$$\mathrm{RCF} = 0.000012 \times \mathrm{R\,(cm)} \times \mathrm{N\,(r.p.m.)}^2.$$

(c) That the cells remain undisturbed during the slow-freezing process. Attempting to freeze cells in a freezer whose doors are constantly being opened and shut will not work. Even minor increases in temperature results in damage to the cell membrane and ultimately in cell death. The key is to freeze in the late afternoon for overnight freezing, when disturbance is likely to be at a minimum.

A procedure for freezing of murine hybridomas is given in *Protocol 7*.

Protocol 7

Freezing of hybridomas for long-term storage

Equipment and reagents

- Sterile 10 ml pipettes (Becton Dickinson, 7551)
- Sterile 50 ml Blue max centrifuge tubes (Becton Dickinson, 2098)
- 200 μl pipette and sterile tips
- 1 ml cryo-tube freezing vials (Life Technologies, 366656)
- Polystyrene box and cotton wool
- Cryostore canes and sleeves (Life Technologies, 378441 and 343958 respectively)
- -80 °C freezer
- Cryogenic storage vessel (CRY-860-030V, model 18HC, Fisher Science UK) containing liquid N_2
- Freezing medium: 92% FBS, 8% (v/v) dimethyl sulfoxide (sterile DMSO in 5 ml glass ampoules, Sigma D2650); mix thoroughly in sterile plastic bottle and freeze at −20 °C in 300 μl aliquots in sterile Eppendorfs
- Ice

Method

1. The day before freezing, subculture cells into two Petri dishes to give a total of 30 ml of actively growing cell suspension in each dish. The dilution of cells used to prime each dish will depend on the growth rate of the cell line in question.
2. Suspend the cells, taking care to dislodge cells that have adhered to the plastic. Carefully transfer the cells from each Petri dish to separate sterile 50 ml centrifuge tubes and centrifuge at 350 g for 5 min (at 4 °C if possible).
3. Remove all of the culture supernatant by aspiration with a pipette, taking care not to disturb the cell pellet. Discard the supernatant and slowly resuspend each pellet in 300 μl of freezing medium removed from the freezer during centrifugation and kept on ice.

Protocol 7 continued

4. Combine the cell suspensions in one of the tubes, mix gently by pipetting, and transfer 150 μl to each of four replicate cryo-tubes. Place the sealed tubes on ice and, immediately prior to placing in the −80 °C freezer, transfer the upright tubes to a polystyrene box lined with cotton wool. Freeze the cell overnight for at least 16 h.
5. Transfer the frozen cells to liquid N_2 by placing in canes with covers.
6. After at least 24 h, remove one of the vials from the cane and quickly thaw by placing at 37 °C or by warming in the palm of the hand. This takes approx. 1-2 min. Do not allow prolonged exposure of the thawed cells to the freezing medium as this will rapidly cause cell death. Once thawed transfer the cells immediately to a Petri dish containing 20 ml pre-warmed 10% TCM.
7. Check the cells to see if they are alive. Dead cells look 'ghost-like' and live cells have clear, distinct, membranes and look shiny. Place the cells in the incubator and check again after an hour or so. Cell division will have started and will be clearly visible. There may also be some cell death. After 24 h the cells may need to be subcultured as described.

Note: freezing multiple copies of the same hybridoma cell line enables the operator to determine the viability of cells in replicate vials. If the cells in one of the vials are viable after several days storage in liquid N_2, then the remaining vials are guaranteed to be viable also.

4.7 Determination of immunoglobulin class/subclass

The most important reason for determining the Ig class of a mAb is to establish that you have a pure cell line either before or after subcloning (Section 4.5). However, determination of mAb class also allows more informed choices to be made as to, amongst others, the appropriateness of secondary conjugates in certain antibody applications. For example, the format of an indirect double-mAb-sandwich ELISA, may require the use of a heavy chain-specific conjugate (see *Protocol 10*). For determination of mAb class in hybridoma supernatants the most appropriate methods are capture ELISAs (18, 21) or antigen-mediated ELISAs. The latter method is described in *Protocol 8*.

Protocol 8

Determination of Ig class/subclass by antigen-mediated ELISA

Equipment and reagents

- Antigen-coated microtitre strips or plates (see *Protocol 1*)
- Rabbit anti-goat IgG (whole molecule) peroxidase conjugate (Sigma A5420)

Protocol 8 continued

- PBS and PBST (see *Protocols 1* and *3*)
- Peroxidase substrate and reaction stop solutions (see *Protocol 3*)
- Mouse monoclonal antibody isotyping reagents: these can be obtained from a number of sources, but the reagents described were purchased as a kit (Sigma ISO-2). Each kit contains 0.2 ml each of goat anti-mouse IgG_1, IgG_{2a}, IgG_{2b}, IgG_3, IgM, and IgA pAbs. For long-term storage, each isotype should be frozen in 20 μl aliquots. They should not be exposed to repeated freeze-thawing.

Method

1. Incubate antigen-coated wells with mAb supernatant for 1 h.
2. Wash four times with PBST (3 min each time).
3. Incubate wells for 30 min with goat anti-mouse immunoglobulin pAbs each diluted 1 in 3000 in PBST.
4. Wash wells four times with PBST (3 min each time).
5. Incubate wells for 30 min with rabbit anti-goat peroxidase diluted 1 in 1000 in PBST.
6. Wash wells three times with PBST and once with PBS (3 min each time).
7. Incubate wells with substrate for 30 min, stop colour reaction, and determine absorbance at 450 nm (*Protocol 3*).

5 Characterization of protein and carbohydrate epitopes

Characterization of mAb epitopes is most easily achieved by chemical or enzymatic modification of antigens. The most commonly used methods are periodate oxidation, protease digestion, and heat treatment of antigens either in solution, immobilized to microtitre wells, or after separation by gel electrophoresis and western blotting (8). These are described in *Protocol 9*. More detailed analyses of fungal antigens using, for example, peptide-*N*-glycosidase (PNGase) or endo-β-*N*-glucosaminidase-H (endo-H) digestion are described elsewhere (8, 30).

5.1 Periodate oxidation

Periodate oxidation involves the simultaneous oxidation and cleavage of carbon to carbon bonds that have adjacent free hydroxyl groups or an aldehyde or ketone group adjacent to a hydroxyl group. The laboratory procedures and the interpretation of the results of periodate oxidation studies are complicated exercises. However, in its simplest form, a decrease in mAb binding to an antigen following treatment with periodate provides good preliminary evidence that the mAb binds to a carbohydrate epitope. Conversely, an increase in binding provides preliminary evidence that the mAb binds to a protein epitope, such that by removing the carbohydrate component of the antigen, you expose the protein epitopes for mAb binding.

5.2 Protease digestion using pronase and trypsin

Modification of antigens by protease to the extent that mAb binding is reduced or eliminated provides strong evidence that an epitope has a protein component. The two proteases most commonly used in the analysis of fungal antigens are a non-specific protease (pronase) derived from *Streptomyces griseus* and bovine trypsin. Trypsin is a pancreatic serine protease with substrate specificity based upon positively charged lysine and arginine side chains.

Pronase treatment is usually performed both on immobilized antigens using a saline pronase solution and on soluble antigen extracts (lysates) using pronase immobilized to agarose beads. The reason for this is that a lack of effect of insoluble pronase treatment on mAb binding, but a reduction in binding by soluble pronase treatment, could result from detachment of glycan from the microtitre plate following digestion of the protein backbone, rather than digestion of the epitope itself.

5.3 Heating

A reduction in antibody binding following heat treatment of soluble antigens by boiling or autoclaving provides additional evidence for recognition of protein epitopes. However, while carbohydrate epitopes can be extremely heat stable, prolonged exposure to high temperatures will also result in degradation. Consequently, the effects of heat treatment on mAb binding should only be used as additional evidence of binding to protein epitopes.

5.4 Mild alkaline hydrolysis (β-elimination)

Mild alkaline hydrolysis results in cleavage of glycans *O*-linked through the β-hydroxy amino acids serine and threonine, but not of glycans *O*-linked through tyrosine or hydroxylysine, or through hydroxyproline, or *N*-linked through asparagine.

Protocol 9

Characterization of epitopes by chemical and enzymatic modifications

Equipment and reagents

- Antigen-coated microtitre strips or plates (see *Protocol 3*)
- PBS and PBST (see *Protocols 1* and *3*)
- 50 mM sodium acetate buffer pH 4.5: 37.0 ml of 50 mM NaOAc (1.7 g $CH_3COONa.3H_2O$ dissolved in 250 ml dH_2O) + 63.0 ml of 50 mM HOAc (0.75 g HOAc in 250 ml dH_2O)
- 20 mM sodium *m*-periodate (Sigma, S1878) dissolved in 50 mM NaOAc buffer pH 4.5
- Trypsin (EC: 3.4.21.4) solution (Sigma, T4424) diluted 1 in 2.5 with PBS to give a solution containing 1 mg/ml
- Trypsin inhibitor (Sigma, T6522) dissolved in PBS to give a solution containing 0.1 mg/ml

Protocol 9 continued

- Insoluble protease XIV attached to 4% crosslinked beaded agarose (Sigma, P4531) and underivatized agarose support (Sigma, A7299)
- 50 mM NaOH: 0.2 g NaOH dissolved in 100 ml dH_2O
- Pronase (protease XIV, EC3.4.24.31, Sigma, P5147) dissolved in PBS to give a solution containing 0.25 U per well coated with 50 μl of antigen solution (equivalent to 9 mg solid per ml of PBS)

A Periodate oxidation of immobilized antigens (17)

1 Incubate antigen-coated wells with 50 μl of periodate solution for the appropriate time at 4 °C in sealed plastic bags. Incubate an equivalent number of control wells with acetate buffer only.

2 Wash wells four times (3 min each) with PBS and assay by ELISA (*Protocol 3*).

Note: the effect of periodate oxidation on epitope binding varies considerably between antibodies. A range of treatment times should therefore be tested for each antibody–epitope interaction (*Table 1*). Typical times vary from 16 h to 1 h or less.

B Protease digestion of immobilized antigens (17, 18)

1 Incubate antigen-coated wells with 50 μl of pronase or trypsin solution for 4 h at 37 °C and 4 °C in sealed plastic bags. Incubate an equivalent number of control wells with PBS only.

2 Wash wells four times (3 min each) with PBS. To trypsin treated wells add 50 μl of trypsin inhibitor solution, incubate at RT for 10 min, and wash wells a further three times with PBS. Assay treated wells by ELISA (*Protocol 3*).

C Pronase digestion of antigen extracts (18)

1 Wash pronase-coated agarose beads three times by repeated centrifugation and resuspension in PBS. Prepare a suspension containing 10 mg of beads per ml buffer and a control consisting of underivatized agarose. To 9 ml of antigen solution add 1 ml of agarose bead or control agarose suspension. Incubate with mixing at 4 °C and 30 °C for 4 h.

2 Remove agarose supports by centrifugation and transfer 50 μl samples of supernatant to microtitre well for immobilization, washing, and assay by ELISA (*Protocol 3*).

D Boiling (18)

1 Prepare antigen extracts in PBS and boil. At 10 min intervals, remove samples, allow to cool, centrifuge at 12 000 g for 5 min.

2 Transfer 50 μl samples to microtitre wells for immobilization, washing, and assay by ELISA (*Protocol 3*). Control wells should contain untreated extract only.

E β-elimination

1 To antigen-coated microtitre wells add 50 μl of 50 mM NaOH. Incubate for 24 h at RT in sealed plastic bags.

2 Wash plates three times (3 min each time) with PBS and assay by ELISA (*Protocol 3*).

Table 1 Epitope characterization by chemical and enzymatic modification

Anastomosis	Periodate oxidation				Pronase digestion		β-elimination			
group	EE1[a]	EE1	EE1	EE1	EH2[a]	EH2	EE1	EE1	EH2	EH2
	0.5 h(–)	0.5 h(+)	1.0 h(–)	1.0 h(+)	4.0 h(–)	4.0 h(+)	24 h(–)	24 h(+)	24 h(–)	24 h(+)
1	1.497[b]	0.708	1.311	0.263	1.065	0.406	1.024	1.115	1.088	0.424
1-IA	1.010	0.341	1.003	0.261	1.104	0.237	1.067	1.153	1.104	0.439
1-IB	1.526	0.677	1.455	0.292	1.119	0.387	1.051	1.150	1.079	0.323
1-IC	1.354	0.848	1.368	0.581	1.368	0.717	1.027	1.142	1.426	0.488
3	1.007	0.266	1.020	0.208	1.226	0.703	1.136	1.160	1.204	1.078
4HG-I	1.541	1.173	1.314	1.112	1.387	0.712	1.125	1.180	1.222	0.799
4HG-II	1.578	1.362	1.319	1.167	1.469	0.604	1.033	1.164	1.180	0.899
4	1.475	1.131	1.247	0.989	1.362	0.744	1.063	1.147	1.142	0.767
4	1.320	1.153	1.381	0.976	1.211	0.871	1.026	1.144	1.124	0.781
7	1.237	0.176	1.123	0.119	0.819	0.208	1.076	0.597	0.808	0.089

[a] Monoclonal antibodies EE1 and EH2 bind to carbohydrate and protein epitopes respectively on the same glycoprotein antigen. They were developed against lyophilized mycelium from an anastomosis group 4 isolate of *Rhizoctonia solani* and are described in Thornton *et al.* (18).

[b] Numerical values are means of two replicate samples and are the absorbances (450 nm) in ELISA after chemical or enzymatic modification of surface washings immobilized to microtitre wells, after different time periods. Figures in parentheses indicate treated (+) or control (–) wells.

Examples of the effect of periodate oxidation, pronase digestion, and β-elimination on the binding of two mAbs to their respective antigens are shown in *Table 1*. Monoclonal antibodies EE1 and EH2 bind to carbohydrate and protein epitopes respectively on the same glycoprotein antigen secreted by actively growing mycelium of the soil-borne plant pathogen *Rhizoctonia solani* (18, 25, 32). The differential sensitivity of the antigens secreted by different anastomosis group (AG) isolates of the pathogen to chemical and enzymatic modification illustrates the need to study a range of incubation periods on mAb binding.

6 Other commonly used immunological assays

A number of immunological assays, other than those described in this chapter, have been developed or adapted for studies of fungi. These assays and their various permutations are described in depth elsewhere (31). Of these assays, the double-antibody-sandwich ELISA (DAS-ELISA) has found widespread application. *Protocol 10* describes a typical indirect DAS-ELISA using a mAb for antigen 'capture' and either a second mAb or a pAb as the 'detector' species. For descriptions and use of double mAb sandwich assays see refs 31 and 32.

Protocol 10

Indirect DAS-ELISA

Equipment and reagents

- Immunlon 2 HB flat-bottomed microtitre plates (see *Protocol 3*)
- Variable volume (50–100 μl) multichannel pipette (Anachem)
- Microplate reader fitted with 450 nm filter
- PBS and PBST (see *Protocols 1* and *3*)
- Goat anti-rabbit IgG (whole molecule) peroxidase conjugate (Sigma, A6154)
- Goat anti-mouse Ig (heavy chain-specific) peroxidase conjugates: anti-mouse IgG (γ chain-specific) peroxidase (Sigma, A3673), anti-mouse IgA (α chain-specific) peroxidase (Sigma, A4789), anti-mouse IgM (μ chain-specific) peroxidase (Sigma, A8786)
- Peroxidase substrate and colour reaction stop solution (see *Protocol 3*)

Method

1. To wells of microtitre plate add 50 μl of 'capture' mAb supernatant. Tap plate gently to ensure even distribution over well surface. Incubate overnight at 4 °C to allow immobilization of antibody. Wash and dry plates (see *Protocol 3*). Antibody-coated plates can be stored at 4 °C in sealed plastic bags.
2. Block wells with 100 μl of blocking solution (*Protocol 3*) for 10 min to 1 h at RT, in a sealed plastic bag.
3. Remove blocking solution and add 50 μl of antigen extract.[a] Incubate for 1 h at RT in a sealed plastic bag.
4. Wash wells four times with PBST (3 min each time).
5. Add 50 μl of 'detector' pAb dissolved in PBST[b] or 'detector' mAb supernatant[c] and incubate for 1 h at RT in sealed plastic bag. Addition of blocker to the pAb diluent is optional.
6. Wash wells four times with PBST (3 min each time).
7. Add 50 μl of anti-rabbit peroxidase conjugate diluted 1 in 5000 in PBST, or anti-mouse Ig (heavy chain-specific)[c] peroxidase conjugate diluted 1 in 2000 in PBST, and incubate for 1 h at RT in sealed plastic bag. Addition of blocker to the PBST is optional.

   Note: the concentrations of secondary enzyme conjugates used should be optimized for each antibody–antigen interaction.
8. Wash wells three times with PBST (3 min each time) and once with PBS. Add 50 μl of substrate to each well and incubate for 30 min at RT. Stop colour reaction and determine absorbance at 450 nm (*Protocol 3*).

[a] For DAS-ELISAs, antigens are usually solubilized in PBST. Regardless of the extraction buffer used, it must contain detergent to prevent non-specific binding of antigen.

[b] The dilutions of pAb used should be optimized for each DAS-ELISA. This is best determined by titre using the principles described in *Protocol 3*. Dilutions are typically 1 in 1000 or greater.

Protocol 10 continued

[c] For double-mAb-sandwich-ELISAs, the second 'detector' mAb must be of a different class to the 'capture' mAb, unless the 'detector' MAb is directly conjugated to an enzyme conjugate. For example, if an IgA immunoglobulin is used as the 'capture' species, then the 'detector' species must belong to the class IgM or IgG (32). The reason for this is that an anti-mouse polyvalent (IgG, IgA, IgM) enzyme conjugate, such as that used in PTA-ELISAs (see *Protocol 3*), will not differentiate between the 'capture' and 'detector' mAbs in the sandwich which will lead to non-specific binding. However, commercial heavy chain (γ (IgG), α (IgA), μ (IgM))-specific enzyme conjugates are available that will discriminate between Ig class.

7 Availability and commercial development of antibodies

A list of diagnostic assays for fungal pathogens including antibody and nucleic acid-based methods has been compiled and can be found on the COST 823 fungus diagnostic database at website www.dsmz.de/COST. Further information on the availability of mAbs to *Botrytis*, *Cladosporium*, *Pseudocercosporella*, *Septoria*, *Penicillium*, *Aspergillus*, and *Humicola* can be obtained from Dr F. M. Dewey at the address given at the COST website.

Over the past decade a number of commercial companies have been established that provide pAb and mAb production to customer-supplied antigens and peptide sequences, as a chargeable service. Antigens are sent either lyophilized, in solution, or as PAGE gel slices. For pAb production the customer can choose their preferred host animal (rabbit, mouse, sheep, goat, hen, or rat). Immunization programmes typically last three months and immunization protocols are tailored to suit the needs of the customer. Pre-immune serum is supplied along with test serum taken after the third immunization. Monoclonal antibody production is modular and proceeds in phases *viz.* immunization, fusion, screening and selection of hybridomas and cloning, and isotyping. Supernatants are supplied before and after the cloning procedure to allow mAb selection. The customer receives the corresponding cloned hybridoma(s) with isotype(s).

Two companies providing such a service are listed below. Further information can be obtained at the respective websites.

1. EUROGENTEC

   European Headquarters: Parc Scientifique du Sart Tilman, 4102 Seraing, Belgium.
   Tel: +32 4 366 01 50
   Fax: +32 4 365 51 03
   E-mail: info@eurogentec.com
   Website: http://www.eurogentec.be
   In the UK:
   Tel: +44 (0)990 084 263
   Fax: +44 (0)990 084 288

2. Immune Systems Limited
PO Box 120, Paignton, Devon TQ4 7XD, UK.
Tel: +44 (0)1803 526 556
Fax: +44 (0)1803 526 776
Website: http://www.IS-L.com

References

1. DeVay, J. E., Schnathorst, W. C., and Foda, M. S. (1967). In *The dynamic role of molecular constituents in plant-parasite interaction* (ed. C. J. Mirocha and I. Uritani), pp. 313–28. American Phytopathological Society, Minnesota.
2. Doubly, J. A., Flor, H. H., and Clagget, C. O. (1960). *Science*, **3**, 229.
3. Kohler, G. and Milstein, C. (1975). *Nature*, **256**, 495.
4. Harlow, E. and Lane, D. (1988). *Antibodies: a laboratory manual*. Cold Spring Harbour Press, NY.
5. Del Sorbo, G., Scala, F., Capparelli, R., Iannelli, D., and Noviello, C. (1993). *Phytopathology*, **83**, 1178.
6. Mutasa-Gottgens, E. S., Chwarszczynska, D. M., Halsey, K., and Asher, M. J. C. (2000). *Plant Pathol.*, **49**, 276.
7. Bermingham, S., Dewey, F. M., and Maltby, I. (1995). *Appl. Environ. Microbiol.*, **61**, 2606.
8. Pain, N. A., O'Connell, R. J., Mendgen, K., and Green, J. R. (1994). *New Phytol.*, **127**, 233.
9. Thornton, C. R. and Dewey, F. M. (1996). *Mycol. Res.*, **100**, 217.
10. Hardham, A. R., Suzaki, E., and Perkin, J. L. (1986). *Can. J. Bot.*, **64**, 311.
11. Callow, J. A., Mackie, A. J., Roberts, A. M., and Green, J. R. (1992). *Symbiosis*, **14**, 237.
12. Pain, N. A., O'Connell, R. J., Bailey, J. A., and Green, J. R. (1992). *Physiol. Mol. Plant Pathol.*, **41**, 111.
13. Mackie, A. J., Roberts, A. M., Green, J. R., and Callow, J. A. (1993). *Physiol. Mol. Plant Pathol.*, **43**, 135.
14. Estrada-Garcia, M. T., Green, J. R., Booth, J. M., White, J. G., and Callow, J. A. (1989). *Exp. Mycol.*, **13**, 348.
15. Wakeham, A. J. and White, J. G. (1996). *Physiol. Mol. Plant Pathol.*, **48**, 289.
16. MacDonald, M. M., Dunstan, R. H., and Dewey, F. M. (1989). *J. Gen. Microbiol.*, **135**, 375.
17. Thornton, C. R., Dewey, F. M., and Gilligan, C. A. (1997). *Phytopathology*, **87**, 123.
18. Thornton, C. R., Dewey, F. M., and Gilligan, C. A. (1993). *Plant Pathol.*, **42**, 763.
19. Xia, J. Q., Lee, F. N., and Kim, K. S. (1992). *Can. J. Bot.*, **70**, 1790.
20. Priestley, R., Mohammed, C., and Dewey, F. M. (1994). In *Modern assays for plant pathogenic fungi: identification, detection and quantification* (ed. A. Schots, F. M. Dewey, and R. P. Oliver), pp. 149–56. CABI, Oxford, UK.
21. Karpovich-Tate, N., Spanu, P., and Dewey, F. M. (1998). *Mol. Plant Microbe Interact.*, **7**, 710.
22. Thornton, C. R. and Gilligan, C. A. (1999). *Mycol. Res.*, **103**, 443.
23. Dewey, F. M., Thornton, C. R., and Gilligan, C. A. (1997). In *Advances in botanical research and plant pathology* (ed. J. H. Andrews and I. C. Tommerup), Vol. 24, pp. 275–308. Academic Press, London, UK.
24. Bermingham, S., Maltby, L., and Dewey, F. M. (1996). *New Phytol.*, **132**, 593.
25. Thornton, C. R. (1996). *Soil Biol. Biochem.*, **28**, 527.
26. Otten, W., Gilligan, C. A., and Thornton, C. R. (1997). *Phytopathology*, **87**, 730.
27. Kennett, R. H., Davis, K. A., Tung, A. S., and Kilman, N. R. (1978). *Curr. Top. Microbiol. Immunol.*, **81**, 77.
28. Goding, J. W. (1980). *J. Immunol. Methods*, **39**, 285.

29. Foung, S. K., Sasaki, D. T., Grumet, F. C., and Engleman, E. C. (1982). *Proc. Natl. Acad. Sci. USA*, **79**, 7484.
30. Bleddyn Hughes, H., Carzaniga, R., Rawlings, S. L., Green, J. R., and O'Connell, R. J. (1999). *Microbiology*, **145**, 1927.
31. Dewey, F. M. and Thornton, C. R. (1995). In *New diagnostics in crop sciences* (ed. J. H. Skerritt and R. Appels), pp. 151–70.
32. Thornton, C. R., O'Neill, T. M., Hilton, G., and Gilligan, C. A. (1999). *Plant Pathol.*, **48**, 627.

List of suppliers

Agar Scientific Ltd., 66a Cambridge Road, Stansted, Essex CM24 8DA, UK.
Agent: Ted Pella Inc., 4595 Mountain Lakes Boulevard, Redding, CA 96003, USA.

Aldrich, Sigma-Aldrich Pty Ltd., PO Box 970, Castle Hill, NSW 2154, Australia.

Anderman and Co. Ltd., 145 London Road, Kingston-upon-Thames, Surrey KT2 6NH, UK.
Tel: 0181 541 0035
Fax: 0181 541 0623

Aurion, Costerweg 5, 6702 AA, Wageningen, The Netherlands.
URL: htpp://www.aurion.nl
Agent: Electron Microscopy Sciences (see below for address).

BAL-TEC, Technotrade International, 7 Perimeter Road, Manchester, NH 03103-3343, USA.
URL: htpp://www.Bal-tec.com
Agent: Balzers, Föhrenweg 16, FL-9496, Liechtenstein.

Beckman Coulter (UK) Ltd., Oakley Court, Kingsmead Business Park, London Road, High Wycombe, Buckinghamshire HP11 1JU, UK.
Tel: 01494 441181
Fax: 01494 447558
URL: http://www.beckman.com
Beckman Coulter Inc., 4300 N Harbor Boulevard, PO Box 3100, Fullerton, CA 92834-3100, USA.
Tel: 001 714 871 4848
Fax: 001 714 773 8283
URL: http://www.beckman.com

Becton Dickinson and Co., 21 Between Towns Road, Cowley, Oxford OX4 3LY, UK.
Tel: 01865 748844 Fax: 01865 781627
URL: http://www.bd.com
Becton Dickinson and Co., 1 Becton Drive, Franklin Lakes, NJ 07417-1883, USA.
Tel: 001 201 847 6800
URL: http://www.bd.com

Bio 101 Inc., c/o Anachem Ltd., Anachem House, 20 Charles Street, Luton, Bedfordshire LU2 0EB, UK.
Tel: 01582 456666
Fax: 01582 391768
URL: http://www.anachem.co.uk
Bio 101 Inc., PO Box 2284, La Jolla, CA 92038-2284, USA.
Tel: 001 760 598 7299
Fax: 001 760 598 0116
URL: http://www.bio101.com

Bio-Rad Laboratories Ltd., Bio-Rad House, Maylands Avenue, Hemel Hempstead, Hertfordshire HP2 7TD, UK.
Tel: 0181 328 2000
Fax: 0181 328 2550
URL: http://www.bio-rad.com

Bio-Rad Laboratories Ltd., Division Headquarters, 1000 Alfred Noble Drive, Hercules, CA 94547, USA.
Tel: 001 510 724 7000
Fax: 001 510 741 5817
URL: http://www.bio-rad.com

Biospec Products, Bartlesville, OK, USA.

Biosupplies Australia Pty Ltd., PO Box 835, Parkwille, Victoria 3052, Australia.
URL: htpp://www.biosupplies.com.au

British BioCell International, Golden Gate, Ty Glas Avenue, Cardiff CF4 5PX, UK.
URL: http://www.british-biocell.co.uk

Citifluor Ltd., Connaught Building, City University, Northampton Square, London EC1V 0HB, UK.

CP Instrument Co. Ltd., PO Box 22, Bishop Stortford, Hertfordshire CM23 3DX, UK.
Tel: 01279 757711
Fax: 01279 755785
URL: http://www.cpinstrument.co.uk

Dupont (UK) Ltd., Industrial Products Division, Wedgwood Way, Stevenage, Hertfordshire SG1 4QN, UK.
Tel: 01438 734000
Fax: 01438 734382
URL: http://www.dupont.com
Dupont Co. (Biotechnology Systems Division), PO Box 80024, Wilmington, DE 19880-002, USA.
Tel: 001 302 774 1000
Fax: 001 302 774 7321
URL: http://www.dupont.com

Eastman Chemical Co., 100 North Eastman Road, PO Box 511, Kingsport, TN 37662-5075, USA.
Tel: 001 423 229 2000
URL: http://www.eastman.com

Electron Microscopy Sciences, 321 Morris Road, Box 251, Fort Washington, PA 19034, USA.
URL: htpp://www.emsdiasum.com
Agent: Science Services Ltd., Greenwood House, 4/7 Salisbury Court, London EC4Y 8BT, UK.

EY Laboratories Inc., 107 North Amphlett Boulevard., San Mateo, CA 94401, USA.
URL: http://www.eylabs.com
Agent: TCS Biologicals Ltd., Botolf Claydon, Buckingham MK18 2LR, UK.

Fisher Scientific UK Ltd., Bishop Meadow Road, Loughborough, Leicestershire LE11 5RG, UK.
Tel: 01509 231166
Fax: 01509 231893
URL: http://www.fisher.co.uk
Fisher Scientific, Fisher Research, 2761 Walnut Avenue, Tustin, CA 92780, USA.
Tel: 001 714 669 4600
Fax: 001 714 669 1613
URL: http://www.fishersci.com

Fluka, PO Box 2060, Milwaukee, WI 53201, USA.
Tel: 001 414 273 5013
Fax: 001 414 2734979
URL: http://www.sigma-aldrich.com
Fluka Chemical Co. Ltd., PO Box 260, CH-9471, Buchs, Switzerland.
Tel: 0041 81 745 2828
Fax: 0041 81 756 5449
URL: http://www.sigma-aldrich.com

FMC BioProducts, 191 Thomaston Street, Rockland, ME 04841, USA.

Hybaid Ltd., Action Court, Ashford Road, Ashford, Middlesex TW15 1XB, UK.
Tel: 01784 425000
Fax: 01784 248085
URL: http://www.hybaid.com

Hybaid US, 8 East Forge Parkway, Franklin, MA 02038, USA.
Tel: 001 508 541 6918
Fax: 001 508 541 3041
URL: http://www.hybaid.com

HyClone Laboratories, 1725 South HyClone Road, Logan, UT 84321, USA.
Tel: 001 435 753 4584
Fax: 001 435 753 4589
URL: http://www.hyclone.com

ICN Biomedicals Australasia Pty Ltd., PO Box 187, Unit 12, 167 Prospect Hwy, Seven Hills, NSW 2147, Australia.

Invitrogen Corp., 1600 Faraday Avenue, Carlsbad, CA 92008, USA.
Tel: 001 760 603 7200
Fax: 001 760 603 7201
URL: http://www.invitrogen.com
Invitrogen BV, PO Box 2312, 9704 CH Groningen, The Netherlands.
Tel: 00800 5345 5345
Fax: 00800 7890 7890
URL: http://www.invitrogen.com

Leica UK Ltd., Davy Avenue, Knowlhill, Milton Keynes MK5 8LB, UK.
URL: htpp://www.leica.com

Life Technologies Ltd., PO Box 35, Free Fountain Drive, Incsinnan Business Park, Paisley PA4 9RF, UK.
Tel: 0800 269210
Fax: 0800 838380
URL: http://www.lifetech.com
Life Technologies Inc., 9800 Medical Center Drive, Rockville, MD 20850, USA.
Tel: 001 301 610 8000
URL: http://www.lifetech.com

Merck Sharp & Dohme, Research Laboratories, Neuroscience Research Centre, Terlings Park, Harlow, Essex CM20 2QR, UK.
URL: http://www.msd-nrc.co.uk

MSD Sharp and Dohme GmbH, Lindenplatz 1, D-85540, Haar, Germany.
URL: http://www.msd-deutschland.com

Miles Inc., Diagnostic Division, Elkhart, IN 46515, USA.

Millipore (UK) Ltd., The Boulevard, Blackmoor Lane, Watford, Hertfordshire WD1 8YW, UK.
Tel: 01923 816375 Fax: 01923 818297
URL: http://www.millipore.com/local/UK.htm
Millipore Corp., 80 Ashby Road, Bedford, MA 01730, USA.
Tel: 001 800 645 5476
Fax: 001 800 645 5439
URL: http://www.millipore.com

Molecular Probes, Inc., PO Box 22010, Eugene, OR 97402-0469, USA.

Nalgene Nunc International, 2000 North Aurora Road, Naperville, IL 60563, USA.

Nanoprobes, 25 East Loop Road, Suit 113, Stony Brooks, NY 11790-3350, USA.
htpp://www.nanoprobes.com

New England Biolabs, 32 Tozer Road, Beverley, MA 01915-5510, USA.
Tel: 001 978 927 5054

Nikon Inc., 1300 Walt Whitman Road, Melville, NY 11747-3064, USA.
Tel: 001 516 547 4200
Fax: 001 516 547 0299
URL: http://www.nikonusa.com
Nikon Corp., Fuji Building, 2-3, 3-chome, Marunouchi, Chiyoda-ku, Tokyo 100, Japan.
Tel: 00813 3214 5311 Fax: 00813 3201 5856
URL: http://www.nikon.co.jp/main/index_e.htm

Nycomed Amersham plc, Amersham Place, Little Chalfont, Buckinghamshire HP7 9NA, UK.
Tel: 01494 544000 Fax: 01494 542266
URL: http://www.amersham.co.uk

Nycomed Amersham, 101 Carnegie Center, Princeton, NJ 08540, USA.
Tel: 001 609 514 6000
URL: http://www.amersham.co.uk

Osmonics Inc., 111 Lindbergh Avenue, Livermore, CA 94550-9520, USA.
URL: htpp://www.osmonics.com
Agent: Porvair Sciences Ltd., Shepperton Business Park, Govett Avenue, Shepperton, Middlesex TW17 8BA, UK.
URL: htpp://porvair-sciences.com

Perkin Elmer Ltd., Post Office Lane, Beaconsfield, Buckinghamshire HP9 1QA, UK.
Tel: 01494 676161
URL: http://www.perkin-elmer.com

Pharmacia Biotech (Biochrom) Ltd., Unit 22, Cambridge Science Park, Milton Road, Cambridge CB4 0FJ, UK.
Tel: 01223 423723 Fax: 01223 420164
URL: http://www.biochrom.co.uk
Pharmacia and Upjohn Ltd., Davy Avenue, Knowlhill, Milton Keynes, Buckinghamshire MK5 8PH, UK.
Tel: 01908 661101 Fax: 01908 690091
URL: http://www.eu.pnu.com

Philips Electronics Australia Ltd., 2 Greenhills Avenue, Moorebank, NSW 2170, Australia.

Polysciences Inc., 400 Valley Road, Warrington, PA 18976, USA.
Agent: Park Scientific Ltd., 24 Low Farm Place, Moulton Park, Northampton NN3 6HY, UK.

Promega UK Ltd., Delta House, Chilworth Research Centre, Southampton SO16 7NS, UK.
Tel: 0800 378994
Fax: 0800 181037
URL: http://www.promega.com

Promega Corp., 2800 Woods Hollow Road, Madison, WI 53711-5399, USA.
Tel: 001 608 274 4330
Fax: 001 608 277 2516
URL: http://www.promega.com

Qiagen UK Ltd., Boundary Court, Gatwick Road, Crawley, West Sussex RH10 2AX, UK.
Tel: 01293 422911 Fax: 01293 422922
URL: http://www.qiagen.com
Qiagen Inc., 28159 Avenue Stanford, Valencia, CA 91355, USA.
Tel: 001 800 426 8157
Fax: 001 800 718 2056
URL: http://www.qiagen.com

RMC-Ventana Medical Systems Inc., 3450 S. Broadmont, Suite 100, Tucson, AZ 85713, USA.
URL: http://www.rmc-scientific.com
Agent: Electron Optical Services Ltd., 52 Higher Urmston Road, Manchester M41 9AP, UK.

Roche Diagnostics Ltd., Bell Lane, Lewes, East Sussex BN7 1LG, UK.
Tel: 01273 484644 Fax: 01273 480266
URL: http://www.roche.com
Roche Diagnostics Corp., 9115 Hague Road, PO Box 50457, Indianapolis, IN 46256, USA.
Tel: 001 317 845 2358
Fax: 001 317 576 2126
URL: http://www.roche.com
Roche Diagnostics GmbH, Sandhoferstrasse 116, 68305 Mannheim, Germany.
Tel: 0049 621 759 4747
Fax: 0049 621 759 4002
URL: http://www.roche.com

Schleicher and Schuell Inc., Keene, NH 03431A, USA.
Tel: 001 603 357 2398

Shandon Scientific Ltd., 93-96 Chadwick Road, Astmoor, Runcorn, Cheshire WA7 1PR, UK. Tel: 01928 566611
URL: http://www.shandon.com

Sigma-Aldrich Co. Ltd., The Old Brickyard, New Road, Gillingham, Dorset XP8 4XT, UK.
Tel: 01747 822211 Fax: 01747 823779
URL: http://www.sigma-aldrich.com
Sigma-Aldrich Co. Ltd., Fancy Road, Poole, Dorset BH12 4QH, UK.
Tel: 01202 722114 Fax: 01202 715460
URL: http://www.sigma-aldrich.com
Sigma Chemical Co., PO Box 14508, St Louis, MO 63178, USA.
Tel: 001 314 771 5765
Fax: 001 314 771 5757
URL: http://www.sigma-aldrich.com

Stratagene Inc., 11011 North Torrey Pines Road, La Jolla, CA 92037, USA.
Tel: 001 858 535 5400
URL: http://www.stratagene.com
Stratagene Europe, Gebouw California, Hogehilweg 15, 1101 CB Amsterdam Zuidoost, The Netherlands.
Tel: 00800 9100 9100
URL: http://www.stratagene.com

Structure Probe Inc., PO Box 656, West Chester, PA 19381-0656, USA.
URL: http://www.2spi.com
Agent: Aztech Trading, 12 Kerman Drive, Loughborough, Leicestershire LE11 5JF, UK.

TAAB Laboratories Ltd., 3 Minerva House, Calleva Industrial Park, Aldermaston, Berkshire RG7 4QW, UK.

United States Biochemical, PO Box 22400, Cleveland, OH 44122, USA.
Tel: 001 216 464 9277

Vector Laboratories Inc., 30 Ingold Road, Burlingame, CA 94010, USA.
URL: http://www.vectorlabs.com
Vector Laboratories Ltd., 16 Wulfric Square, Bretton, Peterborough PE3 8RF, UK.

Whatman International Ltd., Catalogue Sales Department, St. Leonard's Road, Maidstone, Kent ME16 0LS, UK.

Index

The manufacturer's authorised representative in the EU for product safety is Oxford University Press España S.A. of el Parque Empresarial San Fernando de Henares, Avenida de Castilla, 2 – 28830 Madrid (www.oup.es/en or product.safety@oup.com). OUP España S.A. also acts as importer into Spain of products made by the manufacturer.

www.ingramcontent.com/pod-product-compliance
Lightning Source LLC
LaVergne TN
LVHW080035210926
848481LV00011B/3196
* 9 7 8 0 1 9 9 6 3 8 3 7 6 *